Electrical Characterization of Organic Electronic Materials and Devices

Electrical Characterization of Organic Electronic Materials and Devices

Peter Stallinga
Center for Electronics, Optoelectronics and Telecommunications
University of The Algarve

WILEY
A John Wiley and Sons, Ltd., Publication

This edition first published 2009

Registered office
John Wiley & Sons Ltd, The Atrium, Southern Gate, Chichester, West Sussex, PO19 8SQ, United Kingdom

For details of our global editorial offices, for customer services and for information about how to apply for permission to reuse the copyright material in this book please see our website at www.wiley.com.

Library of Congress Cataloging-in-Publication Data

Stallinga, Peter, 1966-
Electrical characterization of organic electronic materials and devices / Peter Stallinga.
p. cm.
Includes bibliographical references and index.
ISBN 978-0-470-75009-4 (cloth : alk. paper) 1. Electronics–Materials. 2. Organic electronics. 3. Organic semiconductors. 4. Electronic apparatus and appliances–Materials.
I. Title.
TK7871.S73 2009
621.381–dc22

2009028761

A catalogue record for this book is available from the British Library.

ISBN: 978-0-470-75009-4 (Cloth)

Typeset in 10.5/13 Sabon by Laserwords Private Limited, Chennai, India

Printed and bound in Great Britain by TJ International, Padstow, Cornwall

Contents

Preface

This book is the summary of more than a decade of investigation of organic electronic materials and devices carried out at the Optoelectronics Laboratory at the University of The Algarve. It is in no way a summary of the literature, though an extensive selection of references is included. (It should be made clear from the beginning that by mentioning a paper in the literature does not imply that I approve of its contents, agree with its conclusions, or even recommend its reading, nor that I fully understood its message; it only serves the purpose of mentioning works on the subject and gives a possible starting point for interested readers. Some of the papers, especially in the introductory parts, I have to admit, I did not even read thoroughly myself; the list of citations at the end is thus a mixture of references and bibliography.)

Every time a concept of electrical behavior came up, I tried to understand it by making calculations, or, more often, by making simulations, until the behavior was in-line with how I understood things to work. Often my understanding of things was wrong and it had to be changed to be in-line with simulation and results in our laboratory and results presented in the literature. Even while writing this book, I changed my thinking substantially in some places, showing that the learning curve is infinite. It also shows that, probably, this book is still full of errors and misconceptions. It is to be considered, still, an unfinished work. It should therefore not be seen as an absolute book, believing everything without questioning. It should rather be seen as a work to stimulate thinking and to help the scientist on the way with proper tools.

I also suffer from the constant need to not only understand things, but to also make the description as simple as possible, to strip away all the unneeded embellishments, and to reduce it to its important core. While obviously missing out on some descriptive power of the ideas, the gain of hopefully making it possible for many more people to understand compensates enough. In my opinion, it is not the task of a scientist to understand nature, it is rather to model it and explain it to others so that they can understand it and further build on it. If not, all gained knowledge will be lost at the end of the career of the scientist.

Never, in any case, did I assume things to be correct in the literature and just conveniently reference them away when encountering similar behavior in our materials and devices. A multitude of papers exist that are far beyond my grasp. After some failed attempts, I ignored these papers (some of them are still mentioned in this book, so that you can take a look yourself). The most remarkable result of this approach is probably the description, from the origin, of the thin-film transistor with, as final end-product, a predicted and implemented transistor based on a metal as the active layer. Looking back at it, I now consider it something completely trivial and expected. Most people will probably agree that it is trivial. However, the same reasoning also led to the accurate description of the contact effects, something that conflicts with a large number of papers in the literature.

It always gave me immense pleasure to work on the subject and to realize that, after all, nature is a fascinating complex world, but when persisting, even for a humble and ignorant human being, it is possible to understand some things. Then, the beauty lies in its simplicity. As a single starting point, I used the idea 'Think as an electron'; when thinking as an electron, the rest will follow. In case of doubt it is best to ask 'What would *I* do if I were an electron?' This is funny, in a way, but in many cases very helpful.

This work, of course, would not have come about without the help of other people. Most important in this respect is Henrique Gomes, my colleague at the Optoelectronics Laboratory. Sometimes, by just saying a couple of phrases, he pointed the way to a better understanding of things, things that would sometimes take me months or even years to figure out in all their myriad implications. Likewise, the comments and observation of Martin Taylor have been extremely helpful. The discussions with Adriano Benvenho were always a good inspiration for a lot of the work. I also thank him for proofreading this book. My students Nelson Pimenta and José Almada are thanked for implementing the transistor admittance spectroscopy system described in Chapter 6.

I hope this text is helpful to you, the reader, too. After all, that is the aim of this work.

Peter Stallinga
Faro, Portugal
February 2009

1

General Concepts

1.1 INTRODUCTION

This book describes the electronic measurements of materials and devices based on amorphous materials, with an emphasis on organic electronic materials. The distinction of what is exactly an organic material is rather vague, but generally speaking it is all those materials that contain carbon. This area is rather vast (see for instance the book on organic chemistry by Wade [1]). For this work the ones that are important are those that are electrically active, basically organic semiconductors. That organic materials can be semiconductors is nothing new. Wallace has shown already in 1947 that graphite is a semiconductor [2]. (Curiously enough, graphene, a monolayer of graphite, is again a hot-topic material [3].) Recently, though, a new wave of research in organic electronics has got under way, powered by the need for cheap electronic devices or other devices that cannot be made by conventional, 'classic' semiconductors and helped by the accidental discovery that organic materials can be as conductive as metals and further stimulated by the continuing advances in technology of fabrication of materials and devices.

Two decades and a Nobel Prize after the start of this new wave of organic electronic research, commercial devices are a reality [4]; manufacturers advertise that their products have light emitting diodes based on organic materials (OLEDs), to name but an example. Like in most technological advances, the understanding of the devices lags behind. There are many issues yet to be solved. The most important may be

Electrical Characterization of Organic Electronic Materials and Devices Peter Stallinga

the electronic processes governing the conduction of charge. On the one hand there exists the idea that the conduction consists of hopping from one molecular orbit to the other. Such hopping models centered around the molecular aspects obviously find many followers among chemists. Conduction is then considered a perturbation to the system. On the other hand there exists the solid-state-physicists approach of treating the materials like any semiconducting material and describing the electronic states of the material in the classical conduction and valence bands. The molecular properties are then treated as a perturbation. The answer will lie probably somewhere in between. However, electronic devices can only be made from semiconductor (conducting) materials. In view of this it is most obvious that existing working devices are best described by classic semiconductor theory, at least as a starting point, and changes should only be introduced into the theory when needed. However, organic electronic materials research evolved from chemistry and the publications by chemists far outnumber the publications of semiconductor physicists. Many papers describe the problems encountered when fabricating the materials, whereas the electronic parameters, such as band gap, effective mass, or electron affinity are difficult to encounter. The electrical characteristics of the resulting materials often serve as a by-product of the chemical research, without even trying to understand or describe them well. As an example, often the mobility is presented as a parameter to describe the quality, the 'success', of the new proposed production route, or to compare one material with another [5]. Review articles often present tables with best mobilities, for instance to show the evolution over time [6–8]. However, just presenting the mobility value without specifying the conditions is as meaningless as presenting a current without specifying the applied bias. In view of this, it is often difficult to compare the results presented by different research institutes. The current book is an attempt to explain the mechanisms involved in the electrical characteristics of organic materials and the measurement techniques used to determine the relevant parameters. The starting point is the theory of the 'old school' of semiconductor devices which is very well described by the classical book of Sze [9], a book which will be cited many times throughout this work. The reader is advised to obtain a copy of this book. More recent books include novel devices, such as thin-film transistors (TFTs), for example the book by Shur [10]. There also exist good books about the measurement techniques in general. The book by Blood and Orton may serve as an example [11]. The current work describes how and when the ideas of these books, written for classical semiconductors, can be applied to organic materials and when they need to be changed or adjusted before application.

As a first observation, a significant difference can be pointed out between classical and organic semiconductors. In classical semiconductors, namely silicon, germanium and GaAs, with the exception of diamond, the band gap, the distance between the conduction band and the valence band, is rather small, in the order of 1 eV, see Table 1.1. In most organic semiconductors, this band gap is much larger and is in the order of 2–3 eV. As a direct result, the optical processes have a larger range and blue photovoltaic devices are feasible. However, the large band gap also allows for the existence of deep levels. These deep levels can make the electronic processes very slow. To give an example, the 'lifetime killer' impurity of gold in silicon introduces a near midgap level with a depth of approximately 0.6 eV. This midgap level of gold in

Table 1.1 (a) Physical properties of some classical semiconductors. (b) Band gaps of some organic semiconductors

(a)

Property	Si	GaAs	C
Atoms/molecules (cm^{-3})	5.0×10^{22}	4.42×10^{22}	1.8×10^{23}
N_C (cm^{-3})	2.8×10^{19}	4.7×10^{17}	10^{20}
N_V (cm^{-3})	1.04×10^{19}	7.0×10^{18}	10^{19}
ε_r	11.9	13.1	5.7
Electrons m^*/m_e	0.26	0.067	0.2
Holes m^*/m_e	0.16	0.082	0.25
Electron mobility ($cm^2\ V^{-1}\ s^{-1}$)	1500	8500	4500
Hole mobility ($cm^2\ V^{-1}\ s^{-1}$)	450	400	3800
Energy gap (eV)	1.12	1.424	5.47
Electron affinity (V)	4.05	4.07	

(b)

Material	Band gap (eV)
PEDNT	0.95
PEDOT	1.6
Polythiophene	2.1
PPV	2.4
PPP	3.0
Polypyrrole	3.2
Pentacene	1.82
Tetracene	3
Sexithiophene (T6)	2.87
Alq_3	2.7

silicon, the fastest possible electron–hole recombination path in silicon, has a recombination time in the order of microseconds. The lifetime of thermal excitation from a level is exponentially dependent on the ratio of level depth and average thermal energy ($kT = 26\,\text{meV}$ at room temperature) [9]. Thus, a typical organic wide band gap material, with a band gap of 2.5 eV has a midgap level that is about 0.6 eV deeper. This translates to a factor $\exp(600/26) = 10^{10}$ in the time constant when assuming the prefactor in the capture cross-section to be constant. In other words, a midgap trap in silicon has a relaxation time in the order of microseconds, whereas in an organic material it causes transients in the order of tens of kiloseconds. Other levels suffer from this effect as well, albeit to a lesser extent. Since many impurity levels derive their electronic structure from the band states, a widening of the band gap might cause a deepening of the impurity levels. This makes organic materials notoriously slow. Ring oscillators have been fabricated working only in the kHz range, where for silicon they can easily be made operating at GHz frequencies. However, the application of organic materials lies not in the area of fast electronics but rather in the area of cheap electronics, large-area electronics [6], or electronic devices that cannot be made or are difficult to make by classical materials. One has to think here about things like printable [12–17], ink-jet printable [18, 19], and flexible electronics [15, 20–23] or electronic textiles [24]. In this respect fully organic devices stable in air are interesting [25]. Further, imagine a display on curved surfaces, etc. Or the color-switchable light emitting device [26]. The ultimate goal is to reach the single-molecule electronics level [27]. With this goal in mind, a wave of research was focused on nanoscale technologies, such as incorporating carbon nanotubes [28, 29] or nanowires [30].

Since, in contrast to inorganic materials, the number of organics is sheer infinite, an organic material can be synthesized that meets any need, for instance, for sensors [31–36]. Immediately it was obvious that organic materials are highly compatible with biological materials and are thus ideal partners in bioelectronics, including biodetectors (transducers) [37–39] or signal processing in biomolecular electronics [40, 41]. DNA or its constituent nucleic acids themselves can easily serve as the active material in (thin-film) transistors [41–43]. Interestingly, in these cases the behavior of the devices is adequately described by the ideas presented here.

Another fundamental difference is the difficulty of making organic materials with the same level of purity and crystallinity as, for instance, silicon. As discussed by Caserta *et al.*, the treatment of the material by a band model might seem inadequate for such noncrystalline materials. However, as stated by them by referring to papers of Ioffe and Regel and Gubanov, 'a periodic electric field of the lattice is not essential for the occurrence of typical semiconducting properties and the band model may be applied also in the case in which there is a loss of periodicity of the lattice' [44]. Long-range order is not needed to give the material semiconducting properties. The edges of the bands are not well defined, though. In other words, noncrystalline (disordered) materials can be treated by semiconductor (band) theory with nondiscrete bands.

Together with the noncrystallinity comes a large density of trap states and they severely change the electronic behavior, as will be shown throughout this book. Trap states are deep localized states that can capture carriers that would otherwise contribute to conduction. It is irrelevant if the trap states originate from the disorder of the material or from impurities. Once captured by a trap, these charges are unavailable for conduction (in other words, the mobility is zero). There exists also models to include hopping or tunneling conduction via traps [45]. These models do not need any band states and are therefore more closely related to insulating materials than semiconducting materials. As a rule of thumb, it can be said that when the (intrinsic) mobility is below $10^{-4}\,\mathrm{cm^2\,V^{-1}s^{-1}}$ the conduction is most likely to be of hopping type and above this limit it is more likely to be of band type. The reason for this is that if conduction is large, the time spent on an individual site is short, and this causes a wide band because of Heisenberg's Uncertainty Relation [46], $\Delta t \times \Delta E \geq \hbar/2$. However, if wave function overlap is small the wave functions more resemble molecular orbits with well defined, sharp energies. At the same time, the lack of overlap causes low conductivity and the charges spend a long time on the molecules between hops.

The question is if it is really necessary to make them so pure. Again, organic materials are not planned to compete with silicon in performance, but mostly in price. The enhanced performance of the purity will not balance against the extra cost associated with the purification.

Finally, there is a fundamental difference between organic and classical materials in that the latter are more mechanically and chemically stable. In this sense, stability means that the materials do not show phase transitions and that the conduction is by electrons and holes only. In some organic materials, the conduction is by ionic transport, this being a much slower and less controlled process. Phase transitions can also play an important role. Recently it has been pointed out that the omnipresent phase transition in organic materials (see for example the work on sexithiophene [47]) at around 200 K is due to water [48], an impurity very difficult to avoid in the low grade fabrication techniques such as spin coating.

1.2 CONDUCTION MECHANISM

As to the basic conduction mechanism, there is a debate between 'free' carriers (electrons in the conduction band and holes in the valence band) on one side and polarons [49] (or bipolarons [50]) on the other side. A polaron is a charge together with its local lattice deformation and polarization, as shown in Figure 1.1.

It seems most obvious, once the theory from semiconductors has been borrowed, to start assuming free carriers and only introduce polarons when it is not possible to explain the data otherwise. In this work the entities responsible for charge transport are considered to be free carriers, for the sake of simplicity, rather than to make a bold statement about the validity of this assumption. The conduction band, as used in this work might, in fact, be a polaronic band.

Another on-going debate is the distinction between band conduction and hopping conduction (Figure 1.2). In band conduction, the charge is highly delocalized and can travel freely. Charges, however, spend most of their time on localized deep states from where they cannot contribute to current, thus reducing the average mobility. One example of this is Poole–Frenkel conduction, which also includes a field-dependence of mobility, or multi-trap-and-release (MTR), which is equal to Poole–Frenkel with the field dependence removed.

In contrast, in hopping conduction models only localized states exist or play a role, for instance because the delocalized bands are too far

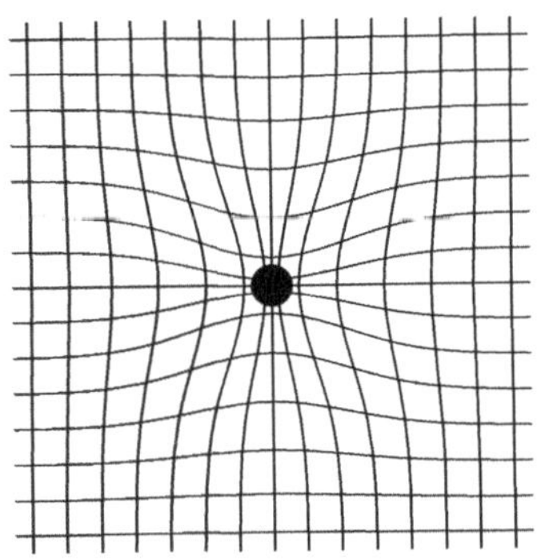

Figure 1.1 A polaron: a charge with its associated lattice distortion

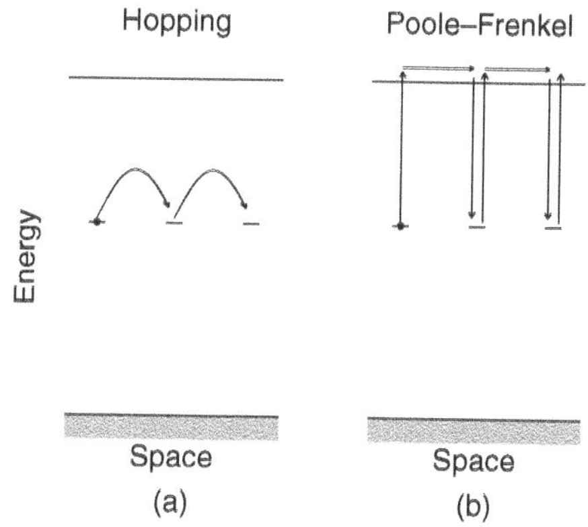

Figure 1.2 Distinction between hopping conduction (a) and Poole–Frenkel conduction (b). In the former charges occasionally jump ('hop') from trap to trap. In the latter carriers spend most of their time trapped but occasionally are excited to the delocalized (conduction or valence) band from where they can contribute to current

away or the temperature too low to allow for thermal excitation of trapped carriers [51]. The charges spend all of their time on these states. Transport of charge occurs by instantaneous hops between these states. This can also cause a field and temperature dependence of the effective mobility, as shown by the simulations of Bässler [45]. Ambegaokar *et al.* [52] extended the work of Mott on hopping/percolation theory that predicts a dependence of conductivity on temperature of the form $\sigma \propto \exp[-(T_0/T)^{1/4}]$. Often, the percolation theory of Shklovskii and Efros is applied to organic materials [53].

Comments by Rakhmanova and Conwell, however, state that Poole–Frenkel conduction seems more likely for organic materials [54]. Also, Waragai *et al.* reject hopping conduction for their transistors because of the unrealistic values found for the parameters [55]. Similarly, Nelson *et al.* reject the idea of hopping conduction for their pentacene transistors [56]. Poole–Frenkel conduction is more adequate for when the devices start having substantial currents and conduction is no longer a perturbation.

Other conduction mechanisms contain a two-band formalism; one conduction band and one defect band [57] (where the designation 'band' for the defect states is somewhat misplaced; they are comparable with the hopping states described above). When the defect band is dense enough, hopping can occur from defect to defect and the associated current can be substantial [58]. This can be nearest-neighbor hopping, or variable-range hopping (VRH) [59, 60].

In this work, a Poole–Frenkel formalism is used. It is assumed that all conduction is through the conduction (and valence) bands with additional efficient traps, deep localized states; to contribute to current, a charge trapped on a deep level has first to be promoted to a conductive band to contribute to current, i.e., no direct hopping from trap to trap is possible. The interesting thing is that the defect band can, in principle, also originate from the polarons. In other words, the traps can be self-traps. A charge can create a lattice distortion that makes it so immobile as to effectively be locked in place, thus behaving like a trap. It would be difficult to explain, though, how it can be that with advances in production techniques, the devices seem to conduct better. After all, a polaron is an intrinsic particle and should not depend much on the quality of the material. However, traps related to impurities such as lattice defects will disappear when going from amorphous to crystalline materials. This difficult discussion is not addressed in this book; no bold statements are made about the exact conduction mechanism. As will be shown, maintaining a two-band formalism with only one band contributing to current, and without stating what the bands represent exactly, will not take anything away from the descriptive power. The discussion about the conduction mechanism is deferred. Many theories exist that are consistent with measured data, yet, the reverse logic cannot be applied, namely that the data are proof of the theory.

As an example of the difficulty of describing amorphous materials consider the analysis of the parameter 'mobility', μ. Mobility is probably the most important parameter in electronic materials, as it is directly linked to the switching speed of the devices, because charges either have to drift or diffuse in and out of the active zone of the device and these are both governed by mobility (diffusion via Einstein's Relation). Switching times are proportional to $L^2/\mu V$, where V is the relevant voltage and L is the relevant device dimension, for instance the channel length in transistors [61, 62]. Therefore, most research is aimed solely at increasing the mobility. Organic materials are often labeled as low-mobility materials. However, we have to be very careful when making such statements. Mostly, the definition of mobility is the basis for the measured low mobility. When we define the mobility as the factor that links current to the product of density of charge (qp, where p is the hole density) and the electric field (E),

$$J = \mu_p q p E \tag{1.1}$$

where J is the current density, it is assumed that we know the density p very well. This assumption is not always true. It is, for instance, often assumed that p is constant, and that all dependencies of the current on temperature are described by a temperature dependent mobility, $\mu(T)$. A more correct picture is one in which the mobility is constant and the charge density is temperature dependent,

$$J(T) = q\mu_p p(T)E. \tag{1.2}$$

Moreover, in the same way, if a material has remarkably low conductivity, this is often not because of low intrinsic mobility, but rather because of unexpected low density of free carriers, for instance because most charge is located on traps and cannot contribute to current. This can be modeled by an effective mobility in which the contributions to current of free and trapped charge are distinguished. The mobility is defined as the proportionality between current and the product of field and *total* charge, $J = \mu_{\text{eff}}\rho E$. This can be written as

$$\mu_{\text{eff}} = \frac{p\mu_p + p_T\mu_T}{p + p_T} \tag{1.3}$$

where p is the density of free charges (holes) in the valence band and p_T is the density of trapped charges, μ_p is the intrinsic mobility of free holes in the valence band and μ_T is the mobility of trapped charges. The designation 'trapped' immediately tells us that $\mu_T = 0$. In this way,

$$\mu_{\text{eff}} = \frac{p}{p + p_T}\mu_p \tag{1.4}$$

or, in other words, the intrinsic mobility μ_p is multiplied by the fraction of total charge that is in mobile states. Moreover, since this fraction can be strongly dependent on factors such as temperature or bias, as will be shown in this book, so will be the effective mobility. In conclusion, saying that organic materials have low mobility is inaccurate. One should rather say 'organic materials are trap-ridden'. Moreover, the unfounded assumption of knowledge of the density of charge is often the reason why different measurement methods result in different values for mobility. As examples are the time-of-flight (ToF) and field-effect transistor (FET) mobilities. These two types of measurement normally give widely different results. These measurements will be discussed in this book. As known from elementary physics, not all current is caused by drift. In cases where a gradient of density exists of any quantity, diffusion will try to equalize the distribution and remove the gradient. If these diffusing entities are charged, this results in a current. For holes,

$$J_p = -qD_p \nabla p \tag{1.5}$$

where D_p is the diffusion coefficient of holes. An important law in physics is Einstein's Relation, that couples the diffusion coefficient to the mobility,

$$D_p = \frac{kT}{q}\mu_p \tag{1.6}$$

where the proportionality factor is called the thermal voltage, known from electronics, and is equal to 26 mV at room temperature. The total hole current in the one-dimensional case, with current along the direction x, thus becomes

$$\begin{aligned} J_p &= -qD_p\frac{\partial p}{\partial x} + q\mu_p E p \\ &= q\mu_p\left(pE - \frac{kT}{q}\frac{\partial p}{\partial x}\right) \\ &= \mu_p p\frac{\partial E_F}{\partial x} \end{aligned} \tag{1.7}$$

where the last step is easily demonstrated if we assume that the carrier concentration depends exponentially on the Fermi level, $p = \exp[(E_V - E_F)/kT]$ and $\partial E_V/\partial x = q\partial V/\partial x = -qE$. The interesting fact is that for many disordered materials, the Einstein Relation is violated and the relation between diffusion coefficient and mobility can be much larger than 26 mV, even way beyond the small correction factor of $1 + 0.35p/N_V$ as given by Sze for classical semiconductors [9]. As will be shown, this can have tremendous impact on the behavior of the devices.

1.3 CHEMISTRY AND THE ENERGY DIAGRAM

Where the basic ingredient of organic materials is the carbon atom, the basic ingredient of organic semiconductors is conjugation. Conjugation is a chain of carbon atoms with alternating single and double bonds, as shown in Figure 1.3. This has two important results, namely the

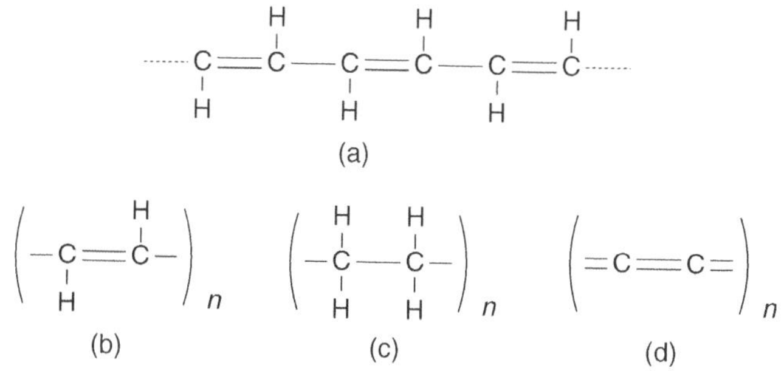

Figure 1.3 (a) The principle of conjugation in organic electronic materials shown in Lewis structures. Conjugation consists of a chain of carbon atoms with alternating single and double bonds. Note that any of the hydrogen atoms can be replaced by a functional side group, for instance to make the compound soluble. (b) The simplest organic polymeric semiconductor, polyacetylene. (c) The nonconjugated equivalent, polyethylene. The difference in room-temperature conductivity between polyethylene and polyacetylene is about seven orders of magnitude. (d) Polyallene with only double bonds also has relatively low conductivity

opening of a band gap, a splitting of the energy levels, in the range of semiconductors, and the delocalization of charge in these levels. Figure 1.4 shows examples of organic semiconductor materials in which the chain can easily be recognized.

The four electrons on each carbon atom in the chain can be considered to reside in sp^2 hybridized orbitals (3) and in a p_z orbital (1). The three sp^2 electrons are used to form covalent bonds via σ molecular orbitals to neighboring carbon atoms in the chain on either side and to the sidegroup (for instance a simple hydrogen atom). The remaining electron in the p_z orbital is then used in a covalent bond via a π molecular orbital with a neighboring carbon atom in the chain on one side only. The result is a chain of alternating single (σ only) and double (σ and π) bonds. Figure 1.5 shows the energy diagram of an interaction between two carbon atoms forming a double bond. After filling the levels from low to high (using Hund's Rules for the spins [63]) it can be recognized that four electrons (two from each carbon atom) are used in bonding, two in σ_- molecular orbitals, and two in π molecular orbitals. The remaining four electrons are in nonbonding (nb) orbitals and are still available for bonding to the rest of the chain and the ligands. The basic feature is the splitting between the π and π^* molecular orbitals caused by the interaction between the p_z atomic orbitals. Interactions between p_z orbitals further away in the chain cause additional, smaller splittings of the levels, as schematically indicated in Figure 1.5(b). A highest occupied

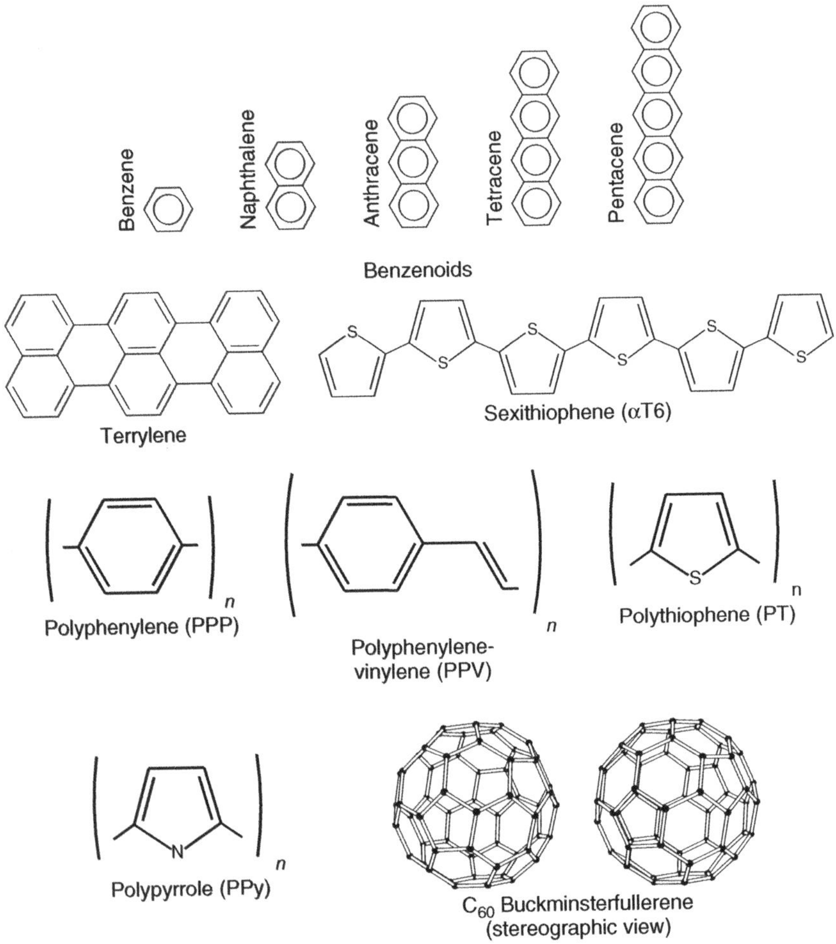

Figure 1.4 Examples of organic semiconducting materials, divided into oligomers and polymers. In each one a chain of alternating single and double bonds can be recognized as in Figure 1.3. The benzenoids (for example, tetracene) are pure carbon-hydrates with fused hexagon structures, such fused rings giving rigidity to the molecules. For phenylenes, the rings are separated, in PPP by a single bond and in PPV by a short conjugated link of two carbon atoms. Both are shown here in the para form (with the links in opposing sites of the benzene ring). The basic structure in thiophenes is a pentagon ring with in each ring one carbon atom replaced by a sulfur atom. Likewise, in pyrroles a carbon atom is replaced by a nitrogen atom. More complex structures include terrylene and buckminsterfullerene (shown in stereographic view), the latter being a purely carbon molecule in which each atom is bound to three neighbors, i.e. has one double bond. For all but buckminsterfullerene, the convention of organic chemistry is used whereby carbon atoms are not shown and neither are hydrogen atoms nor bonds with hydrogen atoms

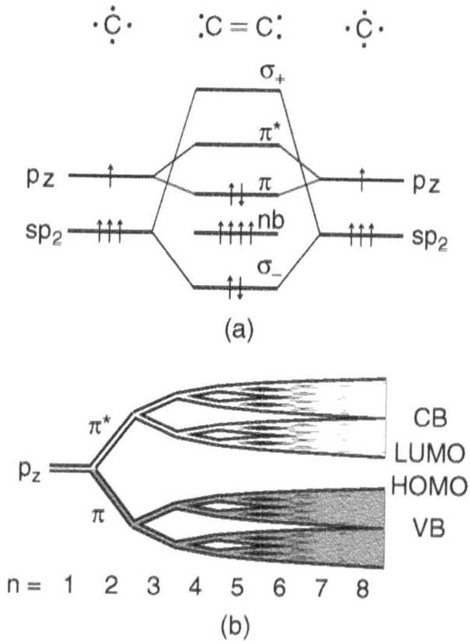

Figure 1.5 (a) Energy diagram of two interacting carbon atoms. sp_2 and p_z atomic orbitals of the two individual carbon atoms combine to form π, σ and nonbonding molecular orbitals. (b) A band structure starts emerging with a narrowing band gap when the conjugation length of alternating single and double bonds is increased. A HOMO and LUMO can be recognized that is part of the valence band (VB) and conduction band (CB), respectively

molecular orbit (HOMO, π) and lowest unoccupied molecular orbital (LUMO, π*) can be recognized.

The formation of a chain of single and double bonds thus causes an energy structure with a HOMO and a LUMO level, with a splitting ('band gap') in the range of semiconductors. From Figure 1.5 it is also clear that a material with only single σ bonds, like polyethylene, will have a much wider band gap and will not easily fall in to the category of semiconductors. The same goes for polyallenes with only double bonds (Figure 1.3d). To result in a semiconductor material, conjugation is essential.

For polymeric organic electronic materials, the size of the chain of alternating single and double bonds is, in principle, infinite. For oligomeric organic materials, the chain is limited to the molecule. Figure 1.6 shows for the specific case of benzenoids how the increasing chain length causes a narrowing of the band gap E_g [64]. For polymer chains, a similar effect of the HOMO–LUMO splitting as a function of chain length has been calculated [65]. In any case, often the term 'conjugation length' is used to designate the length, or average length,

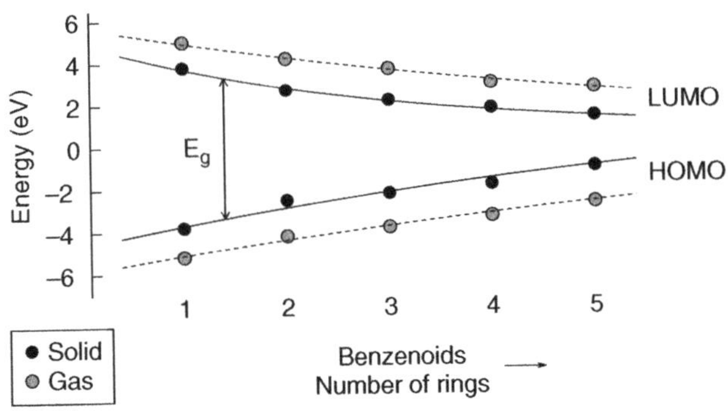

Figure 1.6 Band gap E_g as a function of molecular size in benzenoids. For increasing conjugation length, from benzene to pentacene, the gap narrows. This effect is caused by increased dislocation of the charges and is further amplified when the molecules are close to each other in solids. After Karl [64]

of uninterrupted chain. Figure 1.6 also shows how the band gap is further narrowed when the molecules are placed in the solid state form caused by van der Waals interactions between molecular orbitals on near molecules.

For many molecules, the place of the double and single bonds is not uniquely determined, but has two distinct configurations with exactly the same energy. In these cases, resonance of the two forms causes a dislocation of the electrons; it is no longer possible to identify the place of the single and double bonds, see Figure 1.7; the electron wavefunction becomes one continuous cloud. For other materials, for instance the linear chains with an odd number of atoms, the different forms have slightly different energies, but the difference is small enough to allow for this dislocation effect. In any case, this dislocation effect is essential for conduction. A charge injected into a π molecular orbit can relatively easily migrate along the chain, as demonstrated by Figure 1.8. Often the description 'electron-relay' is used to visualize the current. This seems quite adequate as the passing of charge is similar to a long line of people passing each other buckets of water in order to put out a fire.

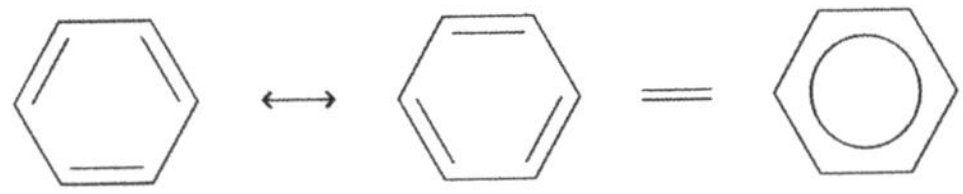

Figure 1.7 Resonance between the two equal-energy configurations of single and double bonds causes dislocation of the charges. In aromatic compounds this is shown by a ring in place of indicating the location of the double bonds

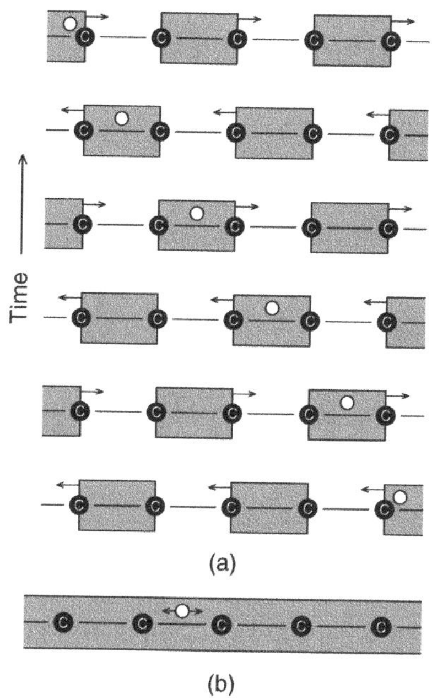

Figure 1.8 (a) Schematic time sequence showing how resonance between two configurations of alternating single (indicated by lines) and double bonds (gray rectangles) gives a way for a charge (indicated by a white circle) injected into a π molecular orbit to travel along the chain. The π molecular orbits are represented by gray planes. The arrows indicate how the positions of double bonds change in a transition from one state to the other. On account of the high rate of transitions between the two resonant states, it is better to think of the system as a single, continuous, highly delocalized wavefunction as in (b)

For limited conjugation lengths, macroscopic conduction is then made possible by further overlap of molecular orbits with neighboring molecules, causing so-called Davydov splitting of the molecular levels. Figure 1.9 summarizes the intramolecular interactions causing molecular orbits and inter-molecular interactions further dispersing the energy levels, and delocalizing the electrons. The interactions cause the wavefunctions with maximum same-sign overlap to have lowest energy and those with proximity of opposite-sign parts to have highest energy. Half of the electronic states are occupied (each p_z orbital contributes one electron and in each electron wavefunction fit two electrons, with opposing spin directions) and in this way a HOMO and LUMO can be identified. When the splitting between these two levels is relatively small, *and* a large delocalization of the electrons occurs, a semiconductor results. In this case it is better to use the jargon of semiconductor physics

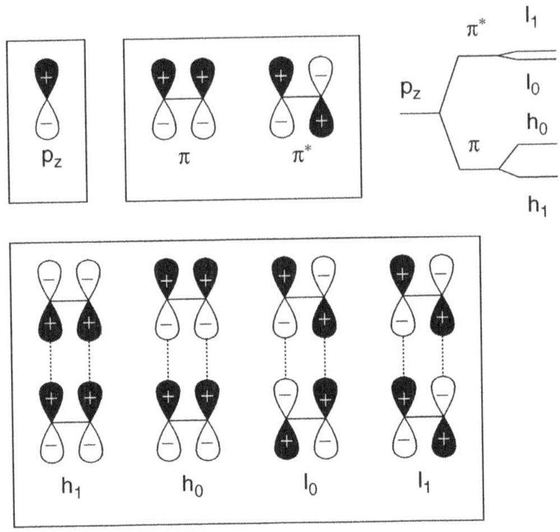

Figure 1.9 Intramolecular interactions between p_z electrons causing molecular orbits π and π^*. Additional (intermolecular) interactions cause a further spreading of the electronic levels. The interactions make the wavefunctions with maximum same-sign overlap to have the lowest energy, in this case h_1 and the wavefunction with maximum opposite-sign overlap to have the highest energy, in this case l_1. Half of the levels are occupied and this causes the distinction between HOMO and LUMO (in this case h_0 and l_0, respectively)

and replace the chemistry words 'HOMO' and 'LUMO' by 'valence band' and 'conduction band', respectively. In this language, we can say that an electron is promoted from the top of the valence band (VB) to the bottom of the conduction band (CB). Both the electron in the CB (LUMO) and the missing electron, or 'hole', in the VB (HOMO) can nearly equally well contribute to current. In practice, most organic semiconductors conduct better via holes because of a higher efficiency of trapping (immobilizing) electrons.

To summarize, the overlap of atomic and molecular wave functions causes a dislocation of electrons and the formation of valence and conduction bands. For materials on the insulating side of the spectrum, the overlap is minimal and the molecular orbitals (LUMO/HOMO) are good designations, whereas for the semiconducting materials, the overlap is strong and it is better to talk about 'conduction band' and 'valence band'. The time a charge spent on an individual molecule is minimal and the accompanying level broadening ΔE is large, further blurring the molecular levels. In these cases, electronic band structures can be calculated, see for instance the calculations on crystalline pentacene [66]. (Note that materials do not have to be crystalline to have semiconductor

Table 1.2 Comparison of room temperature conductivity for some electronic materials

Material	Conductivity ($S\ cm^{-1}$)
Copper	10^{6}
Silicon	10^{-5}
Polyacetylene	10^{-8}
Glass	10^{-12}
Polyethylene	10^{-15}
Diamond	10^{-42}

properties.) From the opposite point of view, materials can be studied on the basis of LUMO/HOMO levels. See for instance the determination of the molecular levels in tetracene by scanning tunneling microscopy [67].

Table 1.2 compares the conductivity of some pure materials.

As mentioned already above, since this book is about semiconducting organic materials, it is most logical to use the semiconductor physics point of view. Roughly speaking, it can be said that materials for optical devices (light emitting diodes, LEDs, etc.) are more insulator-like, whereas materials for switching devices (FETs, etc.) are more semiconductor-like. Moreover, experience has shown that for LEDs, the polymers, for example PPV, are more suitable, whereas for FETs, the oligomeric materials, for example T6, are better [68–71]. This finds its basis on the requirement for transistors of high mobility and thus a need for regularly stacked units, more easily obtained with small molecules. Purity is not a big issue for these devices since it does not limit the mobility. However, for photovoltaics, purity is essential since impurities create nonradiative recombination centers.

Here semiconductor physics is used as a starting point for all materials, but every now and then the insulator theory is mentioned, for example the Fowler–Nordheim tunneling current for insulating layers.

1.3.1 Energy Diagram of Crystalline Materials

One pillar of semiconductor physics is band theory. In the condensed version used to describe the behavior of electronic devices, the *entire* energy diagram of crystalline material can be considered to effectively consist of N_V states at a discrete energy E_V, named the valence band and N_C states at E_C, the conduction band, even while knowing that the real band diagram is much more complicated. (In reality the density of states is closer to a square-root function of energy [72].) The basic underlying idea is that only the states close to the band edges are significant for the electrical behavior. The difference between this effective conduction band and valence band is then called the band gap, $E_g = E_C - E_V$. At low temperatures, the valence band is completely full, occupied with N_V electrons, whereas the conduction band is completely empty. Although the valence band has many electrons, none of them can contribute in this situation to current. This is because for every electron with velocity $+v$ there always exists an electron occupying a state with an exact opposite velocity $-v$, even when an electrical field is present. For low temperatures, the average velocity and thus conductivity of the material is therefore zero and this distinguishes semiconductors from metals. Conductivity in semiconductors can be achieved by promoting electrons from the valence band to the conduction band or by directly injecting electrons in the conduction band or removing them from the valence band. In either case, this allows for a mismatch of electrons with opposite speeds and conduction can occur. Moreover, it is customary to name missing electrons in the valence band 'holes'. In this way, either electrons in the conduction band or holes in the valence band can contribute to current.

There are basically four ways of creating electrons in the conduction band or holes in the valence band, see Figure 1.10:

1. **Chemically:** The idea is to deliberately contaminate the material with chemically foreign dopants which either easily donate or accept electrons from the host material (in chemistry jargon these

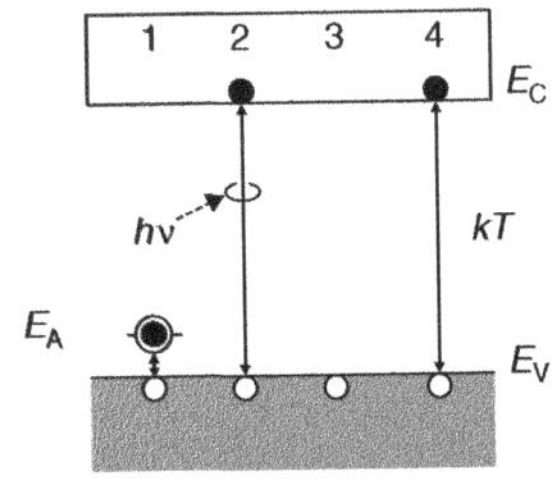

Figure 1.10 Four ways of creating free charge carriers (holes ○ or electrons •) contributing to external current: (1) chemically, by adding acceptors to the material that can capture electrons from the valence band, thus creating free holes ○ and ionized acceptors, N_A^-; (2) optically, the energy of an absorbed photon can be used for the creation of electron–hole pairs; (3) electrically, by changing the bias it can become energetically favorable to inject free holes; and (4) thermally, very similar to optically, but the energy comes from the phonons of the crystal

agents are called reducers and oxidizers whereas in solid-state physics terms they are called donors and acceptors, respectively). Doping in organic materials is not an important issue. This is because for the typical organic devices, LEDs and TFTs, doping is not an important parameter. For both these types of devices, purer materials work better. This is in comparison with some devices of classical semiconductors, such as pn-junction diodes and bipolar transistors, where doping is essential and device-dominating. Doping is not well studied. Often iodine is used when doping is needed; see the work of Minakata [73]. Also, oxygen (and water or air) seem to introduce electrically active levels either in the form of traps or doping [74, 75]. These effects of doping seem to be reversible [76].

2. **Optically:** Electron–hole pairs can be created by the absorption of photons. This is the area of photovoltaics and is not the main subject of this book. However, it can be said that the maximum absorption of light occurs for photon energies close to the band gap. This makes organic materials especially sensitive in the range of visible light. Likewise, recombination of electrons and holes causes the emission of photons and for organic materials blue LEDs are feasible because of their large band gap.
3. **Electrically:** Electrical bias can modify the energetic band diagram in such a way as to allow for the presence of free carriers in the bands. The FET may be used as an example. Even intrinsic materials can be induced into conduction by the presence of an electric field. Likewise, with enough voltage, even the most insulating materials can be made to conduct current. For LEDs, this is the preferred method of carrier generation. Doping of the materials can result in nonradiative recombination paths that can kill the luminescence.
4. **Thermally:** This way of creating free charges in the bands is very similar to the optical method. With high enough temperature, the lattice vibrations can thermally excite electrons from the valence band to the conduction band.

Of these effects, the optical and thermal ones create an equal amount of free electrons and holes. The chemical and electrical ways can create a mismatch of carriers. The chemical way creates uniquely one type of carrier, either electrons or holes. In principle this also applies to electrical creation of carriers. The electrical way can create both carriers only at opposing electrodes in, for instance, LEDs. Moreover, the electrical creation of free carriers is the only way where charge neutrality is not maintained locally (throughout the entire device, charge neutrality is always conserved, of course).

To find the densities of free electrons and holes, the Fermi–Dirac distribution is an important function for semiconductor physics.

$$f(E) = \frac{1}{1 + \exp\left[(E - E_F)/kT\right]} \tag{1.8}$$

specifies the fraction of states at energy E to be occupied and ranges from 0 for $E \gg E_F$ to 1 for $E \ll E_F$. In this, the Fermi level E_F is a parameter to describe the distribution. Often the Boltzmann approximation of this function can be used,

$$f(E) \approx \exp\left[(E_F - E)/kT\right]. \tag{1.9}$$

This is valid when the levels under consideration are far above the Fermi level, $E \gg E_F$.

Applying this to the band diagram of an intrinsic material without impurities, and imposing charge neutrality where the density of electrons n is equal to the density of holes, p, it is easy to show that the density of free electrons (and holes) are strongly temperature dependent [9]

$$p = n = \sqrt{N_V N_C} \exp\left(-\frac{E_g}{2kT}\right) \tag{1.10}$$

and this defines the intrinsic electron density n_i. Whatever happens, the product of p and n will always be equal to n_i^2 in thermal equilibrium. Note also the factor 2 in the denominator in the exponent; the activation energy of density (and thus current, $I \propto n + p$) of intrinsic material is *half* the band gap.

When impurities are added to the energy diagram, they can be either donors, acceptors or traps. The difference between them is that donors can supply one or more electrons to the conduction band, acceptors can steal one or more electrons from the valence band (or donate holes to the valence band) and traps can capture free carriers from a band. As an example, a hole trap can be either neutral or positively charged, whereas an acceptor, that also communicates with the valence band, can be either neutral or negatively charged. This is a small difference that has immense impact on organic materials, materials that are notoriously trap ridden.

When acceptors are added to the band diagram, using the same technique of demanding charge neutrality, $p - n - N_A^- = 0$, it can be shown that there are three regimes in the temperature dependence. Whilst it is difficult to determine the free charge density analytically, it is easy to show it graphically, as presented by Sze after Shockley [9]. The idea is that the hole density p and the sum of electron density n and negatively charged acceptor density N_A^- are plotted as a function of Fermi level. The Fermi level and accompanying hole density can then graphically be found by the crossing point of these two functions. From this method, it can easily be determined that for normal conditions the hole density is equal to the acceptor density. Figure 1.11 shows an example of a silicon crystal with an acceptor density $N_A = 10^{16}$ cm^{-3}. For normal operating temperatures, the hole density is equal to N_A. When the temperature is increased, the slopes of the plots, equal to $1/kT$, are decreased. This moves the crossing point horizontally, indicating a change of Fermi level, but a constant density of holes. It is thus said that the device is working in saturation, since the hole density is independent of temperature. When the temperature is further increased, there comes a point where the crossing point rises above $p = N_A$. In this

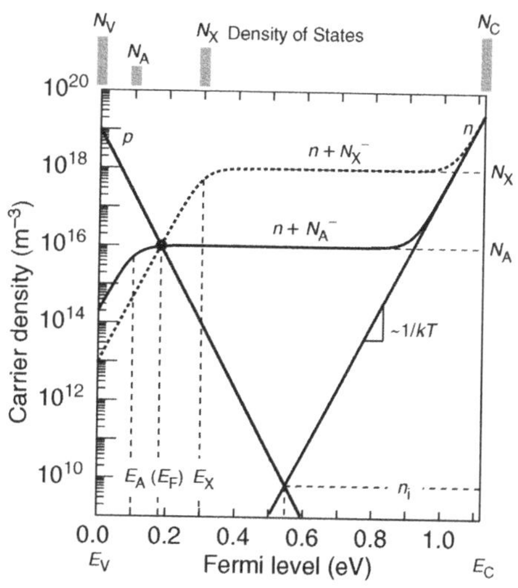

Figure 1.11 Graphical method of finding the Fermi level (E_F) and hole density p as described by Sze after Shockley [9]. The density of holes and the sum of electron and ionized acceptor densities are plotted. Imposing charge neutrality then gives the crossing point of the two curves. A typical example for an acceptor level depth of 100 meV and concentration $N_A = 10^{16}$ cm^{-3}. The plot is for a temperature of 300 K and around these temperatures, the hole density (and thus the conductivity) is independent of temperature. The dotted line shows the situation for an abundant deep acceptor X. This results in equal hole density and thus conductivity

case, the device works in the intrinsic regime. The transition temperature is easily calculated by setting the intrinsic density of Equation (1.10) equal to the acceptor density, namely

$$T_{\text{is}} = \frac{E_{\text{g}}}{2k \ln\left(\frac{\sqrt{N_{\text{C}}N_{\text{V}}}}{N_{\text{A}}}\right)}. \quad (1.11)$$

Above this temperature, the density is as in Equation (1.10). However, when the temperature is lowered, the slopes increase and the Fermi level drops below the acceptor level E_{A}. This implies that the electrons are released from the acceptors and return to the valence band, there recombining with the holes, or alternatively, holes return back to their source, the acceptors. This is called the freeze-out regime; the transition temperature is found with

$$T_{\text{fs}} = \frac{E_{\text{A}} - E_{\text{V}}}{2k \ln\left(\sqrt{\frac{N_{\text{V}}}{N_{\text{A}}}}\right)} \quad (1.12)$$

below which the hole density is equal to [72]

$$p = \sqrt{N_{\text{V}}N_{\text{A}}} \exp\left(-\frac{E_{\text{A}} - E_{\text{V}}}{2kT}\right). \quad (1.13)$$

Once again, the activation energy of density is half the energetic separation of the levels involved, in this case the acceptor depth. Analyzing the above, it can be verified that this is because the charges have opposite sign; a charge jumping from a (neutral) acceptor to the valence band contributes twice, once as free hole and once as ionized acceptor. This causes the square-root dependency implying a factor 1/2 in the exponent, just like the factor 1/2 was introduced in the equation for pure materials because of a charge contributing twice, once as free hole and once as free electron. For trap levels this factor 1/2 will not appear because the sign of the charge is equal in both states, be it on the trap or in the band; creation of one charge will annihilate the other. When traps start playing a role, the activation energy of current therefore is the trap depth, without the factor 1/2.

Figure 1.12 gives an example for p-type silicon with an acceptor density of 10^{15} cm^{-3} and depth of 20 meV. (To be able to show the activation energy in the intrinsic regime, the band gap was lowered to 200 meV.)

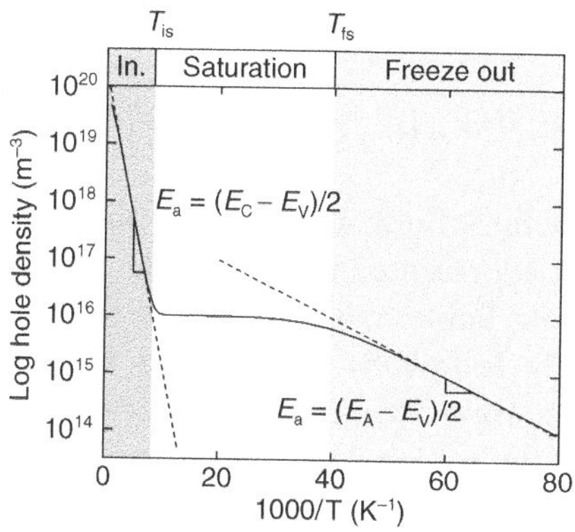

Figure 1.12 Temperature dependence of the hole density showing the three regimes of operation, namely 'intrinsic', 'saturation' and 'freeze-out'. Silicon and like materials normally operate in the saturation regime. Organic materials, due to their high density and depth of impurities, can work in the freeze-out regime

It has to be pointed out that, although the acceptor and donor levels can be much deeper in organic materials compared with silicon, due to the larger band gap, the question is if they really are. In effective-mass theory, the impurity levels form hydrogen-like levels, the deepest one, the S shell, having an ionization energy equal to [77]

$$E_H = \frac{\pi m^* q^4}{8h^2\varepsilon_r^2\varepsilon_0^2} = \frac{m^*}{\varepsilon_r^2} \times 13.6 \text{ eV} \tag{1.14}$$

where h is Planck's constant, ε_r and ε_0 are the relative dielectric constant and the permittivity of vacuum, respectively, and m^* is the effective mass of an electron in the conduction band for a donor level or the effective mass of a hole in the valence band for an acceptor level. For silicon, $\varepsilon_r = 11.9$ and m^* is $0.33m_e$, where m_e is the mass of an electron (in vacuum). This makes a donor or acceptor level of the order of 25 meV. A typical organic semiconductor has $\varepsilon_r = 5$, without further knowledge of the effective mass, a donor or acceptor level is expected to be about 140 meV deep, nearly an order of magnitude deeper than in silicon. This is also the order of magnitude where the semiconductor can be in the freeze-out regime at room temperature, as demonstrated by Equation (1.12).

While it is often difficult to calculate analytically, numerically finding the densities of holes or electrons is trivial. It implies finding the zero of the function of total charge as a function of the Fermi level

$$\rho(E_F) = q\left[p(E_F) - n(E_F) - N_A^-(E_F)\right]. \tag{1.15}$$

The preferred method for finding the zero of this function is the bisection algorithm [78] that is presented in Table 1.3.

This is a fast and reliable 'failsafe' way of finding E_F and the charge densities then easily follow by substituting the found Fermi level in the appropriate equation. Many scientific calculating programs, like Matlab, have zero-finding packages included, but even for the novice programmer the above algorithm is not difficult to implement. The numerical method further has the advantage that the full Fermi–Dirac function can be used instead of the Boltzmann approximation. Most figures and simulations in this work were made in this way.

1.3.2 Energy Diagram of Amorphous Materials

One of the advantages of organic electronics is that they are envisaged to be used for cheap electronics in areas where the cost is more important than the performance. For such applications, single crystals are prohibitively expensive and low-cost applications invariably use lower grade materials. In fact, even polycrystalline materials are too expensive and mostly amorphous materials are used. Also, in some cases this is the only option. When TFTs are made by depositing a certain material on top of a lattice-mismatching insulator layer, invariably the first layers are amorphous and it is exactly that layer that is important in the electrical behavior of TFTs, as will be shown. As a result, TFTs are amorphous devices.

Table 1.3 Finding the Fermi level by the bisection algorithm [78]

Step	Action
1	Choose limits x_1 and x_2 of range of possibilities for E_F
2	Calculate ρ at $E_F = x_1$ and $E_F = x_2$
3	Calculate middle value $x_c = (x_1 + x_2)/2$ and $\rho(x_c)$ at this value
4	If $\rho(x_c)$ and $\rho(x_1)$ of same sign: $x_2 = x_c$ else $x_1 = x_c$
5	If $\lvert x_2 - x_1\rvert > \delta x$ (resolution) then go to step 3
6	$E_F = (x_2 + x_1)/2$

Such materials are full of unterminated 'dangling' bonds. These bonds create deep electronic levels. Moreover, the surroundings of these dangling bonds are not constant throughout the device. The same applies to the conduction levels of conduction and valence bands. As a result, instead of the approximation of discrete levels at the edge of the conduction and valence bands, a more adequate description is one in which a distribution for conduction and band states exists, as described for instance by Bässler [45]. Just as the discrete levels for crystalline materials are a good approximation, so the distributions can be well approximated by two exponentially decaying functions, one for the conduction states ('valence' and 'conduction' bands) and one for the trap states, where the conductive states are also often called 'tail states'; see the work of Shur and Hack where this is applied to TFTs of amorphous silicon [79]. Interestingly, Tessler and Roichman have argued that an exponential distribution of states is not compatible with a Gaussian distribution of states [80]. If true, the exponential distributions seem more adequate since this picture is remarkably good at describing the features of organic electronic devices, as will be shown throughout this book.

As an example, for positive charge states we can use

$$
\begin{aligned}
N_V(E) &= N_{V0} \exp\left(-E/kT_1\right) \\
N_T(E) &= N_{T0} \exp\left(-E/kT_2\right)
\end{aligned}
\qquad (1.16)
$$

for the conductive band tail states and trap states, respectively. Each distribution has two parameters, namely the density at a reference energy (here at $E = 0$) and the slope of the distribution in a logarithmic plot, defining a 'temperature', for example, N_{V0} and T_1. Figure 1.13 schematically draws this energy diagram on a logarithmic scale. Between the tail states of the conduction and valence bands, a large density of trap states exist. In many cases, these trap states can outnumber the conduction states and in the extreme case all charge will eventually wind up trapped. However, any charge must start in a conductive state and we can expect strong transient behavior. Moreover, the sheer abundance of these trap states can cause a symmetry in trapping and detrapping times, compared with the situation in crystalline materials, where trapping is normally fast and detrapping slow (one of the assumptions in deep-level transient spectroscopy, DLTS). Long trapping times cause transient behavior in a long time domain. If we assume that the trapping time is a function of the trap depth, for instance $\tau_T \propto \exp(E_T/kT)$, then a complicated convolution of trapping times and trap densities causes an intricate transient behavior. As will be shown in Chapter 4, the

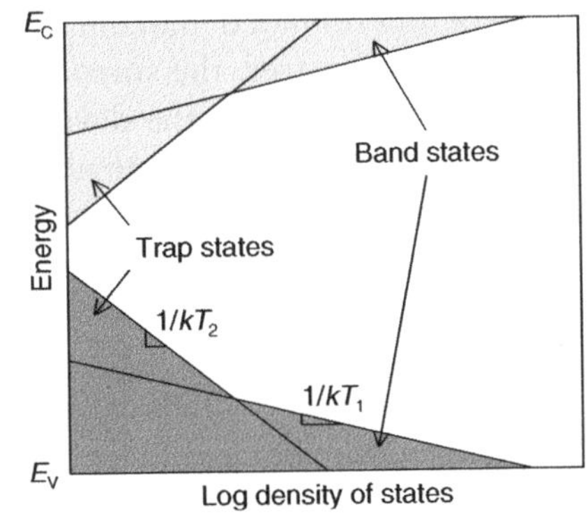

Figure 1.13 Energy diagram for amorphous material showing the conductive 'tail states' and nonconductive trap states, both exponentially distributed in energy. This can in fact readily describe all of the characteristics of organic electronic devices

distributions given here result in a power-law current [$J(t) \propto t^{-\alpha}$] when the traps are abundant enough to be able to capture all charges.

The effects of traps on the electrical characteristics of devices can be summarized as:

- Low conductivity, which is also possible to analyze as low mobility.
- Thermal activation of current.
- Long-lived transient effects. For exponentially distributed density of states (DOS), power-law transients, are expected in constant bias.
- In admittance, a nonconstant Mott–Schottky plot, otherwise used to profile the acceptor concentration.
- TFTs have nonlinear output and transfer curves and a thermal activation energy depending on bias (Meyer–Neldel Rule).
- Anomalous ToF transients.

In the following chapters, conduction processes and device structures will be discussed in a dual approach. On the one hand, measurement techniques and devices are presented as characterization tools. One can think of a Schottky barrier for the determination of acceptor levels in the material. On the other hand, the devices can also be treated as final products and the discussion is then focused on what processes are governing and limiting the performance of these devices.

Electronic devices can roughly be divided into two-terminal devices, such as pn-diodes and Schottky barriers and three-terminal devices, such as TFTs. These devices can then be studied by various measurement techniques that also fall into two categories, namely DC and AC.

1.4 DISORDERED MATERIALS AND THE MEYER–NELDEL RULE

One peculiar aspect of organic electronic materials that is often encountered is the fact that the activation energy of current is dependent on the bias conditions. This is a form of the more general Meyer–Neldel Rule (MNR) that was first reported in 1937 [81], and that can be summarized as follows:

1. The activation energy (the slope in a Arrhenius plot) of a process depends on a certain parameter P. For instance, the activation energy of the current through a transistor depends on the gate bias.
2. The curves in the Arrhenius plot pass through or converge to a common point, at the so-called isokinetic temperature T_{MN} (sometimes used to define a Meyer–Neldel energy $E_{MN} = kT_{MN}$). At this temperature, the process is therefore independent of the parameter.

Another way of presenting this is in the form of an equation, for instance for the process of conduction:

$$\sigma = \sigma_0 \exp\left[-\Delta E(P)\left(\frac{1}{kT} - \frac{1}{kT_{MN}}\right)\right]. \qquad (1.17)$$

Figure 1.14 gives a general example of the MNR. Note that measurements at the isokinetic temperature are not observed. This is represented in Figure 1.14 by the dashed lines; the isokinetic temperature can only be found by extrapolation. For normal MNR, this temperature is positive, but for the anti-MNR, the isokinetic temperature is negative and lies to the left of zero on an Arrhenius plot.

The MNR observation is truly cross-disciplinary and finds even examples outside the electronic domain. It is probably best known for the observation in diffusion processes (where the MNR is also known as 'the compensation effect') [82]. In the electronics world, the MNR is frequently observed in low-conductivity disordered materials. Examples are porous and amorphous silicon [83–85], microcrystalline silicon films [86], polysilicon [87], ionic conductivity [88, 89], glassy materials [90] and organic materials [91–95]. The common factor to all these materials is the existence of a large density of localized states with high activation energy (traps). When barriers exist that are distributed in energy, the MNR can also emerge, as shown for example by Dyre [96].

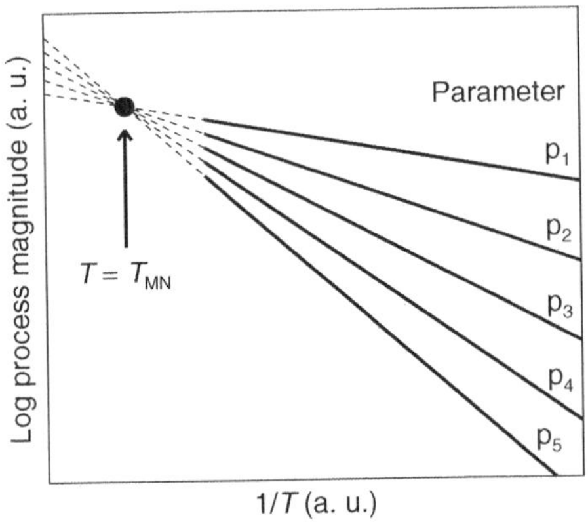

Figure 1.14 Schematic representation of the general form of the Meyer–Neldel Rule (MNR) in Arrhenius plots. The activation energy of a process (for instance conduction) depends on a parameter (for instance bias) in such a way that the Arrhenius plots all pass through or point at a common point. The temperature at this point is called the isokinetic temperature (T_{MN})

In general, when multiple processes exist with different activation energies and depending differently on a parameter, then the MNR results. A good example is the dual-process transport mechanism [97] or a simple silicon p-n diode that has different current regimes [98].

It seems beyond doubt that somehow the disordered character of the materials is an origin for the MNR, even though the MNR is just a phenomenological observation. Many people have established the link between the behavior of the devices in various measurement techniques. For example, the correlation between the MNR and the transient behavior has been reported [99]. In Chapter 6 it is shown that, when an abundant trap state exists that is distributed exponentially in energy, the MNR naturally emerges for the mobility of carriers that have an activation energy depending on the gate bias. Since, as will be shown, traps are also responsible for the anomalous transient behavior, the link between the MNR and the transients is, in fact, the existence of a highly abundant trap state.

1.5 DEVICES

Electronic devices come in numerous types, each with its own application. They are too numerous to present them all here. The focus here is on devices mostly for materials science and device characterization and then mainly on amorphous materials, such as most organics. The range of devices is therefore limited to three types: (1) resistors; (2) Schottky barriers and MIS diodes; and (3) TFTs. Note that, for instance, bipolar diodes, or junction transistors are absent. This is mostly because their implementation with organic electronics are very sparse, if not nonexistent. Note also that the metal-oxide-semiconductor (MOS) FET is absent here (apart from a brief introduction in Chapter 6, for comparison with the TFT). The technology for organics is not at a level to easily implement MOS-FETs. Moreover, a TFT is a much better testbed for electronic characterization of materials. Other interesting devices, such as the metal-base transistor [100, 101], vertical-stack transistor [102], a space-charge-limited transistor or static induction transistor similar to the vacuum-tube triode [103, 104] or magnetically active devices [105], are also left untreated.

1.5.1 Resistor

The most elementary of all devices is the resistor. It consists of a bar of homogeneous material with two contacts at the extremes in one direction. Injection of carriers is assumed to be unhindered and the current only limited by the resistivity of the material. Moreover, the amount of surplus charge in the device is considered nil for any bias, so that the capacitance, defined as the derivative of the charge–voltage function, is equal to zero. Therefore, the only electronic device parameter of a resistor is its resistance R, which is defined by Ohm's Law

$$R \equiv \frac{V}{I}. \tag{1.18}$$

The reciprocal of this resistance is called conductance G and is a more adequate entity in electrical characterization. Both device parameters resistance and conductance can be expressed in their material parameters, resistivity ρ and conductivity σ, according to

$$R = \rho\frac{d}{A} \tag{1.19}$$

$$G = \sigma\frac{A}{d} \tag{1.20}$$

where A is the cross-section area and d the length of the device, respectively, see Figure 1.15.

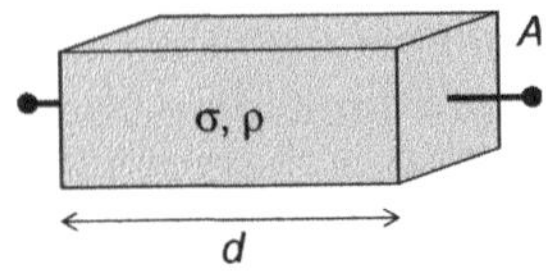

Figure 1.15 Simple resistor. A bar of material with length d and cross section area A

Because current is proportional to charge density, field and mobility, low-mobility materials need large fields to have any substantial currents. Moreover, often large amounts of surplus charge can exist inside the material. Devices thus easily enter into the so-called space-charge-limited-current (SCLC) regime, where the device is no longer homogeneous, with charge density and field a function of space. Devices then acquire a capacitance and a conductance that is bias dependent. This will be explained in Chapters 2 and 3, treating the DC and AC properties of two-terminal devices, respectively.

Resistors are purely symmetric devices and for this reason they cannot show any rectification (when the polarity of the bias is reversed, the currents are of the same magnitude). Any symmetric two-terminal device can then be called a resistor.

Because of its homogeneity and absence of contact effects, the resistor is the preferred element when determining the mobility in a ToF experiment, as described in Chapter 5. For the same reason, this structure is also adequate for describing LEDs. In this case two different metals are used for the opposite electrodes to ensure ohmic contacts for one type of carrier at each end. This makes these devices asymmetric and rectifying. However, the analysis in the forward direction can be done with the resistor structure.

1.5.1.1 Ohmic Contacts

One of the main problems of resistor devices is how to avoid turning them into Schottky diodes. When a metal is used for the contacts, these tend to form rectifying contacts that hinder charge injection. There are basically two ways of avoiding contact effects for the resistor devices:

- work function matching;
- impurity diffusion.

Impurity diffusion means causing a very high density of defect levels close to the contacts. Although this does not remove the (Schottky) barrier, the extent of the depletion zone can be so much reduced that charges can easily overcome this zone by tunneling directly to the appropriate band. Moreover, the defect levels themselves can act as jumping points for the charges to travel easily to and from the bands. In fact, with a huge defect density, these defects can form their own bands through which conduction can take place. This technique is quite common in the semiconductor industry to connect the metallic tracks to the active components with minimum contact resistivity.

Work-function matching is often used in organic semiconductor devices. The idea is to select a metal for the contacts that does not form Schottky barriers for the charges of interest. This implies that the Fermi level in the metal is aligned with the band of interest. Since the Fermi level is linked to the vacuum level in the metal by means of the work function ϕ_m (and the conduction band in the semiconductor by the electron affinity χ, and in turn the valence band is linked to the conduction band by the energy gap E_g), it means selecting a metal that has the appropriate work function. Finding a good hole-injecting metal is often not a problem. The sum of electron affinity and band gap of the organic semiconductors lies in the range of work functions of most noble metals. Gold is a common choice for ohmic hole-injecting contacts. If transparent electrodes are needed, normally ITO (indium–tin oxide) is used. However, finding a good metal for electron injection is more difficult, since for that a low-workfunction metal is needed. These metals then suffer from the side effect of being highly reactive. The use of calcium for electron-injecting contacts may serve as an example [106]. For this reason, these effects were one of the first to be studied in OLEDs [107]. Figure 1.16 shows an example of an LED where both electrons and holes are easily injected into the device at opposite ends.

1.5.2 Schottky Diode

Another two-terminal device is the Schottky diode. This consists of a Schottky barrier and, in series, a purely resistive layer called the 'bulk'. The result is a rectifying element with current in one polarity ('forward') of the bias much larger than in the other polarity ('reverse'). Because of this asymmetry, the electrodes can be distinguished and are called 'anode' and 'cathode', where the anode is the one that has positive bias in forward polarity. This is for historical reasons.

A Schottky barrier is created from the intimate contact of a metal and a semiconductor. As in the rest of this chapter, the semiconductor is considered to be of p-type, but the ideas are equally well applicable to

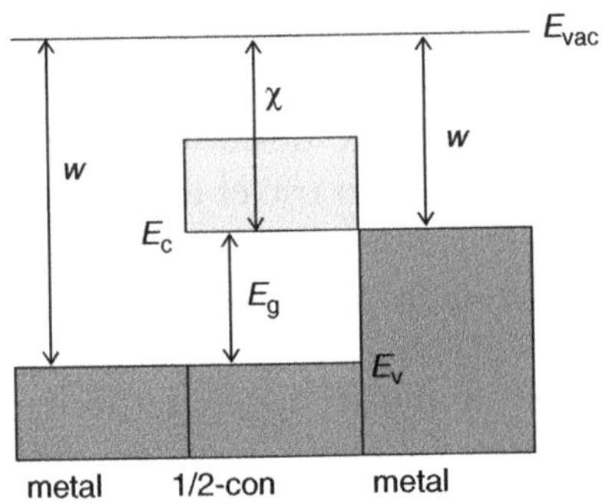

Figure 1.16 Diagram showing how an appropriate choice of work function for the metallic contacts can avoid barriers. The left electrode metal has a high work function and this facilitates hole injection, whereas the right electrode has a low work function which is good for electron injection

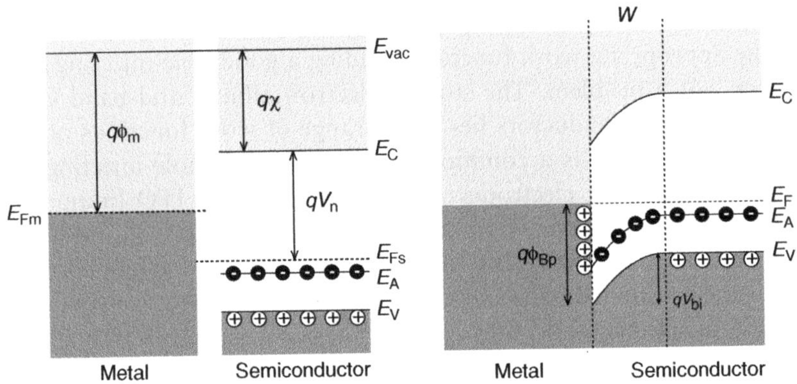

Figure 1.17 Energy diagram of a metal–semiconductor structure (Schottky diode) before and after contact

n-type materials, with the correct changes of sign, etc. Figure 1.17 shows the space–energy diagram of a metal–semiconductor structure. Because of a difference in Fermi level in the materials, E_{Fm} in the metal and E_{Fs} in the semiconductor, the system can gain energy by transferring charge. In the case of Figure 1.17 it implies electrons jumping from the metal into the semiconductor and consequently a build-up of negative charge in the semiconductor stored as immobile charge of ionized acceptors N_A^-. Alternatively it can be seen as free holes sucked out of the semiconductor. The parameters that describe the barrier are:

Barrier height, ϕ_{Bp}: This is the barrier as seen by the (majority) carriers, coming from the metal. This barrier is independent of bias. It is the difference between the top of the electron seas in the metal and the semiconductor before contact, $E_{Fm} - E_V$. Using the material parameters, this can be expressed as

$$q\phi_{Bp} = (q\chi + E_g) - q\phi_m \tag{1.21}$$

where ϕ_m is the workfunction of the metal, χ is the electron affinity of the semiconductor and E_g is its electronic band gap.
Built-in voltage, V_{bi}: This is the barrier as seen from the side of the semiconductor and is the band bending the holes have to overcome to reach the metal. It is the difference in Fermi level before contact, thus

$$V_{bi} = \chi + V_n - \phi_m \tag{1.22}$$

where qV_n is the Fermi level depth relative to the conduction band.
Depletion width, W: This is the extent of the zone in which the ionized acceptors are uncompensated by holes. This zone is therefore depleted of free carriers. In a simplified approach, the end of the depletion zone can be considered abrupt and it is possible to calculate the extent of this zone.

Figure 1.18 shows the rectangular distribution of charge and the resulting electrical field and bending. The local potential can be calculated using Maxwell's equations. The most important for our calculations is Gauss's Law, which states that the divergence of displacement ($D = \varepsilon E$) is equal to the charge density

$$\nabla \cdot D = \rho \tag{1.23}$$

where ρ is the charge density. For the one-dimensional case of the Schottky diode, this reduces to Poisson's equation; considering that $E(x) = -\mathrm{d}V(x)/\mathrm{d}x$ this becomes

$$\frac{\mathrm{d}^2 V(x)}{\mathrm{d}x^2} = -\frac{\rho(x)}{\varepsilon} \tag{1.24}$$

With boundary conditions $V(\infty) = 0$, $E(\infty) = 0$ the solution, for a rectangular distribution $\rho(x) = -qN_A$ for $0 < x < W$ (compensated by

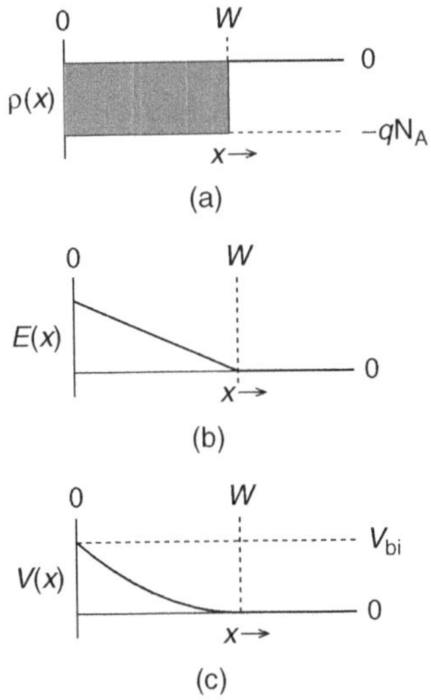

Figure 1.18 Distribution of (a) charge, (b) electric field and (c) potential on the semiconductor side of a Schottky barrier. Not shown is a thin layer of positive charge in the metal that maintains overall charge neutrality; this is important when solving Poisson's equation and deriving the equations given here

an infinitesimal layer of positve charge in the metal at $x = 0$, maintaining overall charge neutrality) and zero elsewhere, is

$$E(x) = \frac{qN_A}{\varepsilon_s}(W - x) \tag{1.25}$$

$$V(x) = \frac{qN_A}{2\varepsilon_s}(W - x)^2 \tag{1.26}$$

where ε is the permittivity of the semiconductor. In the absence of external bias, the potential at $x = 0$ (and in the metal) and total band bending is equal to the built-in voltage V_{bi}. When a bias V is applied, the band bending changes resulting in a total potential at the interface $V(0) = V_{bi} - V$, where the sign is used that a forward bias, that reduces the band bending and increases the current, has positive sign. Thus, the extent of the depletion width is depending on the bias,

$$W(V) = \sqrt{\frac{2\varepsilon_s(V_{bi} - V)}{qN_A}}. \tag{1.27}$$

Finally, the total space-charge in the semiconductor side of the device, the integral of the charge distribution in Figure 1.18, is the acceptor density multiplied by the depletion width

$$Q(V) = \sqrt{2q\varepsilon_s N_A (V_{bi} - V)}. \tag{1.28}$$

1.5.3 MIS Diode and MIS Tunnel Diode

A metal–insulator–semiconductor (MIS) diode is a Schottky diode with an insulator separating the metal and the semiconductor. This insulating layer effectively blocks any DC current; the electronic behavior is therefore only visible in AC techniques such as admittance spectroscopy as treated in Chapter 3.

The MIS diode can be seen as a precursor to the FET. The difference between them basically being that a transistor has two electrodes ('source' and 'drain') connected in-plane in the semiconductor. The potential at the metal side of the device then controls the charge and drain-source conductance in the semiconductor side of the insulator. As such, the MIS-diode is a valuable device for studying and characterizing the electrical behavior of (field-effect) transistors.

When the insulator layer is thin, with the device approaching the Schottky-diode structure, substantial leakage currents can exist that are comprised of charges tunneling through the insulator. The direct effect of these currents is that such an MIS tunnel diode, although close to being a Schottky diode, is no longer a single-carrier-type device (hole-current or electron-current only), but substantial minority-carrier currents can exist. These effects are best visible in the admittance data and for that reason the thin-insulator MIS diode with minority-carrier injection, including the DC behavior, is discussed in Chapter 3. These minority carriers are essential for the working of optoelectronic devices and such interfacial insulating layers are thus not necessarily detrimental to the device.

1.5.4 Thin-film Transistor

A TFT is basically a resistor. The important device dimensions are its length L and width W (the thickness is not an important parameter!). Nearby is placed a third electrode called the 'gate', resulting in a three-terminal device. Figure 1.19 shows an example of a TFT defining the parameters. Ideally, there is never current flowing from the gate to the resistor, and the only function of the gate is to modulate the charge density in the resistor. Once a substantial amount of free charge is induced in the resistor, it is said that a 'channel' is formed. With a

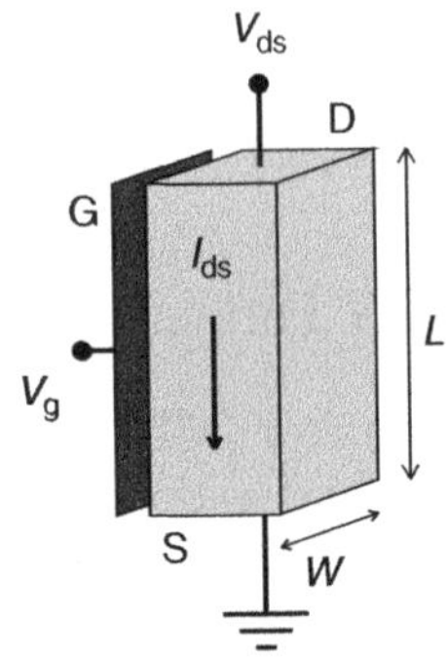

Figure 1.19 A TFT is a resistor with electrode drain (D) and source (S) and nearby a third electrode called the gate (G). The gate modulates the charge-carrier density in the resistor and thus its resistance

large density of free charge, the resistor becomes conductive and large external currents can be observed for relatively small biases applied.

Although the device is still fully symmetric, the electrodes of the resistor are often called 'drain' and 'source' to denote their functionality in an electronic circuit. The source (often connected to ground) is that electrode that is emitting the majority carriers. For instance, in a TFT with a p-type (hole) channel, the source is at a relative positive bias. Holes are emitted from this source, travel through the resistor, and sink into the drain on the other side.

The output current I_{ds}, is a function of both drain-source voltage (V_{ds}) and gate-source voltage (V_{gs}). Since the source is often connected to ground, the subscript 's' can be omitted. (In this work, V_{ds} and V_g are used.) There exists therefore two type of scans: Either the gate is kept constant and the drain is varied, resulting in so-called 'output curves', or simply 'I–V curves', or the drain is kept constant and the gate varied resulting in 'transfer curves'. The first parameter is the conductance, which is defined as the derivative of the output curve,

$$g \equiv \frac{\partial I_{ds}}{\partial V_{ds}} \tag{1.29}$$

which depends on the gate and drain-source bias, as will be shown in Chapter 6.

The transistor has, apart from resistance and conductance of the channel, an important parameter called the trans-conductance, which is defined as the ratio of changes of the output current to the changes of the gate bias,

$$g_m \equiv \frac{\partial I_{ds}}{\partial V_g}. \tag{1.30}$$

However, the single most important parameter of electronic materials is the mobility μ. As will be shown, the mobility is proportional to the trans-conductance. When determined in this way, it is called the field-effect mobility μ_{FET} to distinguish it from mobilities measured with other techniques.

The TFT is treated in Chapter 6. It is there argued that, although the behavior of the TFT is very similar to the better-known MOS-FET, it does deserve its own model. The basic difference is that a MOS-FET is a three-dimensional device, whereas the TFT should be treated as purely two-dimensional. See Figure 1.20 and Table 1.4 for a summary of measurement techniques presented in this book.

Device structure	Technique	Shallow level position	Shallow level density	Deep level position	Deep level density	Carrier mobility	Conduction model	Interface states	Difficulty cost
Resistor	I-V	+	+			+	++		1
	Hall					++			8
	ToF					++	+		8
Schottky	I-V	+	+				+		1
	C-V		+++					+	3
	Admitance Spectroscopy	+	+	+	+	+		+++	4
	TSC			++	+++			+	5
	DLTS			+++	+			+	6
Transistor	I-V					+++	++	++	2
	Admitance Spectroscopy					++	++	+	4

Figure 1.20 Summary of devices and measurement techniques presented in this book and their usefulness in obtaining information (+ = 'can be done', ++ = 'good', +++ = 'excellent')

Table 1.4 Summary of measurement techniques and information obtained

Structure	Measurement	Principal information	Chapter
Resistor	I–V	Transport model	2
	I–V (SCLC)	Mobility	2
	I–T	Shallow level	2
	Hall	Mobility	2
	ToF	Mobility	5
Schottky barrier	I–V	Transport model	2
	Admittance spectroscopy	Shallow and deep levels, barrier height, ε	3
	DLTS	Deep levels	4
	TSC	Deep states	4
TFTs	I–V	Mobility, transport model	6
	Transfer	Mobility, transport model	6

1.6 OPTOELECTRONICS/PHOTOVOLTAICS

The most prominent application of organic materials is in the area of photovoltaics or optoelectronics, meaning the conversion of electricity to and from light. They are commonly known as LEDs, that produce light, and solar cells that convert solar radiation into electric power. While the link between the optical aspects and the electrical characteristics is important and deserves its own approach, these devices are treated in this book as normal devices and only the electrical behavior is described. Other books exist that focus on the optical or optoelectrical properties. Yet, it is relevant to summarize the important aspects of photovoltaics.

A solar cell is an asymmetric device that converts solar radiation to electrical power. The energy of an incoming photon is absorbed by an electron which is thus promoted from the valence band to the conduction band, leaving behind a hole in the valence band, see Figure 1.21. Because of the asymmetry of the device and the associated internal electrical fields, the electron–hole pair is dissociated and the electron and hole drift towards opposite electrodes. Here they are collected and contribute to an external current.

Contrasting, an LED is a device that converts electrical energy into light and this is the oldest semiconductor device; already more than a century has passed since Round noticed light emission from a semiconductor diode, although the credit of inventing the LED probably should go to Losev [108]. LEDs based on organic materials (OLEDs) were first studied in the 1990s [109, 110], with as research aim the development of large-area displays. See the review article of Friend *et al.* for a summary

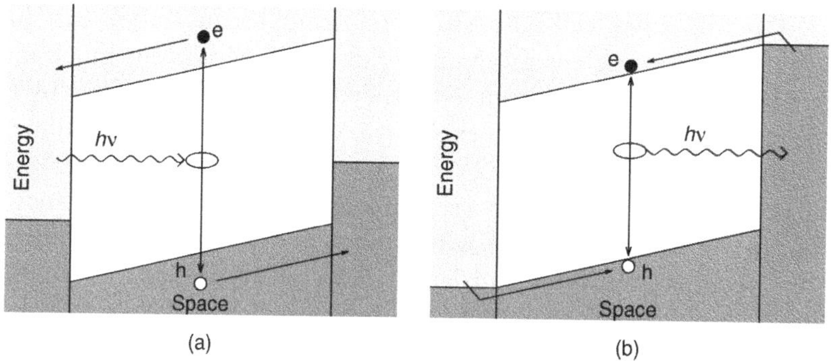

Figure 1.21 The basic working of a solar cell (a) and LED (b). Solar cell: A photon with sufficient energy ($h\nu$) is absorbed and an electron–hole pair is created. The internal electrical fields dissociate the electron–hole pair and the electron and hole each drift to their own electrode. At the electrode they are collected and can contribute to an external current. LED: Electrons and holes are injected into the conduction and valence band, respectively. Where they meet they can recombine, liberating energy in the form of a photon

of the state of polymer OLED research at the end of the 20th century [111], or of Forrest [112], or the work of Blom *et al.*, for the device physics of OLEDs [113].

In an LED an electron and a hole can recombine and the liberated energy emitted in the form of a photon. In this case the device does not need to be asymmetric. However, to control the injection of carriers into the device, and to create a well needed carrier balance, the work function of the electrode materials plays an essential role. Therefore, practical devices are often asymmetric, with one electrode adequate for hole injection into the valence band and the counter electrode having a work function adequate for electron injection into the conduction band. To optimize the light production, a balance between the two types of carriers is needed. In some cases, in order to reach this balance, (partially) blocking one type of carrier may even increase the luminescence efficiency of the device [114]. Therefore, the choice of electrodes is essential. In fact, the same applies to solar cells when it comes to harvesting the carriers.

Of course, the most important aspect of photovoltaic devices is how efficient they convert energy from one domain to the other. As an example, the solar cell quality is described in terms of efficiency, which is the ratio of outgoing electrical power to the incoming solar radiation power. Commercial (silicon) solar cells can reach an efficiency of about 15%, where organic devices fall about one order of magnitude below that. Organic solar cells are still interesting because they can have unique properties such as mechanical flexibility. Moreover, much more interesting than the quantum efficiency is the cost-per-watt. In this respect, silicon is notoriously expensive, even in the polycrystalline or amorphous state.

Figure 1.22 shows three possible reasons for a lower efficiency of LEDs: (a) When the material is not pure, nonradiative recombination centers such as impurities and lattice defects can reduce the quantum yield. It is obvious that making the material better (purer and more crystalline) will increase the yield. (b) When the work function of the metal of one of the electrodes is not adequate for injection of that type of carrier, the result is that the current consists only of one type of carrier and no electron–hole recombination can take place. A large current without light emission results. (c) When the mobility of one type of carrier is much larger than the mobility of the other type, the electron–hole recombination takes place close to an electrode. In these cases the luminescence is quenched [115].

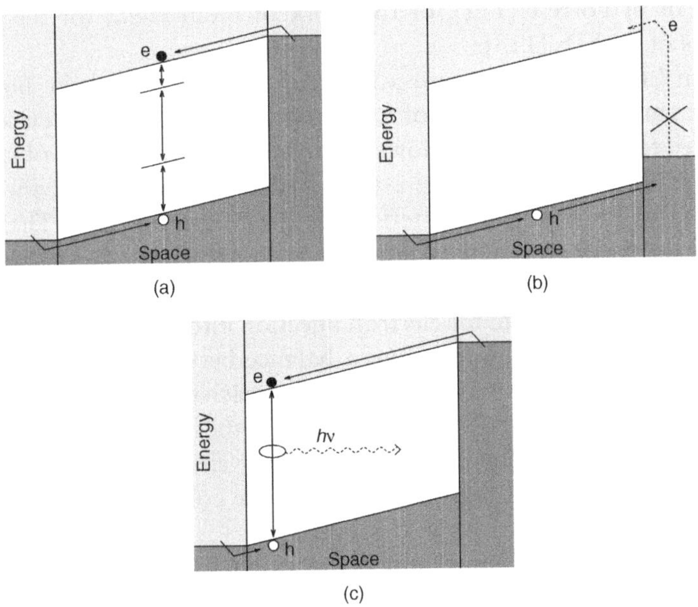

Figure 1.22 Three possible reasons for low luminescence intensity: (a) nonradiative recombination paths via defects; (b) mismatched carrier injection due to wrong choice of work function (resulting in current but no light); (c) mismatched carrier mobility (resulting in luminescence at electrode)

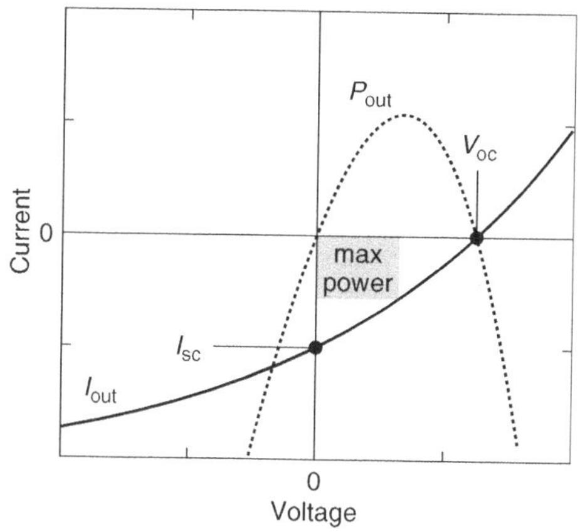

Figure 1.23 The filling factor as a rapid evaluation tool of a solar cell. The I–V curve under illumination (solid curve) defines an open-circuit voltage (V_{oc}) and a short-circuit current (I_{sc}). The output power ($P = -V \times I$) has a maximum somewhere in between. This maximum can also be found by fitting a maximum-area rectangle under the curve in this quadrant of the I–V plot. The filling factor is the ratio of the area of this rectangle to the total area under the curve with positive power

A rapid way of determining the efficiency of a solar cell on basis of the I–V curves is by measuring the filling factor. Figure 1.23 gives an example. The I–V curve of the device under illumination is the solid trace and is labeled I_{out}. When the terminals are short-circuited an optical current can be measured (I_{sc}). An external bias of V_{oc} has to be applied to annihilate this current. This is also the voltage that results at the terminals when they are left in open-circuit. This defines a region where the current has opposite sign compared with the external bias, and thus a negative power consumption ($P = V \times I$). The maximum power output ($P_{out} = -P$) occurs for a voltage somewhere between 0 and V_{oc}, depending on the exact shape of the I–V curve. With power being the product of voltage and current, the maximum power can easily be found by fitting a rectangle with maximum area in the quadrant positive-V-negative-I. To quantify the quality of the device, we can use the filling factor, which is the ratio of the area of this maximum-power rectangle to the total area under the curve. Silicon-based devices have filling factors of about 40%. In principle, the filling factor is independent of the illuminated power, as long as the device output is linear with the illumination power. This because, when the illumination is increased, the output current and power increase, but also the short-circuit current increases and the filling factor is invariant.

When using the filling factor (ff), the power conversion efficiency η can be calculated according to

$$\eta = \frac{J_{sc} V_{oc} ff}{P_o} \times 100\% \tag{1.31}$$

where P_o is the optical power absorbed by the device [116, 117].

One of the controlling parameters of the efficiency of a solar cell is the band gap of the material. This is best explained by Figure 1.24. After generation of hot carriers by the absorption of an energetic photon, three processes start: carrier cooling, carrier drift, and carrier recombination. In normal circumstances, the carrier cooling time is much faster than the carrier drift and collection time. Therefore, the energy that comes out of a single electron–hole pair is, at best, the energy gap of the semiconductor. In even worse cases, the electron–hole recombination time is also faster and no current comes out of the device. Moreover, if the diffusion time is shorter than the drift time, the carriers come out equally on both sides. This shows the necessity for a built-in field as in Schottky diodes or pn-junctions.

In normal circumstances, the carrier cooling time is ultra fast and the energy that can be extracted per electron–hole pair is the energy gap. For a well designed device, nearly all of the carriers make it to electrodes, and the limiting factor is thus the band gap. Larger band gaps have the advantage that the remaining energy extracted per electron–hole pair is larger. However, the trade-off is the reduced number of absorbed

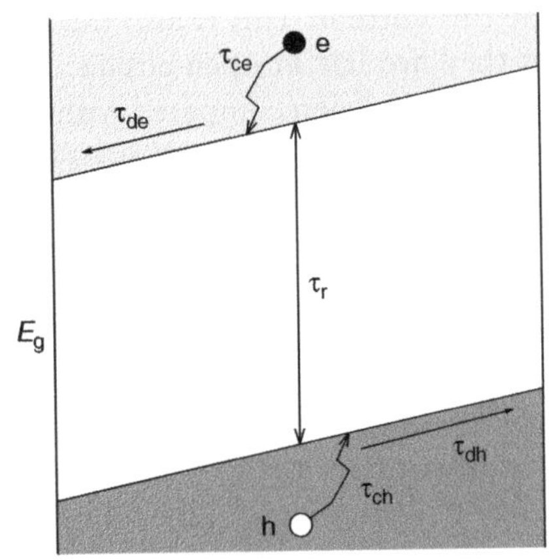

Figure 1.24 Schematic showing the important carrier processes in solar cells after generation of hot carriers. Carrier cooling (τ_c) showing in multi-steps, carrier drift (τ_d), and carrier recombination (τ_r)

photons, since photons with energy below the band gap are not absorbed, and there are fewer electron–hole pairs to start with. Figure 1.25 shows two examples of conversion curves. To find out which of the curves is most efficient, the curves have to be convoluted with the solar spectrum. Schockley and Queisser have calculated that the maximum efficiency for such a single band gap material under solar spectrum illumination is of the order of 30% for a band gap of about 1.1 eV [118]. Interestingly, this is close to the band gap of silicon and that is probably the reason why silicon is the most widely used material for solar cells. This together with the old adage that states that 'everything that *can* be done with silicon *will* be done with silicon'.

Another way of measuring the quality of a photovoltaic device is the quantum efficiency. This is defined as the number of particles (quanta) coming out divided by the number of particles going in. For instance the number of electron–hole pairs generated per incoming photon in solar cells, or the number of photons per electron–hole pair in LEDs. The external quantum efficiency (EQE) is then defined by the external observables (current and photon flux), for instance by the electron–hole pairs that make it to the outside and result in an external current. For a

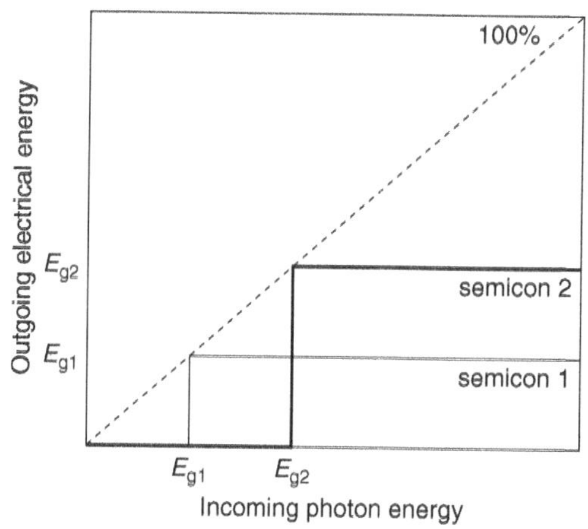

Figure 1.25 Schematic showing the effect of the band gap on the overall efficiency of a solar cell. Photons with small energies are not absorbed and do not contribute to electrical power. Above this threshold, the energy extracted is always equal to the band gap because cooling of the carriers is fast and this rapidly removes the excess energy. Two plots are shown for two different band gap materials. These have to be convoluted with the solar spectrum to find out which one is more efficient. For comparison, the plot for a hypothetical 100% efficient solar cell is also drawn (dashed line)

solar cell and LED, respectively:

$$\mathrm{EQE_{SC}} = \frac{I/q}{\text{photons s}^{-1}} \tag{1.32}$$

$$\mathrm{EQE_{LED}} = \frac{\text{photons s}^{-1}}{I/q} \tag{1.33}$$

where I/q is the number of charged particles per second. The measurement of photon flux is straightforward for solar cells, since lasers are often used. For the luminescence of LEDs, normally Lambertian integrating spheres have to be used. For LEDs the radiant efficiency can be defined as the product of external quantum efficiency (photons per electron) and ratio of photon energy (eV) and electron energy (volt).

A lot of research effort is spent on optimizing these parameters since the goal for the final devices obviously is to make them as energy efficient as possible.

It is not intended here to describe the optoelectronic properties of these devices. They are treated like any other device. The aim of this book is to describe how to apply measurement techniques to extract information from the devices. As an example, an LED can be subjected to a ToF experiment explained in Chapter 5 in order to extract the carrier mobility parameter. The reader is referred to other books when information about the optical or electro-optical properties is sought.

2

Two-terminal Devices: DC Current

2.1 CONDUCTANCE

Conductance G is a device parameter that specifies the ratio of current and voltage,

$$G \equiv \frac{I}{V} \tag{2.1}$$

and is thus the reciprocal of the more familiar resistance. In a homogeneous device, without contact effects or other limiting barriers, the conductance is a scaled version of the material parameter conductivity σ,

$$G = \sigma \frac{A}{d} \tag{2.2}$$

where A is the area and d is the length of the device. It has the unit siemens and is the most elementary characterization parameter of an electronic device. In spite of its simplicity, in practice the behavior of conductance and conductivity can be complicated, since it can be influenced by various effects, that can make it depend on bias, probing frequency, and history, apart from the obvious dependence on temperature.

At this point it is therefore interesting to discuss the various conduction models. By this is meant the steady-state (DC) current. This to distinguish it from, for instance, current transients after a bias or illumination pulse.

As a starting point, it can be said that the current density J is comprised of a drift current, which is proportional to the local field $E_x = -dV/dx$, local free charge density p and mobility μ_p, and a diffusion component, which is proportional to the gradient of charge density dp/dx. Most organic materials are p-type which means that the density of electrons is so small as to make the electron current insignificant. Even in cases where a large electron density can be invoked, the mobility of electrons is normally comparatively lower and electron current remains small. Later, some examples of ambipolar devices will be shown. Assuming only holes contribute to the current:

$$J = -q\mu_p p \frac{dV}{dx} - qD_p \frac{dp}{dx} \tag{2.3}$$

where the mobility μ_p and diffusion coefficient D_p are, under normal conditions, linked via the Einstein Relation, $D_p = \mu_p \times kT/q$. The density of holes can be temperature dependent, as described in Chapter 1. Often the diffusive currents can be ignored (especially in the types of devices presented in this book) and the current proportional to the charge density

$$J = q\mu_p pE \tag{2.4}$$

The effects of reduced (free) charge density can also often be modeled in a reduced effective mobility, which is the convention in organic literature, as discussed in Chapter 1.

The question now boils down to determining the carrier densities. As shown in Chapter 1, for intrinsic materials the carrier density is equal to

$$p = \sqrt{N_V N_C} \exp\left(\frac{-E_g}{2kT}\right) \tag{2.5}$$

and this defines the intrinsic density n_i. Whatever happens, the product of p and n will always be equal to n_i^2 in thermal equilibrium. This is useful to remember. No donors, acceptors or traps can change this. For a system with acceptors, it was shown that nearly all acceptors are ionized and

$$p \approx N_A^- \approx N_A \tag{2.6}$$

under normal conditions. However, when the acceptor level is deep, or the temperature is low and freeze-out takes place, this is no longer valid. In that case, the fraction of ionized acceptors and the free-hole density and thus conductivity depend on the temperature,

$$p = \sqrt{N_A N_V} \exp\left(-\frac{E_A - E_V}{2kT}\right) \tag{2.7}$$

The activation energy of this charge is the acceptor depth divided by two. The factor two stems from the fact that a charge 'counts' twice. For instance a positive charge coming from a neutral acceptor will create two charges, a free hole and a negatively ionized acceptor.

The current density is proportional to this charge density, the mobility μ and the local electric field E, $J = qp\mu E$. Assuming the mobility to be independent of temperature, or only slightly dependent (as in $\propto T^{3/2}$), and the hole density and electric field to be constant over the entire length of the sample, it is easily shown that the current is temperature activated and an Arrhenius plot of the logarithm of constant-bias current through a resistance vs. reciprocal temperature is a straight line, the slope of which reveals *half* the acceptor depth. The comparison with silicon is important. For that material, the temperature dependence reveals information about the factor limiting the mobility, for instance scattering mechanisms, because freeze-out is not easily obtained, whereas in wide band gap materials it can reveal the impurity level depth. In this respect, traps behave similar to dopants. An abundant deep trap can trap free charges in the same way as dopants can recapture them from the bands. In the presence of deep traps, the device can be in the freeze-out regime, with the density of charge on the traps and free charge in the bands, strongly depending on temperature, the activation energy of current revealing the trap depth (without the factor two!). The current is only proportional to the free charge, and thus becomes temperature dependent. Imagine a system with a constant amount of positive charge, ρ_0 (for instance coming from fully ionized acceptors, or injected from an electrode), with part of the charge trapped on deep traps, p_T, and part in the valence band, p; $\rho_0/q = p + p_T$. The current is then given as

$$J(T) \propto q\mu_p p(T) = \mu_p \frac{p(T)}{p(T) + p_T(T)} \rho_0. \tag{2.8}$$

This can also be modeled by a temperature-dependent mobility, as is convention in the literature,

$$J(T) \propto \mu_p(T)\rho_0 \tag{2.9}$$

where knowledge of total charge ρ_0 is often assumed. The effective mobility is then given as

$$\mu_p(T) = \frac{p(T)}{p(T) + p_T(T)} \mu_{p0} \tag{2.10}$$

Examples of this will be given for Poole–Frenkel conduction.

2.1.1 Ohmic Conduction

In ohmic conduction, for simple resistances, ignoring the contact effects (interface regions), at low currents, the quasi-static approximation is valid. In this case, the density of holes is considered homogeneous, making the second term in Equation (2.3) disappear. If also the field is constant over the entire sample and the mobility assumed independent of temperature, the current density is directly proportional to the hole density of Figure 1.12.

$$J = q\mu_p(T)\frac{V}{d}p(T). \tag{2.11}$$

Thus, simple measurement of mobility can be performed in resistors. Accurate knowledge of the hole density is inherently assumed in these measurements to find the mobility, or vice versa. Examples of this will be given in the current work, namely in the framework of traps.

2.1.2 Poole–Frenkel

The effective mobility can be severely reduced by the presence of traps in the material. (Hole) traps are localized deep electronic states that can be positively charged by the capture of free holes. The levels can thus be labeled as T^0/T^+, denoting the two possible charge states. Once a charge is bound to a trap, it becomes immobile (localized) and does not contribute to the current anymore. The current J and effective mobility μ_p are then determined by the fraction of charge that is in the band states. Moreover, since the charges can be thermally excited from the traps into the valence band, the fraction of free holes and thus the current becomes strongly temperature dependent. Poole and Frenkel have shown how the thermal excitation from the traps can be (electrical) field assisted, or, in other words, the thermal barrier for excitation is lowered in the presence of an electrical field, as is demonstrated in Figure 2.1. The result is that when a field is present, the fraction of charge in the valence band

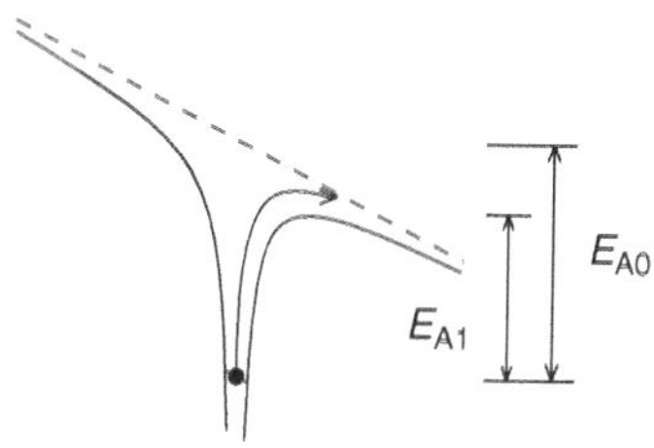

Figure 2.1 A local field at the position of a trap will add a potential to the energy diagram that lowers the effective thermal activation barrier from E_{A0} to E_{A1}. According to the Poole–Frenkel model, a hole is more easily excited into the valence band and a larger fraction of charge is thus found to contribute to current

increases and the effective mobility with it. The Poole–Frenkel current then becomes [9]

$$J = J_0 V \exp\left(\frac{-E_a + \sqrt{qE/\pi\varepsilon}}{kT}\right) \tag{2.12}$$

where the electrical field is considered constant $dV/dx = V/d$, making J_0 device geometry dependent. Equation (2.12) can be summarized in an effective mobility

$$\mu_{PF} = \mu_0 \exp(\sqrt{E/E_0}) \tag{2.13}$$

where μ_0 and E_0 are sample and temperature dependent. This equation is often used in the literature and seems to work very well for most electrical data of organic materials.

The above is the classic form of Poole–Frenkel conduction as found in the literature and is based on a system with discrete, abundant traps apart from the discrete conduction and valence band states used in band theory. For organic semiconductors the trap states or the band states or both can be distributed in energy as shown in Chapter 1. It is therefore useful to present a more general analysis of Poole–Frenkel conduction. Very similar to the discussion of the thermally activated mobility in TFTs (Chapter 6) it will now be shown how the current can have a power-law temperature dependence when energy states are distributed in energy. For this, a system is analyzed with a constant positive charge (ρ_0), for instance coming from shallow acceptors in the saturation regime ($N_A^- = N_A$) or charge injected by electrodes, and consisting of free holes, p, and trapped holes, N_T^+. (Free electrons are neglected in this analysis.)

The total charge is thus

$$p + N_T^+ = \rho_0/q. \tag{2.14}$$

The balance between trapped and free charge is assumed to be in thermal equilibrium. Assuming the mobility of trapped charge to be zero, the conduction is only caused by free holes, $J = q\mu_0 pV/d$. It is now interesting to look at some specific cases.

1. **Abundant discrete trap.** To start, we repeat the calculation for an abundant discrete trap. The trap is in fact much more dense than valence band states, $N_T \gg N_V$, and $N_T^+ \gg p$. If the Fermi level is far enough away from the trap level and valence band level, Boltzmann distributions can be used to find the density of free and trapped holes. As a function of Fermi level and temperature they become, respectively

$$p = N_V \exp\left(\frac{E_V - E_F}{kT}\right) \tag{2.15}$$

$$N_T^+ = N_T \exp\left(\frac{E_T - E_F}{kT}\right). \tag{2.16}$$

The Fermi level can be eliminated from these equations when substituted into Equation (2.14); Assuming $p \ll N_T^+$,

$$\begin{aligned} p &= p\frac{\rho_0/q}{p + N_T^+} \\ &\approx \frac{p}{N_T^+}\rho_0/q \\ &= \frac{N_V}{N_T}\exp\left(-\frac{E_T - E_V}{kT}\right)\frac{\rho_0}{q}. \end{aligned} \tag{2.17}$$

Under constant-charge conditions, the current is thus of the form

$$J \propto \mu_0 \frac{N_V}{N_T}\exp\left(-\frac{E_T - E_V}{kT}\right) \tag{2.18}$$

which is Poole–Frenkel without the field dependence. The field can lower the barrier for charge emission and effectively lower the trap depth. What is important to note is that the current is temperature activated with an activation energy equal to the trap depth, $E_a = E_T - E_V$.

2. **Exponentially distributed trap.** The next calculation is a system of an abundant trap, distributed exponentially in energy,

$$N_T(E) = N_{T0} \exp\left(\frac{E_V - E}{kT_2}\right) \tag{2.19}$$

where N_{T0} is the density of traps at the valence band level (unit: $m^{-3}J^{-1}$) and T_2 (unit: K) is the parameter describing the distribution, namely $1/kT_2$ is the slope of the distribution in a logarithmic plot. In this case the dependence of trapped charge on Fermi level of Equation (2.16) has to be replaced by an integral over all trap states,

$$N_T^+ = \int_{-\infty}^{\infty} N_T(E)[1 - f(E - E_F)]dE \tag{2.20}$$

where now also the complete Fermi–Dirac statistics are used instead of the Boltzmann approximation. Substitution in the total charge equation and eliminating E_F results in (see Appendix A, with $\rho = \rho_0$)

$$p = N_V \left[\frac{\rho_0}{qN_{TT}(T)}\right]^{T_2/T}. \tag{2.21}$$

with

$$N_{TT}(T) = \alpha(T) N_{T0} \frac{(kT_2)^2}{kT_2 - kT} \tag{2.22}$$

in which $\alpha(T)$ is a slightly temperature dependent factor between 1 and 0.8 in the temperature range 0–T_2. (For $T \geq T_2$ the integral diverges.) The current is proportional to this free charge p and thus any dependence of ρ_0 or N_{TT} on temperature is raised to the power T_2/T; indeed a complicated temperature dependence can result for this case.

3. **Exponentially distributed trap and valence band.** The next calculation includes a valence band also distributed exponentially in energy,

$$N_V(E) = N_{V0} \exp\left(\frac{E_V - E}{kT_1}\right) \tag{2.23}$$

where N_{V0} is the density of levels at $E = E_V$ and T_1 is the parameter describing the distribution, similar to T_2 above. In this case the dependence of p on Fermi level of Equation (2.15) has also to be replaced by an integral over all valence band states,

$$p = \int_{-\infty}^{\infty} N_V(E)[1 - f(E - E_F)]dE. \tag{2.24}$$

Substitution in the total charge equation and eliminating E_F results in (see the Appendix A, with $\rho = \rho_0$)

$$p = N_{VT}(T)\left[\frac{\rho_0}{qN_{TT}(T)}\right]^{T_2/T_1} \tag{2.25}$$

with

$$N_{VT}(T) = \beta(T)N_{V0}\frac{(kT_1)^2}{kT_1 - kT} \tag{2.26}$$

in which β is a function similar to α, but scaled with T_1 instead of T_2. The current is proportional to p and thus, in this case, any dependence of ρ_0 or N_{TT} on temperature is raised to the power T_2/T_1 which can be any number.

2.1.3 Tunneling

When the conductivity of the material is further reduced, the conduction process is Fowler–Nordheim tunneling [107, 119–121], also known as 'field emission'. In this regime, the current is strongly bias dependent, but the influence of temperature is nonexistent.

$$J_{FN} \propto \left(\frac{V}{d}\right)^2 \exp\left(-\alpha\frac{d}{V}\right) \tag{2.27}$$

where α is a factor that depends on things like the barrier height and effective mass of charges. Modifications of the theory exist to explain discrepancies, for instance a temperature dependence, or the Schottky effect that lowers the effective barrier.

Fowler–Nordheim conduction is normally found in nearly insulating materials. For truly insulating materials, no current is possible inside the device, but when the layer is thin enough, tunneling can occur from one electrode to the other [46]. The difference between true tunneling and Fowler–Nordheim tunneling is that in the latter a triangular energy diagram is used and charges are injected into the bands of the insulator, from where they can travel freely, while in the former the carriers directly appear on the other side. Fowler–Nordheim tunneling is often a very adequate description in OLEDs; to optimize the EQE (external quantum efficiency), these materials need to be as pure as possible, a side effect of which is a reduced conductivity. Charges of opposite sign are injected into the active layer of these LEDs by Fowler–Nordheim tunneling and recombine radiatively at the meeting point. On the other hand, tunneling is a good description for injection of carriers at the contacts.

2.1.4 Space-charge-limited Current

Finally, in so-called space-charge-limited current (SCLC) [122], the density of injected free carriers is so high that the material is locally charged by this free charge and the associated electric fields and band bendings are substantial. In this case, the band diagram can no longer be considered independent of the injected carriers; the carriers themselves distort the energy diagram. In SCLC the currents are quadratically dependent on the bias.

To prove this, consider that *all* charge is free charge, $\rho(x) = qp(x)$ and contributing to current, and the diffusion current in Equation (2.3) is negligible, thus

$$J = q\mu_p p(x)E(x). \tag{2.28}$$

The field and charge are related via Poisson's equation,

$$\frac{dE(x)}{dx} = \frac{\rho(x)}{\varepsilon_s} = \frac{qp(x)}{\varepsilon_s}. \tag{2.29}$$

Substituting this into Equation (2.28):

$$\begin{aligned} J &= \mu_p \varepsilon_s \frac{dE(x)}{dx} E(x) \\ &= \frac{1}{2}\varepsilon_s \mu_p \frac{dE^2(x)}{dx} \end{aligned} \tag{2.30}$$

gives as a solution [assuming $E(0) = 0$; i.e., zero field at the injecting electrode]

$$E^2(x) = \frac{2J}{\varepsilon_s \mu_p} x \tag{2.31}$$

or

$$E(x) = \sqrt{\frac{2J}{\mu_p \varepsilon_s} x} \tag{2.32}$$

and the potential is the integral of this (setting the potential to zero at the collecting electrode, $x = d$),

$$V(x) = -\int E(x')dx' = \frac{2}{3}\sqrt{\frac{2J}{\varepsilon_s \mu_p}}(d^{3/2} - x^{3/2}). \quad (2.33)$$

The current can then be found by setting the voltage drop across the device equal to the external voltage, $V(0) = V$:

$$V = \frac{2}{3}\sqrt{\frac{2J}{\varepsilon_s \mu_p}} d^{3/2} \quad (2.34)$$

yielding the SCLC equation [123],

$$J = \frac{9\varepsilon_s \mu_p}{8} \frac{V^2}{d^3} \quad (2.35)$$

also sometimes called Mott's steady-state space-charge-limited conduction model, or simply the Mott–Gurney Law [124]. To achieve this, five basic components are needed:

1. There is only one type of charge carrier present, either electrons or holes, but not both.
2. The material has no free charge by itself; charges are injected into it from one electrode and captured by the other.
3. The carrier mobility μ_p and the dielectric permittivity ε_s are constant throughout the sample.
4. The electric field at the charge-injecting electrode is zero.
5. All current is drift current (diffusion plays no role).

Note also that the device current, as well as the capacitance, and conductance, as will be shown later in Chapter 3, are independent of temperature, as long as the mobility is independent of temperature. In some cases this is advantageous when compared with other types of measurements. The temperature dependence of charge density is eliminated.

The explicit forms of the distribution of space charge, electric field and potential in the SCLC regime are

$$p(x) = \frac{3\varepsilon_s}{4q} \frac{V}{d} \sqrt{\frac{1}{xd}} \quad (2.36)$$

$$E(x) = \frac{3}{2}\frac{V}{d}\sqrt{\frac{x}{d}} \tag{2.37}$$

$$V(x) = V\left[1 - \left(\frac{x}{d}\right)^{3/2}\right]. \tag{2.38}$$

These functions are summarized in Figure 2.2. Note that the highest density of charge is found at the charge-injecting electrode ($x = 0$), but the largest field and voltage drop are found at the counter-electrode. A large charge cloud is present close to the injecting electrode, but there the field is relatively small. This may seem counterintuitive.

The total charge in the device is given by the integral of the charge distribution above and can be calculated as

$$Q = \int_0^d qp(x)\mathrm{d}x = \frac{3\varepsilon_s V}{2d}. \tag{2.39}$$

Interesting for further analysis is the observation that the center of mass of this space-charge distribution lies one-third into the semiconductor,

$$< x_\mathrm{p} > = \frac{\int_0^d xqp(x)\mathrm{d}x}{\int_0^d qp(x)\mathrm{d}x} = \frac{d}{3} \tag{2.40}$$

independent of mobility, bias, temperature, etc. This will come in handy later.

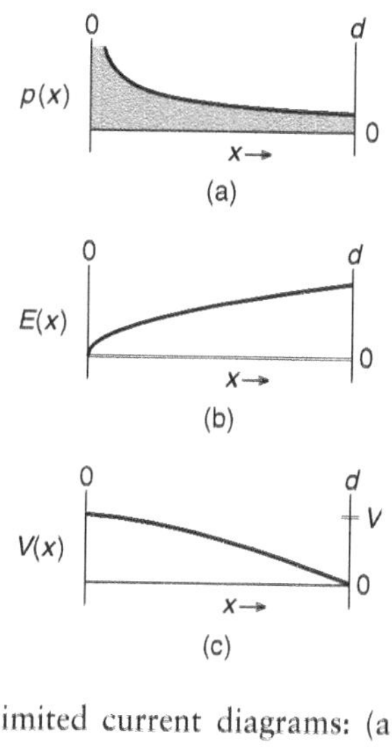

Figure 2.2 Space-charge limited current diagrams: (a) space charge; (b) electric field; and (c) potential

Further observations are needed to better understand the ideas of SCLC. First of all, observe that injection of charge itself is assumed to be no problem. Some work exists to extend the theory to include injection limitations. However, large injection obstacles would turn the device into a normal barrier (Schottky, or tunnel) and remove the validity of the SCLC analysis. It would be barrier-limited instead of space-charge limited. After all, the L in SCLC stands for 'limited' denoting that SC is the limiting factor.

Note also that, according to the above, the entire device is charged, something that is impossible. The charges have to be compensated somewhere to not turn the device into a 'Coulomb bomb'. These charges are on the electrodes. One has to assume that the charge at the injecting electrode is (close to) zero – injection is no problem and holes (and electrons) are abundant in metals – and the collecting electrode has a large compensating negative charge consisting of electrons that have difficulty entering the device because of a high injection barrier, or because the mobility is even much lower than that of the holes and they do not make it very far into the device (during the experiment). In case electrons can also make it easily into the device, it is logical to assume that a space-charge cloud exists at that electrode too, this time made up of electrons. As far as I am aware, no double-SCLC analysis exists in the literature.

Finally, from Figure 2.2 it is obvious that the suppression of the diffusion current is especially dubious at $x = 0$ because there the density of charge (and the gradient with it) go to infinity. However, the SCLC equation is generally accepted in the literature and in many cases can explain experimental data quite well, showing that simplicity often is more important than accuracy.

Again, for low-mobility materials, the mobility μ_p can depend on traps and on the electrical field. For a two-band model, with a conduction band and a zero-mobility trap state, Poole–Frenkel conduction is adequate

$$\mu = \mu_0(T) \exp\left(\alpha\sqrt{E}\right). \tag{2.41}$$

For these field-dependent mobilities, it is not easy to find an analytical solution. A small literature search shows that some people have worked on this subject. Lampert finds an exact solution for a single, discrete trap level [125]. Caserta *et al.* describe a method of extracting the trap density directly from the I–V curves in the SCLC regime [44]. The work of Natali and Sampietro gives an excellent description of how to deal with complicated situations and how to extract mobility information from SCLCs in general [126].

When all traps are filled and converted from the neutral to the charged state, T^0 to T^+, the device enters the trap-free SCLC regime, which is still quadratically dependent on the bias V, but the mobility μ_p is the band mobility. The bias at which this transition occurs is the trap-free limit and to get an idea of the value of this parameter, Poisson's equation can be used. Assuming the only charge in the device to be nothing more and nothing less than all traps charged:

$$V_{TFL} = \iint_0^d \frac{qN_T}{\varepsilon_s} dx = \frac{qN_T d^2}{2\varepsilon_s} \tag{2.42}$$

where N_T is the trap density. Inverting this equation, the trap density can be derived from the trap-free-limit voltage. Figure 2.3 summarizes the SCLCs. The analysis with SCLC and the trap-free limit is quite common for organic materials [127]. Choi and coworkers show how even the distribution of the traps in energy can be derived from the I–V curves in the SCLC regime [128].

2.1.5 Granular Materials; Grain Boundaries

When the material is polycrystalline, at the grain boundaries barriers exist that can modify the simple analysis. Grain boundaries are discontinuities in the crystallinity of the material and cause a large density of

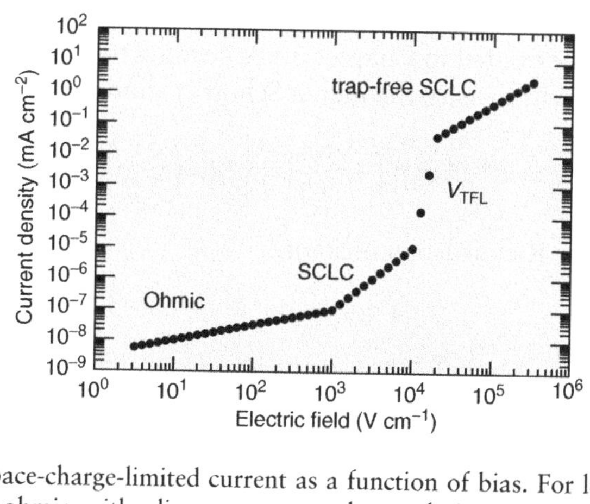

Figure 2.3 Space-charge-limited current as a function of bias. For low biases, the device behaves ohmic, with a linear current–voltage relation. For increased bias, the density of holes becomes larger than the negatively charged acceptors and the device enters the SCLC regime. The effective mobility can still be limited by traps. For a certain voltage all traps get filled and the device enters the trap-free SCLC regime

deep electronic states. These states can cause trapping (charge build-up), a pinning of the Fermi level and band bendings, thus effectively creating a barrier. A junction between the same materials (same Fermi levels) will not by itself cause any barriers; a chemical potential has to exist between regions to cause dislocation of charge and band bendings.

When the grains are large so that they can accommodate all band bendings and a substantial layer of 'metallic' material is in between the two grains, they can be considered back-to-back Schottky diodes. A multitude of such grains can be modeled as a single back-to-back two-Schottky-diode system, the current of which follows

$$J = 2J_0 \tanh\left(\frac{qV}{nkT}\right) \tag{2.43}$$

where J_0 is the reverse-bias saturation current and n is the ideality factor of a single diode. In either bias direction, the current saturates at twice the reverse-bias saturation current of a single diode. Indeed, as shown by Chattopadhyay, the current follows a standard diode equation with an ideality factor equal to $n = 1 + 2kT/E_s$ and a temperature-dependent saturation current [129].

2.2 DC CURRENT OF A SCHOTTKY BARRIER

A Schottky barrier is made by the intimate contact between a metal and a semiconductor, as described in Chapter 1. There are various models to describe the DC conductance of a Schottky barrier. (The admittance behavior will be treated in Chapter 3.) According to thermionic emission theory the current density through a Schottky diode follows [9]

$$J = A^* T^2 \exp\left(-\frac{q\phi_{\mathrm{Bp}}}{kT}\right)\left[\exp\left(\frac{qV}{kT}\right) - 1\right] \tag{2.44}$$

where A^* is the Richardson constant,

$$A^* = 4pqm^* k^2/h^3 \tag{2.45}$$

and ϕ_{Bp} is the barrier height (unit: volt). Alternatively, the current can be following the diffusion theory:

$$J = \left[\frac{q^2 D_{\mathrm{n}} N_{\mathrm{C}}}{kT}\sqrt{\frac{2qN_{\mathrm{D}}(V_{\mathrm{bi}} - V)}{\varepsilon}} \exp\left(-\frac{q\phi_{\mathrm{Bp}}}{kT}\right)\right] \times$$
$$\times \left[\exp\left(\frac{qV}{kT}\right) - 1\right]. \tag{2.46}$$

Finally, a combination of the two theories, the thermionic emission–diffusion theory predicts a current following the equation

$$J = A^{**}T^2 \exp\left(-\frac{q\phi_{Bp}}{kT}\right)\left[\exp\left(\frac{qV}{kT}\right) - 1\right] \tag{2.47}$$

where A^{**} is the effective Richardson constant.

The currents are all of the form

$$J = J_0[\exp(qV/nkT) - 1]. \tag{2.48}$$

In this, n is the ideality factor (≥ 1), parameterizing the deviation from theory. For classic semiconductors such as Si, Ge or GaAs, the value of n is close to 1, while for polymers it is often in the order of 2. Figure 2.4 compares the I–V curve of thermionic emission theory with that of diffusion theory.

From a DC current–voltage plot (I–V plot) we can determine the following:

- The ideality factor n from the slope in forward bias.
- The correct model for conduction
  - In thermionic emission theory the reverse current saturates (levels off).
  - In diffusion theory the reverse current depends on the bias.
- The saturation current J_0 from the reverse-bias saturation current or the extrapolated-to-zero-volt forward-bias current.

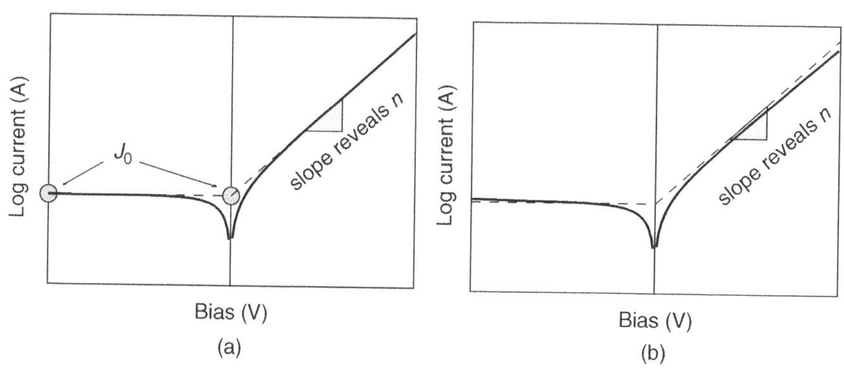

Figure 2.4 I–V curves on a logarithmic scale according to (a) thermionic emission theory [Equation (2.44)] and (b) diffusion theory [Equation (2.46)]. The differences between the two are minimal

- The barrier height (ϕ_{Bp}) when the temperature is varied: A plot of $\ln(J/T^2)$ [or $\ln(JT)$ for diffusion theory] versus $1/T$ in strong forward bias (so that the term '-1' can be neglected in the equations above) results in a straight line with slope $q(V - \phi_{Bp})/k$ from which the barrier height ϕ_{Bp} can be derived. Alternatively it can be done in strong reverse bias, so that the exponent can be neglected and the slope will be $-q\phi_{Bp}/k$.
- The rectification ratio of a device, $r = J(+V)/J(-V)$. An important figure-of-merit since it shows the relative 'leakage' in reverse bias and thus the power efficiency when the device is used in a final circuit.

2.2.1 High-current Regime

For high forward voltages, the current may be limited by the series resistance of the bulk, rather than the Schottky–barrier interface. In that case, the currents follow the laws presented before for homogeneous devices: Ohmic, Poole–Frenkel, space-charge-limited, etc.

In the semi-log I–V curves this is visible as a bending of the curve at strong forward bias, see Figure 2.5.

2.2.2 Displacement Current

As will be shown later, a Schottky diode also has a capacitance. This limits the speed at which the devices can be measured because, apart from the currents described above, there is a current caused by the charging and discharging of the capacitor. This is called displacement

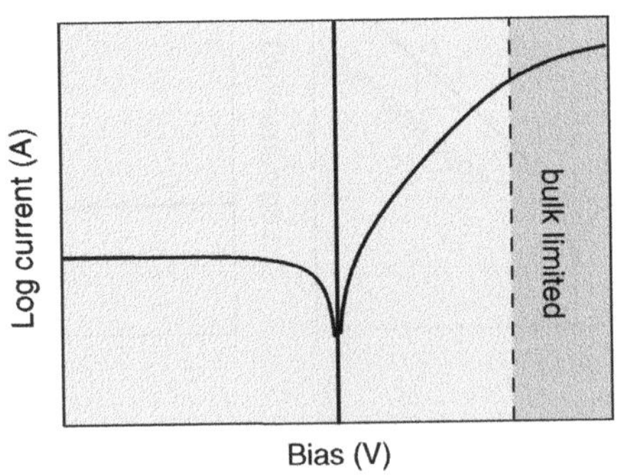

Figure 2.5 I–V curves including the high-current bulk-limited regime, where the currents follow Ohm's Law

current. Given the fact that the charge in a capacitor is the product of capacitance and voltage, the charging current is given by

$$I_{\text{disp}} = \frac{\mathrm{d}Q}{\mathrm{d}t} = \frac{\mathrm{d}}{\mathrm{d}t}(CV) = C\frac{\mathrm{d}V}{\mathrm{d}t} + V\frac{\mathrm{d}C}{\mathrm{d}t}$$
$$= \left(C + V\frac{\mathrm{d}C}{\mathrm{d}V}\right)\frac{\mathrm{d}V}{\mathrm{d}t}. \tag{2.49}$$

The first term in pararentheses is the standard constant-capacitance displacement current, while the second term is added when the capacitance depends on the bias, for instance in a Schottky barrier. With the knowledge of the capacitance of a Schottky barrier to be discussed later [Equation (3.52)], Equation (2.49) can be converted to

$$J_{\text{disp}} = \sqrt{\frac{2(V_{\text{bi}} - V)}{q\varepsilon N_{\text{A}}}}\left[1 - \frac{V}{2(V_{\text{bi}} - V)}\right]\frac{\mathrm{d}V}{\mathrm{d}t} \tag{2.50}$$

and can thus even be used to get information about the interface in devices, namely N_{A} and V_{bi}. However, if unwanted, reducing the scanning speed ($\mathrm{d}V/\mathrm{d}t$) will obviously avoid this problem. Displacement currents can easily be recognized by the large hysteresis. In Equation (2.50), it is clear that for up scanning ($\mathrm{d}V/\mathrm{d}t$ positive) the displacement current is added, while for down scanning ($\mathrm{d}V/\mathrm{d}t$ negative) it is subtracted from the normal I–V curve. Figure 2.6 gives an example of an I–V curve with relatively large displacement currents.

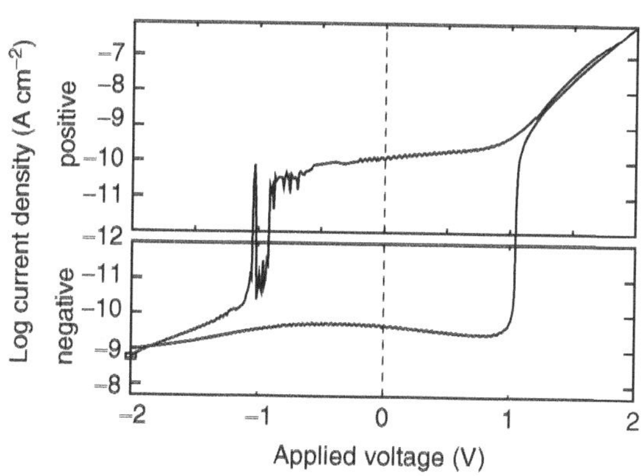

Figure 2.6 Tell-tale signs of displacement current, namely a large hysteresis. The top trace represents the current in an up scan, while the lower trace is a down scan

2.3 DC MEASUREMENTS

The most simple measurement is resistivity of a resistor. For the moment the problems that may exist to inject carriers are ignored. In most cases organic materials have high resistivity (due to low mobility owing to a large density of traps, as discussed above). In these cases, simple two-wire measurements are adequate. Because the currents can be very low for any reasonable bias applied, the quality of the measurements is mainly determined by the sensitivity of the amperimeter. Commercial equipment is available that has a resolution in the ato-amp range.

2.3.1 van der Pauw

When the resistance of the device is low, cable resistances may distort the measurement. In that case it is best to revert to a four-point probe, see Figure 2.7(a). A popular technique is the van der Pauw four-point measurement [130] and commercial equipment is readily available for this type of measurement. The advantage of this scheme is that the effect of cable resistances can be eliminated. The current is programmed by the outside two electrodes (A and D in Figure 2.7) and currents are not altered by cable resistance and this makes the value of the current well known. Since the current cannot disperse anywhere, it also passes between any two cross-sections in between A and D, for instance the points B and C indicated in Figure 2.7. A voltage probe connected to these points (ideally) does not tap the current, and no voltage drop occurs inside the cables connected to B and C. The voltage drop between B and C is therefore accurately known. Since also the current between

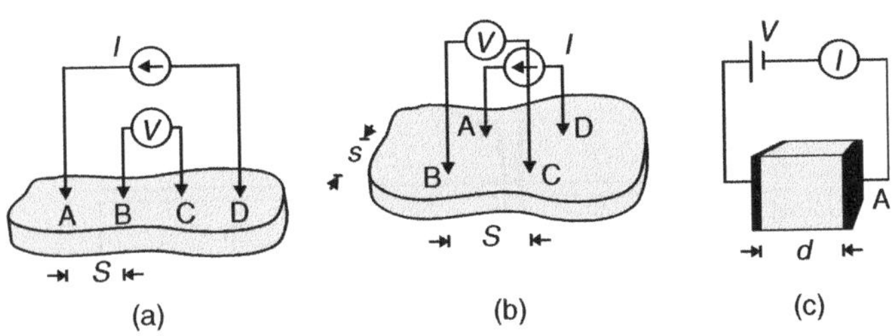

Figure 2.7 (a) Four-point measurement of conductivity with in-line electrodes. A current source forces a current I through the device at points A and D. A voltmeter measures the induced voltage V at two other points, B and C. (b) Four-point measurement with the electrodes in a square configuration. (c) For high-resistance devices the two-point measurement is enough, here shown in a sandwich structure

B and C is accurately known, the resistance can be calculated with high precision, namely

$$R = (V_B - V_C)/I_{AD}. \tag{2.51}$$

To convert this to the material parameters resistivity (ρ, unit: Ω cm) or conductivity (σ, unit: S cm^{-1}) for a slab of material with infinite dimensions, the following relation can be used

$$\rho = R \times 2\pi s \tag{2.52}$$

where s is the distance between two adjacent (equidistant) in-line electrodes. The electrodes can also be placed in a square configuration (see Figure 2.7b) in which case the relation between resistance and resistivity is

$$\rho = R\frac{2\pi s}{2 - \sqrt{2}}. \tag{2.53}$$

For finite samples a correction factor has to be introduced [131].

In cases where the film is very thin and effectively two-dimensional conductivity takes place, the finite-sample correction factor for the square configuration, with the distance between the electrodes s much larger than the film thickness d, becomes

$$\rho = \frac{\pi}{\ln 2} Rd. \tag{2.54}$$

Reversing this equation shows that the measured resistance R is independent of the inter-electrode distance s. It is customary to express the resistance of a thin film in units $\Omega/\square$, which is read as Ω per square, the 'square' being a dimensionless unit but indicating that the resistance was measured in a square configuration. The product R and d is referred to as sheet resistance and Equation (2.54) can be used to convert sheet resistance to resistance and resistivity. It is also obvious that sheet resistance is a device parameter and not a material parameter because it depends on the dimensions (thickness) of the device.

For high resistive bulk measurements, such advanced van der Pauw techniques are not necessary and two-point probes are adequate. In most cases the organic resistor devices are thin films in the sandwich structure (see Figure 2.7c), where the resistivity can then easily be found from the current and voltage by $\rho = A/d \times V/I$, where A is the electrode area and d is the thickness of the organic layer.

2.3.2 Hall Effect

The mobility can also be measured with Hall measurements. For this purpose, the resistor is placed in a magnetic field B_z, see Figure 2.8. The current I_x induced in one direction will cause a charge buildup and a potential in the direction perpendicular to this current and the magnetic field. Measuring this potential V_y yields the Hall mobility. Generally, a charge with (average) velocity $<v_x>$ will feel a Lorentz force equal to

$$F_y = q(E_y + B_z <v_x>). \tag{2.55}$$

In steady state the force is zero. Using the expressions $E_y = V_y/h$ and $<v_x> = \mu E_x = \mu V_x/d$, we arrive at the following Hall mobility

$$\mu_H = \frac{V_y d}{B_z V_x h} \tag{2.56}$$

where h is the height of the sample, the size in the y-dimension, perpendicular to current and magnetic field, and d its length, the size in the current direction x. The problem with Hall measurements is that for small mobilities, like most amorphous materials, the quantity V_y becomes very small, up to the point of being immeasurable. As such, Hall measurements are only adequate for high-mobility materials.

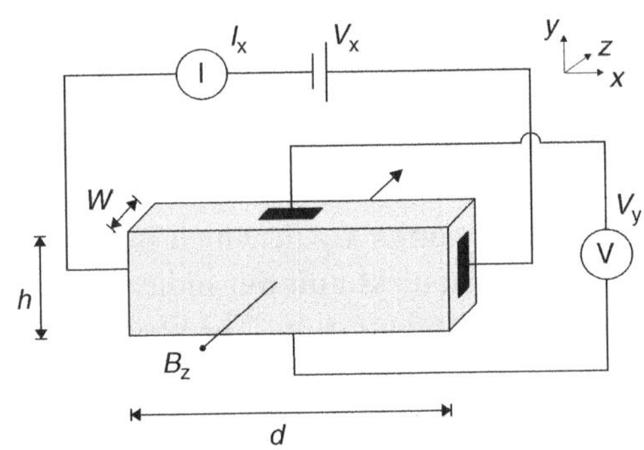

Figure 2.8 Hall measurement. A current I_x is forced through the sample. The magnetic forces working on the moving charges cause a voltage V_y from which the Hall mobility can be extracted

3

Two-terminal Devices: Admittance Spectroscopy

3.1 ADMITTANCE SPECTROSCOPY

Admittance (Y_{dc}) is defined as the ratio of current and voltage,

$$Y_{dc} \equiv \frac{I}{V}. \tag{3.1}$$

This ratio can also be called conductance G and has the unit siemens, which is equal to $1/\Omega$, since the inverse of the above equation yields the resistance. Generally speaking, admittance (and conductance) can be nonlinear, meaning that it depends on the voltage, when the device does not follow Ohm's Law. In admittance spectroscopy we are only interested in the small signal part, which is the *derivative* of the I–V curve.

$$Y_{ac} = \frac{dI}{dV}. \tag{3.2}$$

Note that the DC admittance Y_{dc} of Equation (3.1) is not the same as the AC admittance Y_{ac} of Equation (3.2) at 0 Hz. Y_{dc} is the ratio of I and V, whereas Y_{ac} is the derivative of the I–V curve, see Figure 3.1. In what follows, the admittance is assumed to be the small-signal admittance Y_{ac} and the subscript is dropped, $Y \equiv Y_{ac}$. The DC part of the applied voltage is called 'bias'.

Electrical Characterization of Organic Electronic Materials and Devices Peter Stallinga

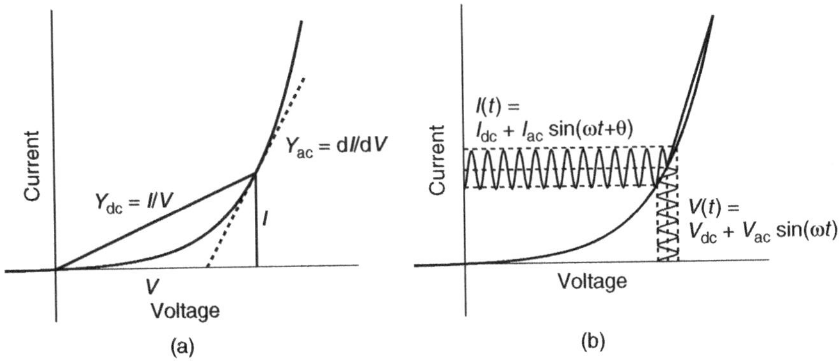

Figure 3.1 (a) Difference between DC admittance and AC admittance. The former is the ratio of current and voltage, while the latter is the derivative of the I–V curve. In the current work, AC admittance is used. (b) Measurement of (AC) admittance by the application of a small AC voltage superimposed on top of a DC bias. The response is a DC current and a phase-shifted AC current. Conductance G corresponds to the in-phase signal, while susceptance B is the 90° out-of-phase signal

In spite of this simple definition of admittance, the relation is quite complicated since

- Current can be a nonlinear function of voltage.
- Current and voltage can have phase shifts and time delays.
- Current can depend on the probing frequency.

Generally speaking, I and V can have different phase. This can easily be described by using complex numbers,

$$Y = G + iB \tag{3.3}$$

where G is conductance and B is susceptance (see Table 3.1 for a list of admittance spectroscopy parameters and their relations). Admittance spectroscopy consists of applying a bias and a small-signal voltage,

$$\begin{aligned} V(t) &= V_{\mathrm{dc}} + V_{\mathrm{ac}}\sin(\omega t) \\ &= V_{\mathrm{dc}} + V_{\mathrm{ac}}\mathrm{Im}(e^{i\omega t}) \end{aligned} \tag{3.4}$$

where V_{ac} is the amplitude of the probe signal, and ω is its radial frequency. The response to this probe signal is a current composed of a

Table 3.1 Admittance spectroscopy device and material parameters and their relations

Parameter	Relation(s)	Name	Unit
Y	$Y = G + iB$ $Y = \mathrm{d}I/\mathrm{d}V$	Admittance	S
$\|Y\|$	$\|Y\| = \sqrt{G^2 + B^2}$	Admittance magnitude	S
Z	$Z = R + iX$ $Z = 1/Y$ $Z = \mathrm{d}V/\mathrm{d}I$	Impedance	Ω
G	$G = \mathrm{Re}(Y)$	Conductance	S
B	$B = \mathrm{Im}(Y)$	Susceptance	S
R	$R = \mathrm{Re}(Z)$	Resistance	Ω
X	$X = \mathrm{Im}(Z)$	Reactance	Ω
C	$Y_C = i\omega C$ $Z_C = -i/\omega C$ $C = Q/V$ $C = \mathrm{d}Q/\mathrm{d}V$	Capacitance (Static; DC) (Dynamic; AC)	F
L	$Y_L = -i/\omega L$ $Z_L = i\omega L$	Inductance	H
L	$L = G/\omega$ $L = 1/\omega R$	Loss	F
$\tan\delta$	$\tan\delta = L/C$ $\tan\delta = 1/\omega RC$	Loss-tangent	-
σ	$(d/A)G$	Conductivity	$\mathrm{S\ m^{-1}}$
ρ	$(A/d)R$	Resistivity	Ω m
ε	$(d/A)C$	Permittivity	$\mathrm{F\ m^{-1}}$
ε_r	$\varepsilon = \varepsilon_r\varepsilon_0$ $\varepsilon_r = K$	Dielectric constant	-
χ	$\varepsilon_r = (1 + \chi)$	Susceptibility	-

DC part and a small signal part that, for small amplitudes V_{ac}, has the same frequency ω, but is shifted in phase by an angle θ.

$$\begin{aligned} I(t) &= I_{dc} + I_{ac}\mathrm{Im}(e^{i(\omega t+\theta)}) \\ &= I_{dc} + I_{ac}\sin(\omega t + \theta) \\ &= I_{dc} + I_{ac}^{0}\sin(\omega t) + I_{ac}^{90}\cos(\omega t) \end{aligned} \tag{3.5}$$

$$I_{ac}^{0} = I_{ac}\cos(\theta) \tag{3.6}$$

$$I_{ac}^{90} = I_{ac}\sin(\theta). \tag{3.7}$$

With the information that

$$\frac{dI}{dV} = \frac{dI}{dt}\frac{dt}{dV} \tag{3.8}$$

$$= \frac{dI/dt}{dV/dt} \tag{3.9}$$

and combining the above information, it is clear that, the admittance parameters are equal to

$$G = I^{0}_{ac}/V_{ac} \tag{3.10}$$

$$B = I^{90}_{ac}/V_{ac}. \tag{3.11}$$

Thus, admittance spectroscopy consists of applying a small-signal voltage and measuring the small-signal response current. The amplitude and phase of the response will yield the conductance and susceptance.

Most relevant for semiconductor electronic materials and device science, are resistance ($R \equiv 1/G$) and capacitance (C). Most measurement equipment will return these values as a function of frequency. This is inspired by the fact that these have most physical meaning. While conductance is rather straightforward, and it is easy to show that the current is in-phase with the applied voltage, capacitance is a form of frequency-dependent susceptance. This is easily shown when we realize that capacitance is, by definition, the capacity to store charge,

$$C \equiv \frac{Q}{V}. \tag{3.12}$$

Or, in other words, how much charge the device can store per unit voltage. Another way of writing this is $Q = CV$. Currents of a capacitor are then charging and discharging the device. With the help of Equation (3.12) this becomes

$$I_C = \frac{dQ}{dt} = C\frac{dV}{dt}. \tag{3.13}$$

When a small-signal voltage [Equation (3.4)] is applied to a capacitor, the current is thus

$$I_C = C\frac{d[V_{ac}\sin(\omega t)]}{dt}$$

$$= \omega C V_{ac}\cos(\omega t). \tag{3.14}$$

Comparing this with Equation (3.5) shows that $I^0_{ac} = 0$ and $I^{90}_{ac} = \omega C V_{ac}$, or, in other words, the conductance and susceptance of a capacitor are given by

$$G_C = 0 \tag{3.15}$$

$$B_C = \omega C. \tag{3.16}$$

The current of a capacitor is fully 90° out-of-phase with the voltage and has a magnitude equal to the product of frequency, capacitance, and probe voltage amplitude.

A useful parameter is loss L. This has the same unit as capacitance, namely farad, and is defined as the conductance divided by the radial frequency, $L \equiv G/\omega$, and thus

$$G_L = \omega L \tag{3.17}$$

$$B_L = 0. \tag{3.18}$$

Loss is therefore, like conductance, in-phase with the probe voltage.

Finally, the loss-tangent is defined as the real part of admittance divided by the imaginary part, and is therefore a dimensionless entity. Combining the above, it can be shown that the loss-tangent is also the ratio of loss and capacitance,

$$\tan\delta \equiv G/B \tag{3.19}$$

$$= L/C = 1/\omega RC. \tag{3.20}$$

Analyzing this, it is clear that δ is the angle the admittance makes with the imaginary axis in a phase diagram, see Figure 3.2. This shows the inheritance of the loss-tangent, namely the parametrization of the 'lossy' aspect of a capacitor; an ideal capacitor has $\delta = 0$. However, an ideal conductor does not store charge and has $\theta = 0$ and $\delta = 90°$. For our research the loss-tangent comes in handy, as will be shown later.

The parameters given until now are all device parameters. In many cases, the material parameters are useful instead, especially for materials scientists. These are called conductivity (σ), resistivity ($\rho = 1/\sigma$) and permittivity (ε). The device parameters can be expressed in these material parameters

$$G = \sigma \frac{A}{d} \tag{3.21}$$

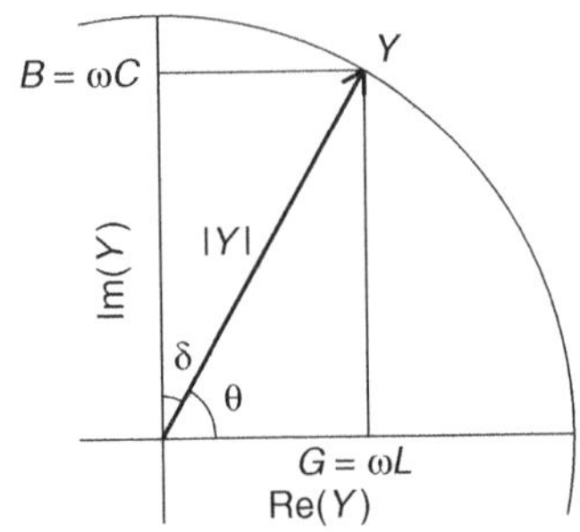

Figure 3.2 Polar representation of admittance (Y) and its components, conductance (G) and susceptance (B). The ratio of these two is the loss-tangent $\tan\delta$, where δ is the angle between the admittance vector and the imaginary axis. θ is the angle between the probing voltage and the response current

$$R = \rho \frac{d}{A} \tag{3.22}$$

$$C = \varepsilon \frac{A}{d} \tag{3.23}$$

where A is the area of the device and d is the length (distance between the electrodes), see Figure 1.15. The permittivity is often expressed in terms of permittivity of vacuum, $\varepsilon = \varepsilon_r \varepsilon_0$, with the dimensionless parameter ε_r called the relative permittivity or dielectric constant, also sometimes called K. Interestingly, for a homogenous device, the loss-tangent is independent of device dimensions and can therefore be classified as a material parameter, as well as a device parameter. Substituting the above definitions, Equation (3.20) becomes

$$\tan\delta = \frac{\sigma}{\omega\varepsilon}. \tag{3.24}$$

A 'lossy' capacitor thus means that the dielectric material is 'lossy'.

Finally, often the parameter 'susceptibility' (χ) is encountered. This is defined as the deviation from unity of the dielectric constant

$$\varepsilon_r = (1 + \chi). \tag{3.25}$$

Note that this is not a device-scaled version of susceptance. A vacuum-filled capacitor, for instance, has zero susceptibility, but nonzero susceptance. To further add to the confusion, often the susceptibility is assumed to have real and imaginary parts, $\chi = \chi' + i\chi''$, thus encompassing both the capacitance and conductance parts of admittance. I leave it here for the reader to work out the relation between the measured entities and χ. In this work χ is not used and is only mentioned for the sake of completeness.

All parameters mentioned in this section are summarized in Table 3.1.

3.1.1 Low-frequency RCL Bridge

On the basis of the above, it is easy to show how an ad-hoc RCL measurement bridge can be assembled from a two-channel lock-in detector. It consists of applying a small AC voltage and determining the in-phase and 90° out-of-phase current amplitudes, I_{ac}^{0} and I_{ac}^{90}, respectively. Figure 3.3 shows this schematically. A sine-wave voltage, based on the internal reference of the lock-in detector is applied to the sample (for simplicity, the bias is not shown). The response current is converted to a voltage signal ('*S*') and multiplied with the square wave reference ('*R*'). In case the current is in-phase with the applied voltage,

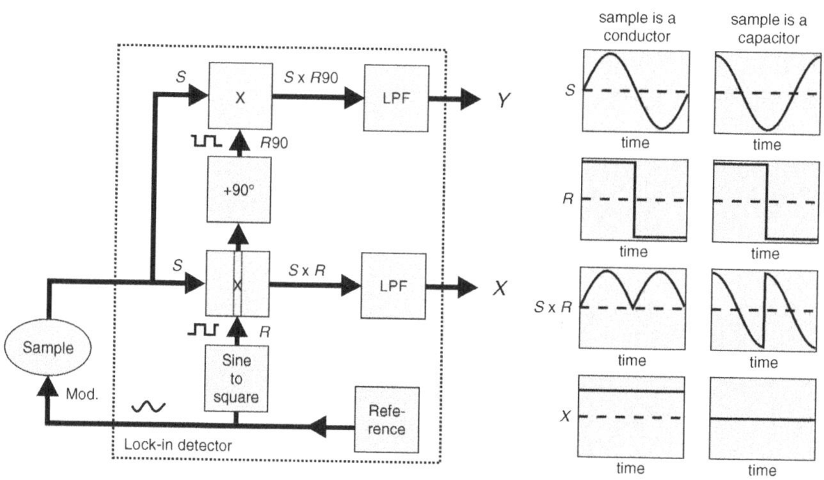

Figure 3.3 Schematic showing how phase-sensitive detection can be used to build an RCL bridge. The reference is used to modulate the signal, which has in-phase and out-of-phase components. When the signal is multiplied by the square-wave reference ('*R*') and (low-pass) filtered, the output signal ('*X*') is proportional to the amplitude of the in-phase component only, as shown. The in-phase part can be associated with conductance (and resistance), while the out-of-phase part represents the susceptance (and capacitance). In a two-channel lock-in detector, as shown, the out-of-phase signal amplitude ('*Y*') can be found by using a 90° phase-shifted reference signal. Thus, *X* is proportional to *G* and *Y* is proportional to *B*, the latter which, in turn, can be converted to capacitance according to Equation (3.16)

the resulting signal ('$S \times R$') is a rectified signal with a DC component ('offset'), and components at 2ω, 4ω, etc. The low-pass filter, consisting of a simple RC circuit, removes the frequency components and passes only the DC part with its slow variations (ideally, everything below ω). With the help of Fourier analysis we can calculate the output signal of the lock-in detector. For an in-phase signal (conductance) with amplitude A:

$$S(t) = S\sin(\omega t) \tag{3.26}$$

$$R(t) = \frac{4}{\pi} \sum_{n=1,3,5\ldots}^{\infty} \frac{1}{n} \sin(n\omega t) \tag{3.27}$$

$$S(t) \times R(t) = \frac{4S}{\pi} \sum_{n=1,3,5\ldots}^{\infty} \frac{1}{n} \sin(n\omega t)\sin(\omega t) \tag{3.28}$$

$$= \frac{4S}{\pi} \sum_{n=1,3,5\ldots}^{\infty} \frac{1}{2n} \{\cos[(n-1)\,\omega t] + \cos[(n+1)\,\omega t]\} \tag{3.29}$$

$$X(t) = \frac{2S}{\pi}. \tag{3.30}$$

For an out-of-phase signal (susceptance):

$$S(t) = S\cos(\omega t) \tag{3.31}$$

$$R(t) = \frac{4}{\pi} \sum_{n=1,3,5\ldots}^{\infty} \frac{1}{n} \sin(n\omega t) \tag{3.32}$$

$$S(t) \times R(t) = \frac{4S}{\pi} \sum_{n=1,3,5\ldots}^{\infty} \frac{1}{n} \cos(n\omega t)\sin(\omega t) \tag{3.33}$$

$$= \frac{4S}{\pi} \sum_{n=1,3,5\ldots}^{\infty} \frac{1}{2n} \{\sin[(n-1)\,\omega t] + \sin[(n+1)\,\omega t]\} \tag{3.34}$$

$$X(t) = 0. \tag{3.35}$$

Thus, the output signal 'X' of the lock-in detector is proportional to the amplitude of the in-phase component. The out-of-phase component is completely filtered off (by adequate selection of the cut-off frequency of the low-pass filter). By using a phase shift of 90° either for the sine wave before it is applied to the sample, or to the reference signal used in the multiplier, the system can be tuned for the 90° out-of-phase signal. Modern commercial (two-channel) lock-in detectors often have the possibility

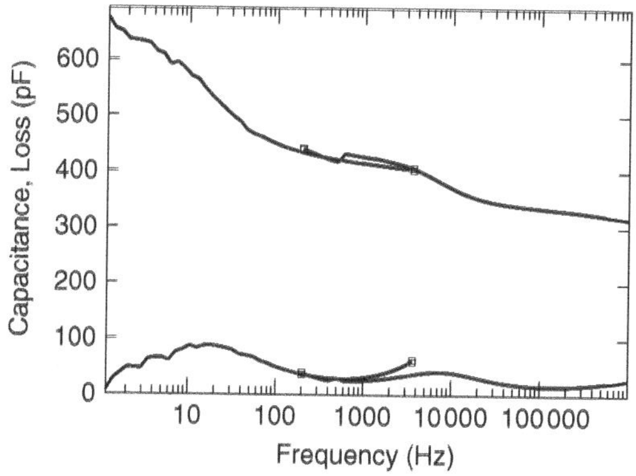

Figure 3.4 Example of a low-frequency RCL measurement based on a Stanford Research Systems SR830 lock-in detector, compared with the mid-frequency range commercial RCL bridge Fluke PM6306

to measure both in-phase and out-of-phase signals simultaneously. As discussed earlier, the in-phase signal corresponds to conductance and the out-of-phase signal corresponds to susceptance. Figure 3.4 shows an example of a low-frequency RCL bridge based on a two-channel lock-in amplifier compared with a commercial medium-frequency RCL bridge.

3.1.2 DC Admittance

DC admittance consists of determining the DC resistance and capacitance. While the former is easily defined by Ohm's Law as the ratio of DC voltage and current, the measurement of DC capacitance is strange, since according to the definition of capacitance [Equation (3.12)], this implies the measurement of charge inside a capacitor for any given voltage. That is not a sinecure. However, in the case where the DC conductance of the device is zero (no steady-state current), the charge inside the device can be found by integrating the current. In this way, the so-called quasi-static capacitance can be found by measuring the additional charge that enters the capacitor after a voltage step,

$$C \equiv \frac{\Delta Q}{\Delta V} = \frac{\int I(t)\mathrm{d}t}{\Delta V}. \tag{3.36}$$

In some cases this can give useful additional information. Commercial equipment is available that performs this type of measurements.

3.2 GEOMETRICAL CAPACITANCE

The geometrical capacitance is formed by the electrodes filled with a material. Any time we have an electronic component and connect it externally, the component electrodes 'feel' each other and this capacitive coupling is unavoidable. The geometrical capacitance for a parallel-plate electrode is given by

$$C_{\mathrm{geo}} = \frac{\varepsilon A}{d} \tag{3.37}$$

where A is the area of the electrodes, d is the distance between them and ε is the permittivity of the material of the sample.

If the device itself is resistive, it is the capacitance that will be measured. In other cases, care has to be taken to not forget about the geometrical capacitance. It depends on the device as to whether and how we have to take this into account.

3.3 EQUIVALENT CIRCUITS

The admittance spectra of electronic devices can often easily be modeled using electronic equivalent circuits. In fact, this is common practice in the literature. A process (also often called the Maxwell–Wagner process) with relaxation time τ can be represented by a serial circuit consisting of a resistance R and a capacitance C, with the condition $\tau = RC$, with C describing the total charge involved in the process ($C = \Delta Q/\Delta V$), for instance the capacitance of the interface, or the charge being trapped on deep states in a cycle of the probing signal. The resistance R has no physical meaning other than to parameterize the RC time τ.

When the admittance of this circuit is measured, this admittance can be translated into a parallel capacitance and resistance (or loss), whose values C_{p}, R_{p} and L_{p} depend on the radial frequency ($\omega = 2\pi f$), according to

$$\begin{aligned} C_{\mathrm{p}}(\omega) &= \left(\frac{1}{\omega^2\tau^2 + 1}\right) C \\ R_{\mathrm{p}}(\omega) &= \left(\frac{\omega^2\tau^2 + 1}{\omega^2\tau^2}\right) R \\ L_{\mathrm{p}}(\omega) &= \left(\frac{\omega\tau}{\omega^2\tau^2 + 1}\right) C \end{aligned} \tag{3.38}$$

where $\tau = RC$. The loss then has a maximum at a radial frequency of

$$\omega_{\text{max}} = 1/\tau \tag{3.39}$$

at which frequency the loss reaches half the value of the low-frequency capacitance and the capacitance itself drops and coincides with the loss at this frequency, $L_{\text{m}} = C_{\text{m}} = C/2$. [See Figure 3.5 for a simulation of the capacitance and loss according to Equation (3.38) with 1 nF capacitance and a cut-off frequency of 1 kHz.] The data are also often represented in a so-called Cole–Cole plot, namely L_{p} vs. C_{p} which is a semicircle for a single simple Maxwell–Wagner process, see the inset of Figure 3.5. Another popular form of representing the data is the loss-tangent ($\tan\delta$), which is defined as the loss divided by the capacitance, and is thus the tangent of the angle in a Cole–Cole plot. In the case of a simple process, the loss-tangent is equal to $\tan\delta = \omega/\omega_{\text{max}}$ and is thus a featureless function, see the dashed curve in Figure 3.5. In other cases, this function can be more informative, as will be shown later.

To summarize, a peak in the loss spectrum at ω_{max} indicates a physical process with relaxation time $\tau = 1/\omega_{\text{max}}$. Moreover, the temperature dependence of the peak will reveal the activation energy of the underlying process if the relaxation time follows Arrhenius behavior, $\tau \propto \exp(E_{\text{a}}/kT)$.

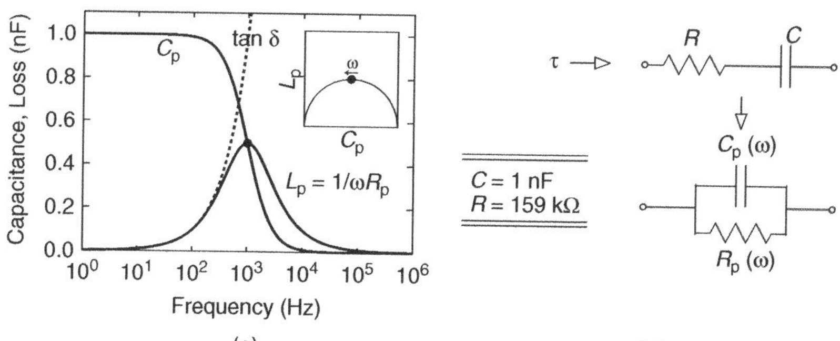

Figure 3.5 (a) Simulation of the spectra of capacitance (C_{p}), loss ($L_{\text{p}} = 1/\omega R_{\text{p}}$) and loss-tangent (dashed) of a Maxwell–Wagner process with relaxation time τ. The inset shows the same simulation in a Cole–Cole representation. For this simulation a capacitance $C = 1$ nF and a cut-off frequency of $f_{\text{c}} = 1/2\pi\tau = 1$ kHz was used. (b) Equivalent circuit of a resistance R and a capacitance C to represent the process with $\tau = RC$. The circuit translates into a parallel circuit with as-measured components R_{p} and C_{p} that depend on the measuring frequency, see Equation (3.38)

Moreover, these effects are cumulative. In other words, when more than one process is occurring in the device, in the loss plots a peak for every process is observed with an amplitude proportional to the charge involved in the process ($C = \Delta Q/V_{ac}$), and a semicircle is added to the Cole–Cole plots. See for example the double-relaxation presented in Figure 3.6. For a continuous distribution of relaxation times, the plots can become very complicated; the loss no longer reaches half-way the low-frequency capacitance and the Cole–Cole plots are depressed semicircles, with the center falling below the horizontal axis $L_p = 0$.

A more general form of Equation (3.38) is then

$$C_p(\omega) = \int \left(\frac{1}{\omega^2\tau^2 + 1} \right) C(\tau) d\tau$$

$$L_p(\omega) = \int \left(\frac{\omega\tau}{\omega^2\tau^2 + 1} \right) C(\tau) d\tau \tag{3.40}$$

where $C(\tau)$ is now the capacitance distribution function (unit: F s^{-1}). The capacitance can directly be linked to the DOS involved in the process, since it represents the capacity to store charge on these states. The relaxation time of the process is often the thermalization time of the states and this, as will be shown in Chapter 4, depends on the energetic depth of the states. These two elements govern the admittance spectra. Figure 3.7 gives an example of a system with charge being trapped on deep states. The DOS of these states is considered to be constant over a

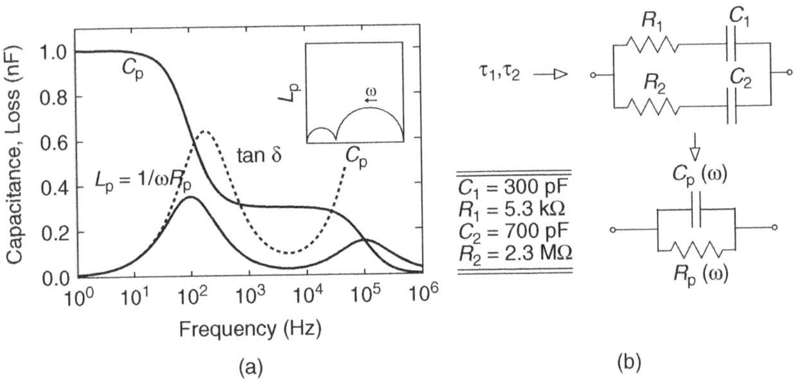

Figure 3.6 (a) Simulation of the spectra of capacitance (C_p), loss ($L_p = 1/\omega R_p$) and loss-tangent ($\tan\delta$, dashed) of a double Maxwell–Wagner process with relaxation times τ_1 and τ_2. (b) The circuit translates into a parallel circuit with as-measured components R_p and C_p that depend on the measuring frequency. The simulation values are shown

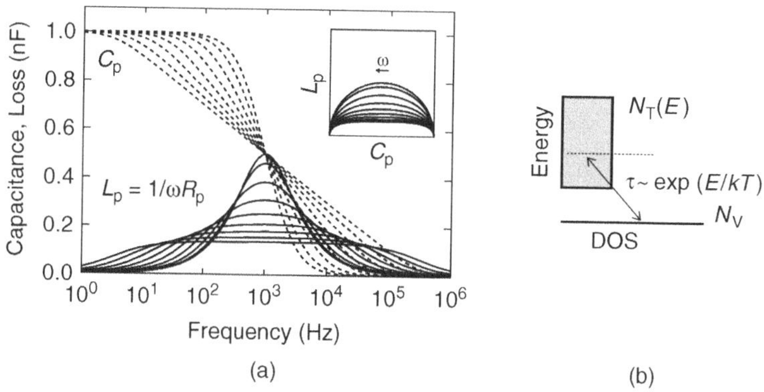

Figure 3.7 (a) Simulation of the spectra of capacitance (C_p, dashed) and loss ($L_p = 1/\omega R_p$, solid) of a system with traps homogeneously distributed in energy and with relaxation times τ exponentially depending on the depth. Each plot represents a certain width of the DOS, resulting in a dispersion in relaxation times, $\tau_{max} : \tau_{min}$ ranging from 1 to 10^5, centered around $\tau_0 = 159\,\mu s$. The Cole–Cole representation (inset) becomes flattened. (b) The associated energy diagram with a rectangular DOS for the electronic states involved

width ΔE centered around a certain value E_0. The thermalization time of the traps is assumed to be exponentially depending on their depth, with the value for τ at the center of the distribution giving rise to a peak in loss at 1 kHz, $\tau(E_0) = 159\,\mu s$.

Thus, the admittance spectroscopy data can give an insight into the DOS diagram of the states involved in the processes. However, there exist more sensitive techniques to determine the deep states; for instance, the transient techniques described in Chapter 4.

There can be other reasons than a limiting relaxation process for a cut-off frequency; for instance, purely electrical reasons. The Schottky barrier might serve as an example. A Schottky diode is normally composed of a depletion region and a bulk region, see Figure 1.17. Each region has its own capacitance and resistance. In this case the resistance has its conventional meaning. The electrical circuit is shown in Figure 3.8. The effective capacitance and conductance of this circuit depend on the measuring frequency, ω:

$$C(\omega) = \frac{R_d^2 C_d + R_b^2 C_b + \omega^2 R_d^2 R_b^2 C_d C_b (C_d + C_b)}{(R_d + R_b)^2 + \omega^2 R_d^2 R_b^2 (C_d + C_b)^2} \qquad (3.41)$$

$$G(\omega) = \frac{R_d + R_b + \omega^2 R_d R_b \left(R_d C_d^2 + R_b C_b^2\right)}{(R_d + R_b)^2 + \omega^2 R_d^2 R_b^2 (C_d + C_b)^2}.$$

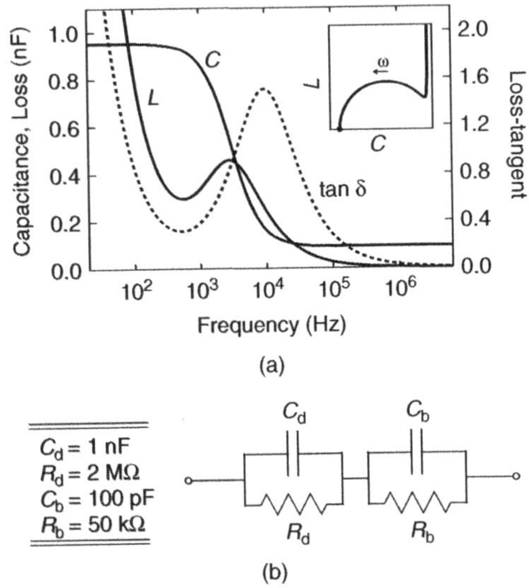

Figure 3.8 (a) Simulation of the measured capacitance (*C*), loss ($L = 1/\omega R$) and loss-tangent (dashed) of a Schottky barrier with a depletion region and a bulk region. (b) Equivalent circuit of the Schottky diode

This shows that, for low frequencies, the capacitance is nearly equal to the depletion capacitance (if $R_d \gg R_b$) and the resistance is equal to the total device resistance ($R = R_d + R_b$). For high frequencies, the capacitance approaches the geometric capacitance, the series-sum of the bulk and depletion capacitances, $C_{geo} = (C_d^{-1} + C_b^{-1})^{-1}$ and the resistance is equal to the bulk resistance when $R_d \gg R_b$ and $C_d \gg C_b$, see Figure 3.8. The Cole–Cole plot (see inset) reveals a slightly lowered semicircle which is also dislocated from the origin. In fact, this dislocation (indicated by a dot) is equal to the geometric capacitance and allows for the determination of the dielectric constant, ε_s, of the semiconductor, since $C_{geo} = \varepsilon_s \varepsilon_0 A/d_s$.

Furthermore, the loss-tangent is equal to

$$\tan\delta = \frac{\omega C_d R_b}{1 + (\omega R_b)^2 (C_d + C_b) C_b}. \qquad (3.42)$$

Assuming $R_b \ll R_d$ and $\omega^2 R_b R_d C_d^2 \gg 1$, the loss-tangent has a maximum at a frequency

$$\omega_{max} = \frac{1}{R_b}\sqrt{\frac{1}{C_b (C_b + C_d)}}, \qquad (3.43)$$

see the dashed curve in Figure 3.8.

The temperature dependence of the maximum in loss-tangent can easily be calculated on the basis of the individual temperature dependencies of resistance and capacitance. The bulk resistance follows a temperature dependence [Equation (2.7)]

$$R_b \propto \exp\left(E_A/2kT\right) \tag{3.44}$$

and the depletion capacitance will be discussed in the next section.

3.4 RESISTOR; SCLC

The AC conductance of a device, being defined as the derivative of the I–V curves can easily be found when the DC characteristics are known. To find the capacitance, the amount of charge in the device as a function of bias has to be found. The capacitance is then the derivative of this Q–V relation.

Thus, for a resistor working in ohmic regime, with a linear I–V curve, the conductance is constant. Since the amount of surplus charge in the device is negligible, the Q–V relation is flat and the capacitance zero. However, any two-terminal device is always composed of two electrodes and this set of electrodes filled with a material has its own capacitance called the geometrical capacitance G_{geo} which is unavoidable. (See Section 3.2 for more details on the geometrical capacitance.)

In the Poole–Frenkel or tunneling regimes, the same applies for the capacitance. Since the amount of injected charge is negligible, the capacitance is zero. The conductance can be found as the derivative of the I–V curves, which are generally speaking highly nonlinear.

More interesting becomes the case of the SCLC regime. The DC characteristics were treated in Section 2.1.4. The AC admittance (capacitance and conductance) of a device in the SCLC regime can also easily be calculated. First, the conductance is defined as the derivative of the I–V curve, see Equation (2.35). The conductance density is therefore

$$G_{SCLC} \equiv \frac{dJ(V)}{dV} = \frac{9\varepsilon_s\mu_p}{4}\frac{V}{d^3} \tag{3.45}$$

and depends linearly on the bias. To find the capacitance density C_{SCLC} we have to find an explicit form of the charge–voltage relation; the (dynamic) AC capacitance is then the derivative of this function. For that we first have to revert to the function of total charge in the device in the SCLC regime given in Chapter 2,

$$Q_{\mathrm{SCLC}} = \frac{3\varepsilon_s}{2}\frac{V}{d}. \tag{3.46}$$

The SCLC capacitance density is therefore

$$C_{\mathrm{SCLC}} \equiv \frac{dQ_{\mathrm{SCLC}}}{dV} = \frac{3\varepsilon_s}{2d} = \frac{3}{2}C_{\mathrm{geo}} \tag{3.47}$$

where C_{geo} is the geometric capacitance density of the same electrodes filled with a dielectric, $C_{\mathrm{geo}} = \varepsilon_s/d$. We could have taken an alternative route and arrive at the same relation if we had used the fact that the center of mass of the space-charge distribution lies one-third into the semiconductor. Effectively, the metal plates have approximated to a distance two-thirds of d, and the capacitance is given by $C_{\mathrm{SCLC}} = \varepsilon_s/(2/3)d$ yielding the same result as given above.

The SCLC capacitance is thus a featureless device parameter that does not yield any information (beyond a possible verification of the assumption of SCLC, or the estimation of the permittivity of the material, ε_s). However, from the ratio of loss and capacitance, the so-called loss-tangent $\tan\delta \equiv G/\omega C$, the permittivity can be eliminated from the equation and the mobility estimated,

$$\tan\delta_{\mathrm{SCLC}} = \frac{\mu_p V}{2\omega d^2}. \tag{3.48}$$

While this easily follows from simple calculations, no reports in the literature are readily found for this sort of measurement.

The above analysis assumes that the device is always in steady state and charge has enough time to arrive there where it is predicted. This implies a low oscillating frequency of the external field. For higher frequencies, the charge does not have enough time to even enter the device. In that case, the SCLC capacitance is zero and the total measured capacitance is expected to return to the geometrical capacitance value. Remarkably, this simple analysis is not in line with predictions and observations of earlier work by Shao and Wright [132] that foresee an increase in capacitance for high frequencies, rather than a decrease.

3.5 SCHOTTKY DIODES

Since conductance does not have any phase shift compared with the probing voltage, it can be found directly from the derivative of the

I–V curve, as shown by Equation (3.2). Since all models of the current through a Schottky barrier are of the form

$$J(V) = J_0\left[\exp(qV/nkT) - 1\right] \tag{3.49}$$

the conductance of a Schottky barrier is, independent of the frequency, of the form

$$G(V) = G_0 \exp\left(\frac{qV}{nkT}\right) \tag{3.50}$$

and as an experimental analytical tool therefore not very interesting. Measurement of G_0 as a function of temperature might be meaningful. However, it does not give additional information compared with the much cheaper and simpler DC technique.

The capacitance of a Schottky barrier can readily be calculated when one realizes that (dynamic) capacitance is defined as the amount of charge that moves in or out of the device when the bias is changed, $C \equiv \mathrm{d}Q/\mathrm{d}V$. The amount of charge in the interface is the shaded area shown in Figure 1.18(a),

$$\begin{aligned} Q &= -qN_A WA \\ &= -A\sqrt{2q\varepsilon_s N_A(V_{bi} - V)} \end{aligned} \tag{3.51}$$

where A is the area of the interface between the metal and semiconductor. The measured (dynamic) capacitance is thus

$$C(V) \equiv \frac{dQ(V)}{dV} = A\sqrt{\frac{q\varepsilon_s N_A}{2(V_{bi} - V)}}. \tag{3.52}$$

Upon closer examination, this is equal to

$$C = A\varepsilon_s/W. \tag{3.53}$$

In other words, the Schottky diode, in terms of capacitance, behaves like a metal-plates capacitor with the electrodes at distance W and filled with a medium with permittivity ε_s. Since W depends on the bias, the capacitance becomes bias-dependent as well [Equation (3.52)] and that distinguishes a static capacitance from the dynamic capacitance described above.

The best way of representing data is always a form in which they fall on a straight line when the theory is correct. For the C–V data of

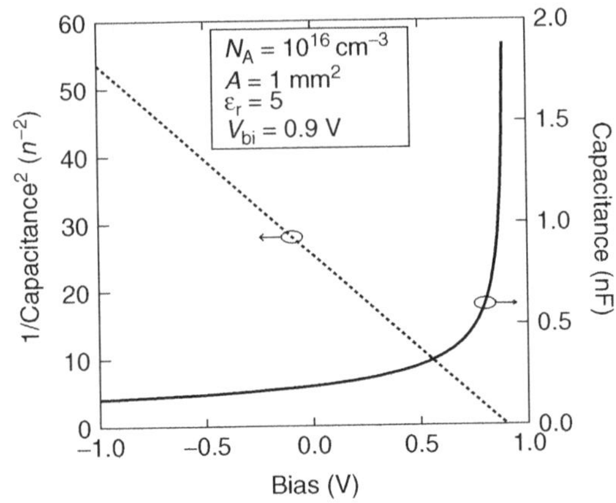

Figure 3.9 Example of a Mott–Schottky representation of the capacitance–voltage measurements of a Schottky barrier

a Schottky barrier, therefore, a so-called Mott–Schottky plot is most adequate, which is C^{-2} vs. V, the slope of which reveals the acceptor concentration,

$$\frac{\mathrm{d}}{\mathrm{d}V}\frac{1}{C^2} = -\frac{2}{q\varepsilon_s A^2 N_A} \tag{3.54}$$

and the intercept with the horizontal axis at $V = V_{bi}$, normally found by extrapolation. An example is given in Figure 3.9.

The capacitance of Equation (3.52) is applicable when the device is in depletion and is therefore also often called depletion capacitance. When the device is placed in forward bias, the capacitance is called diffusion capacitance and is not so interesting from the point of view of a materials scientist. It is more interesting to place the device in strong forward bias and enter the SCLC bulk-limited regime.

Finally, it has to be mentioned that the Schottky barrier analysis is only allowed when the dimensions y, z parallel to the interface can be considered stretching to infinity for the sake of the calculations. In devices that are very reduced in size this is no longer valid. Moreover, other objects present in the vicinity, can destroy the symmetry of the device and undermine the validity of the Schottky barrier analysis. In other words, the Schottky barrier analysis cannot be applied to every metal–semiconductor contact. As an example, the Schottky barrier analysis is not applicable to metal contacts in thin-film FETs, first because the contact is not an area contact but a line contact and secondly, the gate

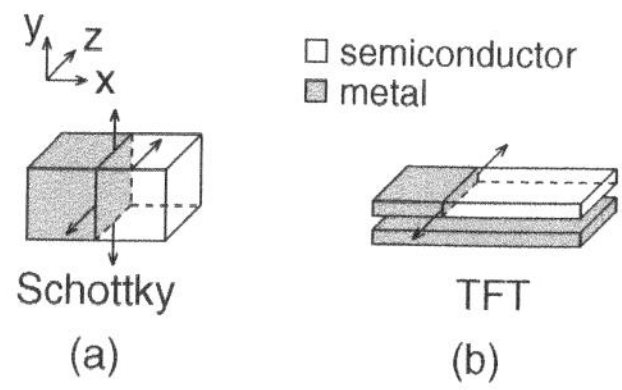

Figure 3.10 Difference between (a) a pure Schottky barrier and (b) a metal–semiconductor contact in a TFT. In (a) the contact can be considered infinite in two directions perpendicular to the interface, as indicated by the arrows. The interface area A then enters as a scaling parameter while in (b) the interface is a line contact with a scaling parameter W. Moreover, the gate metal destroys the symmetry. Obviously, the Schottky barrier analysis cannot be applied to the metal contacts in TFTs

is in very close proximity and further destroys the symmetry. Figure 3.10 shows the difference between a Schottky diode and a metal–semiconductor contact in a TFT. The metal–semiconductor contacts in TFTs will be treated in more detail in Chapter 6.

In some cases, the low mobility of carriers of organic materials may disturb the standard Schottky barrier admittance. In the above analysis it was assumed that, when the bias is changed, the free holes move in and out instantaneously. The capacitance is then the ratio of charge moving in to the change in voltage. For materials with low mobilities, the charges do not have time to respond before a cycle of the probing voltage is completed.

The re-establishment of a new distribution of charge is governed by diffusion. The time it takes for the holes to move in or out of the device and reach the edge of the depletion zone is quadratically dependent on the distance to travel

$$\tau = (\Delta x)^2/D \qquad (3.55)$$
$$= \frac{q}{kT}\frac{\Delta x^2}{\mu}$$

where the Einstein Relation $D = kT\mu/q$ was used. As an example, to travel a distance of 300 nm in silicon takes about 25 ps, because D_n is about 36 cm^2 s^{-1} (and $\mu_n = 1400$ cm^2 V^{-1}s^{-1}). Organic materials can have mobilities as low as 10^{-6} cm^2 V^{-1}s^{-1}. This makes the diffusion coefficient 9 orders of magnitude smaller and the response time of a diode in the order of 25 ms. Only modulation of the external field with a time period longer than this time will cause charges to move in and out of the interface region and contribute to capacitance. In other words, there is a cut-off frequency in the capacitance

$$f_c = \frac{1}{2\pi\tau} \tag{3.56}$$

which can be very low indeed for organic materials. A real-case of an organic MIS diode is given in a subsequent section. An MIS diode is the preferred device architecture when studying these effects, because the DC current is low enough to not interfere with low-frequency admittance spectroscopy.

3.5.1 Schottky Diode; Nonuniform Doping

In the case of a nonuniform doping profile, the relation $Q = qAWN_A$ no longer holds. Instead we have to use $Q = \int qN_A(x)A\mathrm{d}x$. The calculation of the depletion width is also different, but the relation $C = \varepsilon A/W$ is still valid. The end result is that the slope in a Mott–Schottky plot still reveals the acceptor concentration but this time only locally at the end of the depletion width and the slope can therefore change with voltage:

$$\frac{\mathrm{d}C^{-2}}{\mathrm{d}V} = -\frac{2}{q\varepsilon_s A^2 N_A(x)} \tag{3.57}$$

with

$$x = W = \varepsilon_s A/C. \tag{3.58}$$

This enables the determination of the doping profile $N_A(x)$ through the C–V measurement. The depth x is determined by the measured capacitance. The dopant density at that place is then determined by the measured derivative in the Mott–Schottky plot.

3.5.2 Schottky Diode; Adding an Abundant Deep Acceptor Level

It is interesting to see what happens to the behavior of the Schottky diode in the presence of a second, deep level. The band diagram of such a system is shown in Figure 3.11. For simplicity, it is assumed here that a level is fully ionized when it is below the Fermi level and

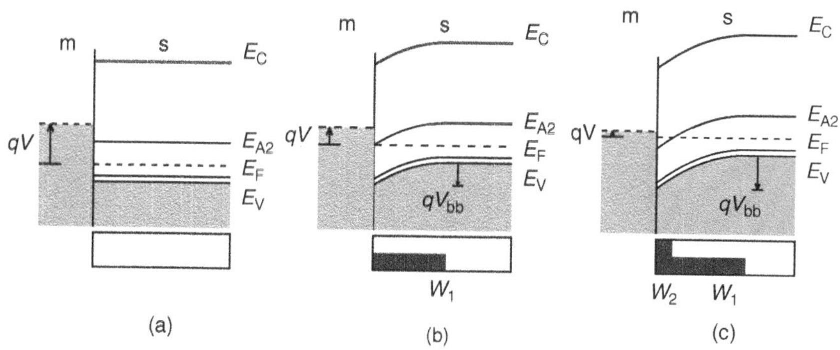

Figure 3.11 Energy diagrams of a Schottky barrier with a second acceptor level, E_{A2}. (The shallow acceptor level is drawn but not labeled.) (a) For $V = V_{bi}$; (b) for $V = V_{A2}$; (c) for $V < V_{A2}$ in which case there is a region with increased density of space charge close to the interface ($x < W_2$). (Negative) space charge density is schematically shown below the images

fully neutral when above it, as if it were at $T = 0$. At small band bendings (large forward biases), the Fermi level is everywhere below the second acceptor level which therefore plays no part in the space charge and the capacitance. The capacitance is given by the regular depletion capacitance, Equation (3.52).

When the bias is lowered, starting from a certain value, the Fermi level crosses the second acceptor level. The bias V_{A2} at which the effect of the second acceptor level is starting is when the band bending is equal to the difference between the deep level and the Fermi level in the bulk (which is denoted ΔV_1 here), thus

$$\begin{aligned} V_{A2} &= V_{bi} - \Delta V_1 \\ &= V_{bi} - (E_{A2} - E_F)/q \end{aligned} \tag{3.59}$$

At this point there exist two distinct regions. One, W_1, with space charge density equal to N_A, and one, W_2, where this density is the sum of the two acceptor densities, $N_A + N_{A2}$. The first region (W_1) has a constant width, independent of bias. This width can be calculated using the theory of the simple Schottky barrier depletion, using $\Delta V_1 = -(E_{A2} - E_F)/q$ for the band bending in this region,

$$W_1 = \sqrt{-\frac{2\varepsilon_s \Delta V_1}{qN_A}}. \tag{3.60}$$

At the edge, the electric field is given by

$$E_{W_1} = \int_{W}^{W-W_1} \frac{qN_A}{\varepsilon_S} dx$$
$$= -\sqrt{-\frac{2\Delta V_1 q N_A}{\varepsilon_S}}. \tag{3.61}$$

Continuing Poisson integration, the electric field in the second region, W_2, can then be found as

$$E(x) = -\int_{W_2}^{x} \frac{q(N_A + N_{A2})}{\varepsilon_S} dx + E_{W_1}. \tag{3.62}$$

Integrating and substitution results in a total band bending

$$V = \Delta V_1 - \frac{q(N_A + N_{A2})}{2\varepsilon_S} W_2^2 - \sqrt{-\frac{2\varepsilon N_A \Delta V_1}{\varepsilon_S}} W_2 \tag{3.63}$$

which can be solved easily, and yields W_2 as a function of bias V. The total depletion width is the sum of the two, $W = W_1 + W_2$.

At first we might think that we can now directly apply the relation between depletion width and capacitance established before [Equation (3.53)] to find the capacitance of this Schottky barrier with deep levels. However, this relation implies intrinsically that all charges taking part in a cycle of the external voltage arrive at the edge of the depletion width W_1. Part of the charge however, arrives much closer to the interface, at the edge of W_2. As an example, for the critical voltage V_{A2} given above, charge arrives right at the interface ($x = 0$). This charge therefore feels an infinite capacitance. For other situations, the capacitance is equal to $C_2 = \varepsilon A/W_2$. The total capacitance is then the parallel sum of the regular capacitance, which can be called C_1, associated with the total depletion width and the capacitance associated with W_2.

$$C_0(V) = \frac{\varepsilon_S A}{W_1 + W_2(V)} + \frac{\varepsilon_S A}{W_2(V)}. \tag{3.64}$$

This, however, is still not the end of the story. The above assumes that thermal equilibrium is established at all times during a cycle, so effectively for $f = 0$ Hz. When the oscillating frequency is too high, the charges do not have enough time to be captured on the deep levels or be emitted from them in region W_2 to establish the thermal equilibrium densities described above. In this case, indeed, all new charge arriving in the device and leaving it in a cycle of the voltage is only at the edge of the total depletion zone, $W = W_1 + W_2$, and the capacitance thus, for high frequencies, is the one that we expected in the first place,

$$C_\infty(V) = \frac{\varepsilon_s A}{W_1 + W_2(V)}. \tag{3.65}$$

This results in dual-slope Mott–Schottky plots, often observed in the literature [133, 134]. Figure 3.12 shows an example of the low and high frequency capacitance for a double-level system.

When the deep states are so slow that they do not even thermalize during the DC bias changes, the effect of the deep levels is only a shift of the capacitance (and conductance) curves along the voltage axis, since a constant density of charge results (via Poisson's Equation) in the addition of a constant voltage. The Mott–Schottky plot only reveals the shallow acceptor concentration.

To summarize, there are three possibilities:

1. The deep-level thermalization is very fast. It is faster than the period of the probing voltage. In this case, the capacitance is high and shows a sharp rise at the voltage where the Fermi level starts crossing the deep level.

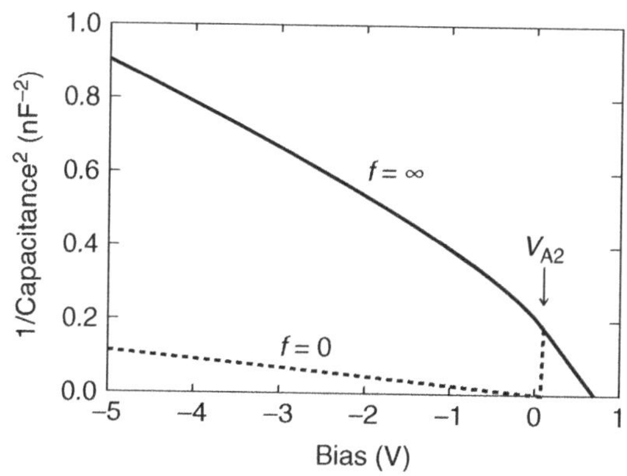

Figure 3.12 Mott–Schottky plot for a double-acceptor-level system as depicted in Figure 3.11. $E_{A2} - E_F = 0.7$ eV, $N_A = 10^{16}$ cm^{-3} and $N_{A2} = 3N_A$, $\varepsilon_s = 11.9\varepsilon_0$. The solid line is for $f = \infty$ where the deep acceptor is not thermalizing within a period of the probing signal and the dashed line is for $f = 0$ with the deep acceptor fully thermalizing

2. The deep-level thermalization time is intermediate. This means slower than the oscillations of the AC probing voltage. In this case, the capacitance is low and shows a change of slope in the Mott–Schottky plots.
3. The deep-level thermalization is slow. This means that the degree of ionization does not even change during a bias scan. In this case, the C–V plot (and the Mott–Schottky plot with it) is shifted along the bias axis, a fact that can be achieved only by a sustained application of the bias, so-called 'stressing' of the device.

A simple and general way of simulating the C–V and Mott–Schottky plot is numerically. Figure 3.13 shows a generic algorithm for determining Schottky diode capacitance. Analytical solutions can also be found in the literature, for example as reported by Taylor and Gomes [135].

3.5.3 Schottky Diode; Minority Levels

Another interesting system is one with, apart from the acceptor (majority) levels, a donor (minority) level. Figure 3.14 schematically shows the energy diagram of a system with two acceptor levels and a donor level. There exist various regions:

I. In the bulk, the Fermi level is assumed to be:

   - above the shallow acceptor level. These acceptors are therefore all ionized (negatively charged);
   - below the deep acceptor level which is neutral;
   - below the donor level which is ionized (positively charged).

   The material is neutral, therefore the free hole concentration in the bulk is equal to $N_{A1} - N_D$.

II. In the normal depletion zone, starting at $x = W$, the free charges have moved out and the space charge is equal to $N_{A1} - N_D$.

III. In the enhanced depletion region, the band bending has caused the deep acceptor level to drop below the Fermi level. These deep acceptors are then also ionized and the space charge density is therefore equal to $N_{A1} + N_{A2} - N_D$.

IV. Finally, in a region close to the interface, the band bending is so large that the donor level drops below the Fermi level. The donors in this region are therefore no longer ionized and stop reducing the space charge which is now $N_{A1} + N_{A2}$.

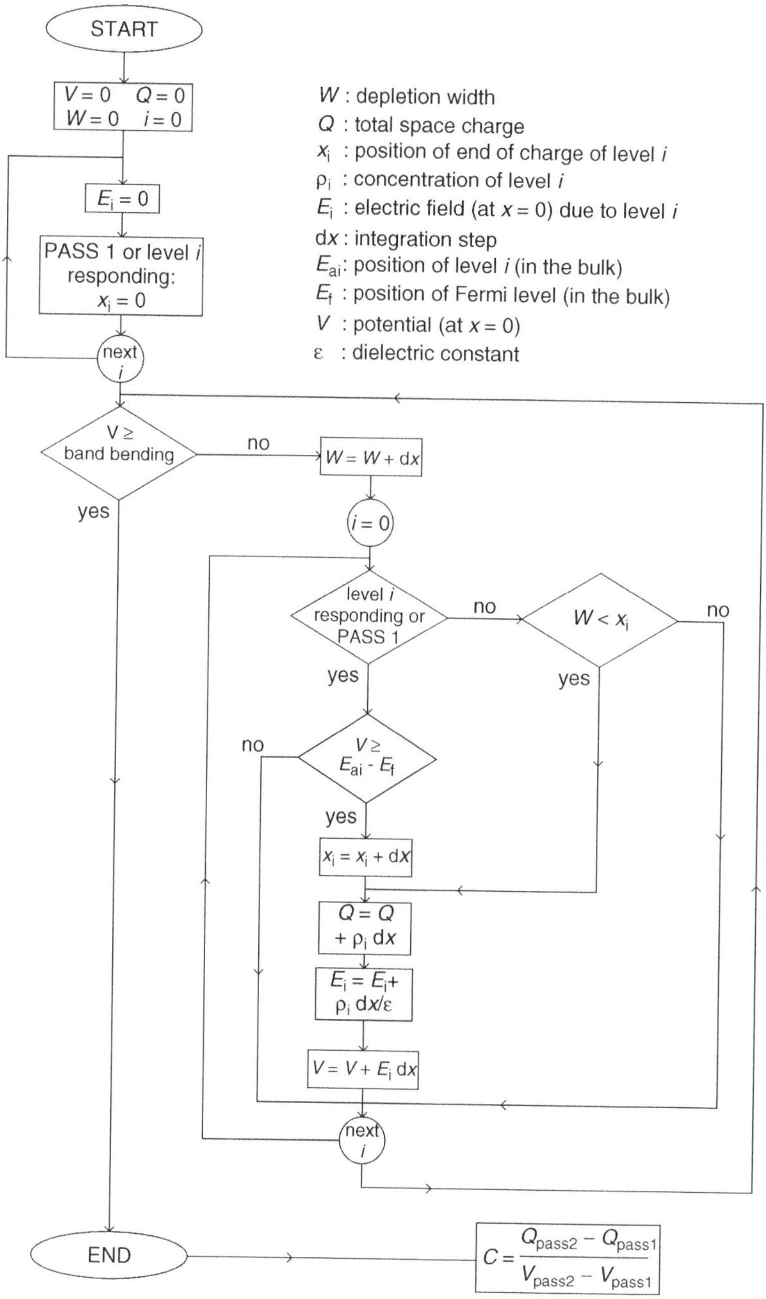

Figure 3.13 Generic algorithm for determining Schottky diode capacitance with several acceptor levels. This algorithm was used to generate Figure 3.12

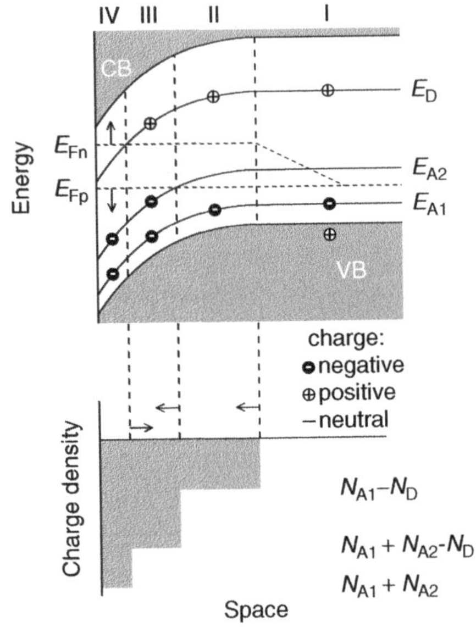

Figure 3.14 Band diagram in the presence of two acceptor levels (a shallow one and a deep one) and a donor level. The donor level communicates with the conduction band and thus thermalizes according to the electron quasi Fermi level

Because the donor level communicates with the conduction band, it thermalizes according to the electron quasi Fermi level, E_{Fn}. Because the quasi Fermi level for electrons and holes behave differently upon changes of bias, so do the regions. For increased forward bias (reducing band bending), the hole quasi Fermi level, E_{Fp}, moves down at the interface, whereas the electron quasi Fermi level moves up, see Figure 3.14. Regions II and III therefore shrink, thereby increasing the capacitance, but region IV grows, thereby reducing the capacitance. This effect is also visible in DLTS, which studies the kinetics of the level filling end emptying. This will be discussed in Chapter 4.

3.5.4 Schottky Barrier; Temperature Dependence

In the classical analysis it is implicitly assumed that all the acceptors are ionized and the charge distribution of Figure 1.18 is caused by the free holes that have moved out of the interface region and left behind uncompensated ionized acceptors with a resulting density of charge equal to N_A. In reality, for organic materials, or in general for low conductivity materials, it is possible that not all acceptors are ionized at room temperature. In other words, the Fermi level is rather close to the acceptor levels in the bulk. However, inside the depletion zone, the acceptor levels, together with the bands, move away from the Fermi level, thus making all the acceptor levels ionized (N_A), even if they are not so in the bulk. In other words, even for acceptors that are not fully ionized in the bulk, all of them contribute to the capacitance, since inside the depletion zone they are all ionized.

As a consequence, changes in temperature, and accompanying changes of the degree of ionization in the bulk and shifts of the Fermi level, therefore have no effect on the depletion capacitance. The capacitance is in first order independent of temperature.

However, this assumes that the deep levels have enough time to thermalize when changes of external voltage occur (either bias, or the oscillating part of the probe voltage in admittance measurements). For deep levels this is not necessarily the case, a fact that was pointed out in the section on capacitance in the presence of deep levels (Section 3.5.2). When the probing frequency is too fast, the capacitance drops to a lower value. This occurs for a frequency that is the reciprocal of the deep-level thermalization time τ. In most cases this thermalization time is thermally activated. We can therefore expect the capacitance to be thermally activated.

In summary if the AC probing frequency is low compared with the deep-level thermalization time, the capacitance is high and independent of temperature. If the probing frequency is of the same magnitude as the thermalization time, the capacitance is strongly temperature dependent and moves from the low-frequency value to the high-frequency value of Figure 3.12 when the temperature is lowered. If the temperature is further decreased so that even the scanning speed is faster than the thermalization time, the capacitance is independent of temperature, but strong stressing effects can be expected, where the C–V and Mott–Schottky curves depend on the scanning speed and even on the history of the device.

3.6 MIS DIODES

MIS devices are MIS layered structures and can be considered Schottky diodes with an insulator layer between the metal and the semiconductor. For ideal MIS diodes, the insulator does not conduct and the only conductivity that can be measured is AC conductance. Even so, the MIS diode is ideal for studying interface states and mobility. Apart from this, the MIS diode is important because it is a precursor to the FET that will be discussed later.

A simple MIS device made of high mobility semiconductor can be considered a capacitor, with the value of the capacitance equal to the oxide capacitance (note, all capacitances given are capacitance densities, with unit F m^{-2})

$$C_{ox} = \frac{\varepsilon_{ox}}{d_{ox}} \tag{3.66}$$

where ε_{ox} is the permittivity of the insulator, and d_{ox} is its thickness. This assumes free charges all the way up to the interface, or in other words, the device in accumulation. The device goes from electron accumulation to hole accumulation. In both cases the capacitance is equal to C_{ox}. In between, there is a transient range where the capacitance reaches a minimum. In this minimum the entire device is free of carriers and the capacitance is the so-called geometric capacitance C_{geo}, which is the parallel-metal-plates capacitance with two dielectrics, or, in other words, it is the series capacitance of the oxide capacitance (C_{ox}) and the semiconductor capacitance (C_s),

$$C_s = \frac{\varepsilon_s}{d_s} \tag{3.67}$$

$$C_{geo} = \frac{1}{C_s^{-1} + C_{ox}^{-1}}. \tag{3.68}$$

For pure materials, where there are no voltage drops in the semiconductor because there is no possibility of space charge inside the semiconductor (no acceptors or donors to ionize); all voltage drop is across the insulator. In this case, the transition from hole accumulation to electron accumulation is instantaneous. The width of the dip in the C–V plot is narrow. The presence of donors or acceptors can change this as will be discussed later.

3.6.1 MIS of Doped Semiconductors

In the traditional case of doped semiconductors, there are several bias regimes, labeled accumulation, depletion, inversion and strong inversion. They will now be discussed. Figure 3.15 shows the energy diagrams for these regimes. In what follows, it is assumed that at zero bias, there is no band bending in the semiconductor (Figure 3.15b), or in other words, $V_{FB} = 0$. In case this voltage is not zero, the bias is effectively offset by this amount, but otherwise the ideas and calculations are the same.

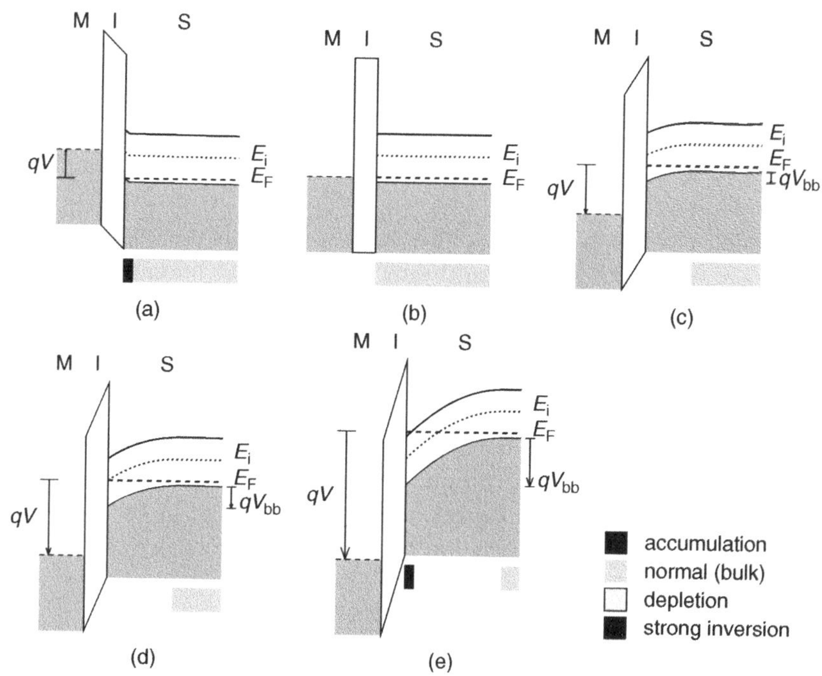

Figure 3.15 Energy diagrams of an MIS device in accumulation (a), flat band (b), depletion (c), onset of inversion (d) and strong inversion (e). Below each diagram the free-charge density is shown schematically. Insulator width and energy levels are not shown to scale

When the device is in **accumulation** (Figure 3.15a), free charge exists all the way up to the interface. The capacitance of the device is thus equal to the insulator capacitance (C_{ox}). At the interface, the free charge density is very high in a very narrow layer. In the rest of the semiconductor, the charge density is equal to the bulk value.

When the device is in **depletion** (Figure 3.15c), a space charge region exists that consists of ionized acceptors (N_A^-) that are not neutralized by free carriers as in the bulk region. Since in this region there is no free charge, it behaves like an insulator of width W whose capacitance is equal to

$$C_D = \frac{\varepsilon_S}{W}. \tag{3.69}$$

The total capacitance is the series capacitance of C_{ox} and C_D. Because the depletion capacitance is bias dependent, so is the total capacitance, see the Equivalent circuit in Figure 3.16. For this reason, an MIS is

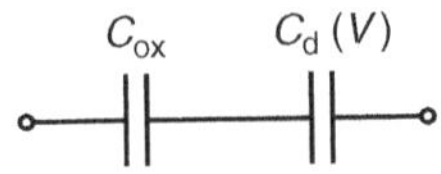

Figure 3.16 Equivalent circuit of an MIS device working in depletion and inversion

also called a varicap. Initially, the depletion width is very small and its capacitance very high. Continuing lowering the bias increases the depletion width and lowers the capacitance.

It is not difficult to calculate the capacitance as a function of bias. We start with assuming a value for the band bending inside the semiconductor, V_{bb}. The way this band bending is formed is equal to the depletion zone in Schottky barriers, and it can thus easily be shown that the depletion width [compare Equation (1.27)], $W = (2\varepsilon_s V_{bb}/qN_A)^{1/2}$. The electrical field is the integral of the space charge in the depletion and reaches a maximum at the interface (with a factor ε_s),

$$\begin{aligned} E_{max} &= \int_0^W qN_A/\varepsilon_s dx \\ &= (qN_A/\varepsilon_s)W \\ &= \sqrt{qN_A 2V_{bb}/\varepsilon_s}. \end{aligned} \tag{3.70}$$

According to Maxwell equations, displacement ($D = \varepsilon E$) is continuous across the interface. On the oxide side of the interface, the electric field is therefore

$$E_{ox} = (1/\varepsilon_{ox})\sqrt{q\varepsilon_s N_A 2V_{bb}}. \tag{3.71}$$

Inside the oxide, this field is constant due to the absence of charges. The total voltage drop in the oxide is the integral of this constant field and is thus $\Delta V_{ox} = d_{ox}E_{ox}$. Together with the band bending in the semiconductor, the total voltage drop across the device, at the onset of strong inversion is

$$V = V_{bb} + (1/C_{ox})\sqrt{2q\varepsilon_s N_A V_{bb}}. \tag{3.72}$$

The first term representing the voltage drop in the semiconductor and the second term is the voltage drop in the oxide. The total (negative) charge density (per m^2) in the semiconductor is the integral of the uncompensated ionized acceptors in the depletion zone,

$$Q = -\int_0^W qN_A \mathrm{d}x = -\sqrt{2q\varepsilon_s N_A V_{bb}}. \tag{3.73}$$

Equations (3.72) and (3.73) can be solved to give the total charge as a function of the total voltage drop. The effective capacitance density as a function of bias is then defined as

$$C \equiv \frac{\mathrm{d}Q}{\mathrm{d}V} = \frac{\mathrm{d}Q/\mathrm{d}V_{bb}}{\mathrm{d}V/\mathrm{d}V_{bb}} = \left(\frac{1}{C_{ox}} + \sqrt{\frac{2V_{bb}}{q\varepsilon_s N_A}}\right)^{-1} = (C_{ox}^{-1} + C_d^{-1})^{-1} \tag{3.74}$$

i.e., the series sum of oxide and depletion capacitance. Since V_{bb} is not a simple function of bias, a rather complicated capacitance–voltage relation results. Some rearranging of terms yields

$$C(V) = \left[\frac{1}{C_{ox}^2} + \frac{2(V_{FB} - V)}{qN_A\varepsilon_s}\right]^{-1/2} \tag{3.75}$$

where also a 'flat-band' voltage was defined that is the bias needed to remove the band bendings, similar to the built-in voltage in Schottky diodes. For this bias, the band bending disappears, the charge arrives at the interface, and the capacitance is equal to the oxide capacitance and equal to the capacitance in accumulation, as expected. For depletion ($V < 0$) the capacitance decreases as it is now the series capacitance of the oxide capacitance and the depletion capacitance, but the Mott–Schottky plot C^{-2} vs. V is no longer linear, unless the oxide layer is thin and the oxide capacitance very large.

In **inversion**, the band bending has increased to such an extent that close to the interface there exists a region where the Fermi level is closer to the conduction carrier band than to the valence band and minority carriers outnumber the majority carriers, hence 'inversion'. Otherwise, the inversion regime is equal to the depletion regime. Space charge is composed of ionized acceptors (the free electrons still play an insignificant role). The capacitance is still the series sum of the depletion capacitance and the insulator capacitance. Figure 3.15(d) shows the onset of inversion, with the Fermi level at the interface equal to the intrinsic level E_i defined as that value of the Fermi level where $n = p$.

For **strong inversion** (Figure 3.15d), the band bending is now so large that at the interface the Fermi level is so close to the conduction band that a narrow layer of electrons (minority carriers) is formed. The bias at which this happens is called the threshold voltage and can be found by substituting $2V_B$ for V_{bb} in Equation (3.72). V_B is the distance from the bulk Fermi level to the midgap level $V_B = [(E_V + E_C)/2 - E_F]/q$.

$$V_T = 2V_B + (1/C_{ox})\sqrt{4q\varepsilon_s N_A V_B}. \tag{3.76}$$

By this definition, V_T will cause at the interface an electron density equal to the hole density in the bulk. To get an idea of magnitude of the above expression and the relative voltage drops in the semiconductor (first term) and insulator (second term) typical values are $N_A = 10^{16}$ cm^{-3}, $2V_B = 1.12$ V (band gap of silicon), $C_{ox} = 173\,\mu F\ m^{-2}$ (200 nm of silicon oxide), $\varepsilon_s = 11.9\varepsilon_0$ (silicon), yields a threshold voltage of 4.68 V, with a voltage drop of 3.56 V in the oxide and 1.12 V in the semiconductor.

From this moment on, the band bending in the semiconductor is constant because any tiny increase in the band bending and associated shift of the Fermi level will bring with it a huge increase in free charge density close to the interface. To a good approximation it can thus be stated that any additional increase in bias is completely absorbed by the insulator. Moreover, for the same reason, all additional charge induced in the device is necessarily close to the interface and the the effective capacitance is thus C_{ox}. Furthermore, the charge induced at the interface from now on overshadows the charge in the rest of the device and the free electron density is linearly proportional to the bias,

$$n = C_{ox}(V - V_T)/q. \tag{3.77}$$

V_T, the threshold voltage, thus represents the bias needed to induce a minority-carrier layer in the semiconductor. This is an important relation that will be used later for metal–oxide–semiconductor field-effect transistor (MOS-FET) devices. Figure 3.17 shows the space charge (ionized acceptors not compensated by free holes plus free electrons) and free charge in an MIS working in strong inversion.

Just before reaching strong inversion, the capacitance is at its minimum. The value of this minimum can be calculated by substituting

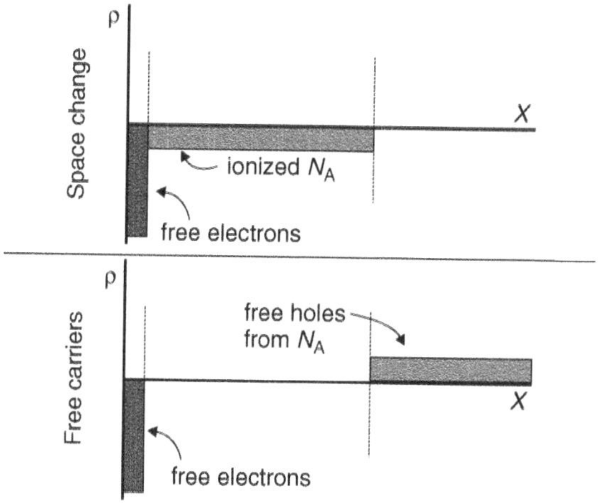

Figure 3.17 Space charge and free charge distribution of an MIS device in strong inversion (not to scale)

$V_{bb} = 2V_B$ into Equation (3.74),

$$C_{min} = \left(\sqrt{\frac{4V_B}{q\varepsilon_s N_A}} + \frac{1}{C_{ox}}\right)^{-1}. \quad (3.78)$$

It is here assumed that the semiconductor layer of the device is thick enough to completely accommodate this depletion. If not, the capacitance settles at the geometric capacitance of Equation (3.68).

Finally, the threshold voltage for an accumulation layer can be calculated to be zero. By definition, for $V = 0$, the hole density at the interface is equal to the hole density in the bulk.

Figure 3.18 summarizes the C–V relation. The capacitance goes from C_{ox} in accumulation (I) to C_{min} in depletion and inversion (II) back to C_{ox} in strong inversion (III).

Again, for high frequencies and low mobility, the charges can have difficulty moving in and out of the device and reaching the interface and establishing new equilibrium within a time period of the external field. This results, as before, in an as-measured capacitance lower than C_{ox} in accumulation and strong inversion. Under such conditions, the spectra of admittance in strong inversion and accumulation will reveal the electron and hole mobility, respectively. In strong inversion, there is another effect that can make the device capacitance frequency dependent. In strong inversion, the capacitance is equal to C_{ox} because all additional charge appears at the interface. However, these charges have to reach

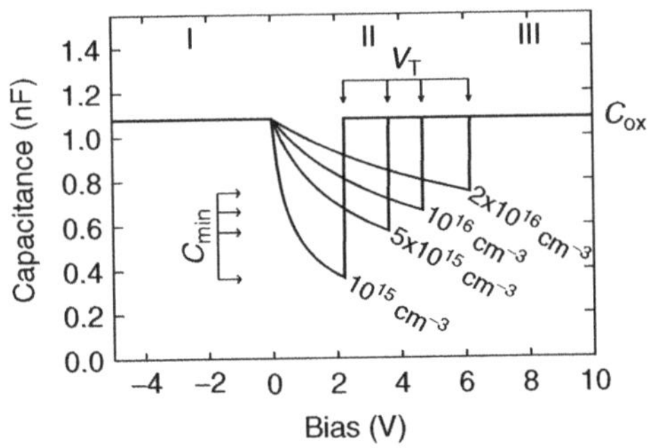

Figure 3.18 C–V plots for an MIS diode for acceptor concentrations as indicated. The three distinct regions are: I, accumulation; II, depletion/inversion; III, strong inversion. For the insulator, a 200 nm layer of silicon oxide was assumed. Area: 2.5 mm × 2.5 mm. The semiconductor layer was assumed large enough to accommodate the entire depletion region in all cases

the interface. They have to pass the depletion width and recombine with majority carriers. In case this generation–recombination process is too slow, effectively the minority band does not exist and the device continues to build up depletion. The entire process can be modeled as a Maxwell–Wagner process but with the infinite-frequency capacitance equal to the depletion capacitance instead of 0. The equivalent circuit is shown in Figure 3.19. Note that the resistance R is not (necessarily) the depletion resistance, but rather is a parameter to model the electron–hole generation–recombination time $\tau = R(C_d + C_{ox})$.

$$C(\omega) = \frac{C_{ox} + \omega^2 R^2 C_d C_{ox}(C_d + C_{ox})}{1 + \omega^2 R^2 (C_d + C_{ox})^2} \tag{3.79}$$

$$G(\omega) = \frac{\omega^2 R C_{ox}^2}{1 + \omega^2 R^2 (C_d + C_{ox})^2}.$$

The loss ($L = G/\omega$) has a maximum at

$$\omega_{Lmax} = \frac{1}{R(C_d + C_{ox})} \tag{3.80}$$

while the loss tangent peaks at

$$\omega_{\delta max} = \frac{1}{R}\sqrt{\frac{1}{C_d(C_d + C_{ox})}}. \tag{3.81}$$

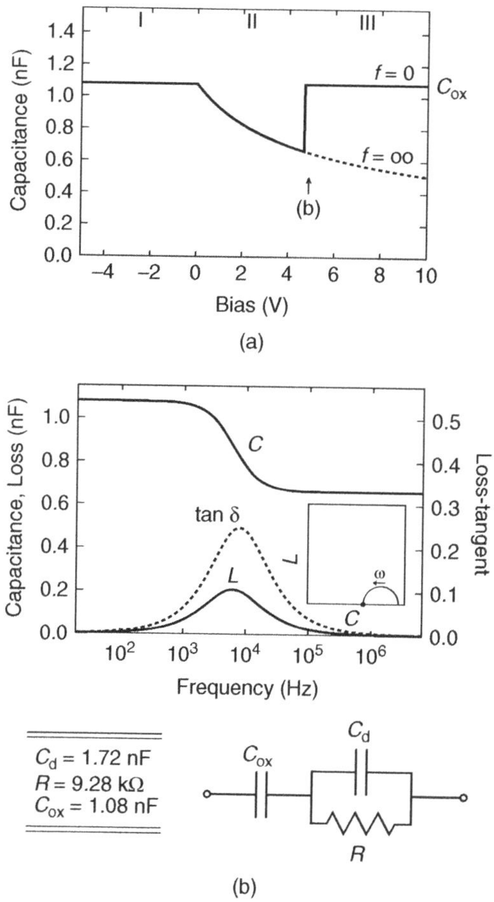

Figure 3.19 (a) Simulation of the measured capacitance (C) as a function of bias for low (solid line) and high (dashed line) frequencies. (b) Simulation of capacitance (C), loss ($L = 1/\omega R$) and loss-tangent (dashed line) of an MIS device in strong inversion for parameters shown

(In these equations, the capacitances are absolute values, with unit F.) Thus, the C–V plots are asymmetric for high frequencies. Figure 3.19 shows C–V plots for low and high frequencies.

3.6.2 MIS with Interface States

Because the DC current in these devices is zero, in MIS diodes it is easy to study interface states. Interface states, or sometimes also confusingly called surface states, are electronic states that are spread normally over a wide energy range, but are in space confined to the interface with the insulator. Interface states are unavoidable due to the discontinuity in the lattice parameters. For Si/SiO_2 interfaces, the density is in the order of 10^{12} cm^{-2} eV^{-1}, but because these interface states are mainly due to unsatisfied dangling bonds, they can easily be passivated by hydrogen attachment. The absence of a high density of dangling bonds in organic materials makes them less likely to present a high density of interface states and densities as low as 10^{10} cm^{-2} eV^{-1} should not be unexpected.

Nicollian and Goetzberger have extensively studied and documented the behavior of the interfaces in MIS systems by admittance spectroscopy [136], and a good summary is also found in the book by Sze [9]. Here the interface states will be briefly discussed and an example given.

Imagine an MIS structure with a bunch of interface states somewhere in the forbidden gap of the semiconductor. Figure 3.20 shows a schematic diagram of an MIS device with interface states for different bias conditions. The dashed line represents the Fermi level that is equal for holes and electrons, due to the lack of DC current. At forward bias, the Fermi level is under the interface states and these are therefore empty (contain no electrons). At strong reverse bias, the interface states are completely below the Fermi level and are therefore completely full and modulating the Fermi level with an external voltage will have no effect on the amount of charge present there; no current will exist that is associated with the movement of charge.

For some biases the Fermi level is resonant with the interface states and they are partly filled. Charge now has an extra path to make it across the device and the DC conductance will increase [137]. Moreover, when, in this case, the Fermi level is modulated (by the AC component of the external voltage) charge will flow into and out of the interface (states). Charge can now reach all the way up to the interface and the capacitance thus increases, in principle, to C_{ox} (see the treatise

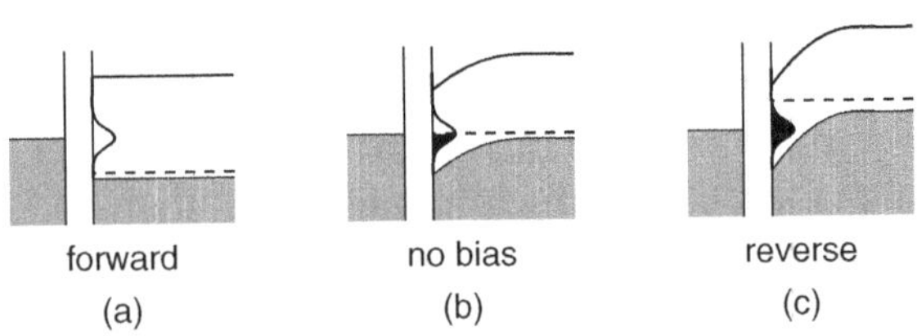

Figure 3.20 Schematic energy diagrams of an MIS with localized electronic levels at the semiconductor–insulator interface for different biases. The dashed line represents the Fermi level. For certain biases, the Fermi level is resonant with the interface states and this opens paths for charge to reach closer to the interface

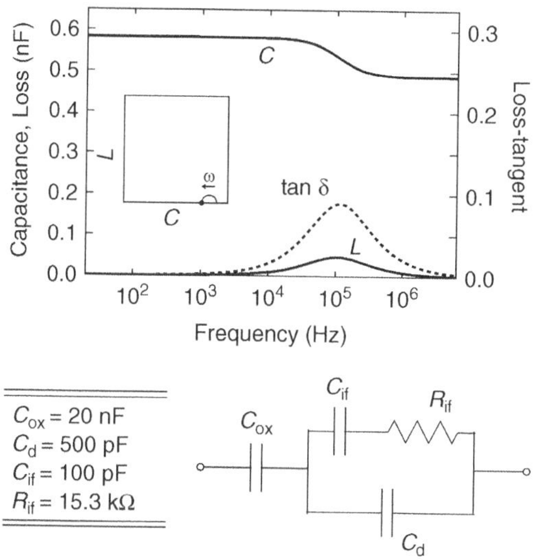

Figure 3.21 Simulation of an MIS with interface states. For low frequencies, the capacitance is $C_d + C_{if}$ while for high frequencies it is C_d (assuming $C_{ox} = \infty$)

on the similar effect of strong inversion). However, the number of interface states is limited and this limits the capacitance. Figure 3.21 shows the equivalent circuit of the device in depletion. In Figure 3.21 C_d is the depletion capacitance, C_{if} is the interface state capacitance, and R_{if} is the parameter that models the thermalization time τ that includes transit through the depletion region and capture and emission rates (see Chapter 4), whichever is the limiting factor; $C_{if}R_{if} = \tau$. This makes the measured capacitance and conductance frequency dependent. The serial subcircuit of C_{if} and R_{if} can be translated into a parallel circuit with frequency-dependent components, as in the case for the Maxwell–Wagner process (see Section 3.3). When the value of the oxide capacitance is very large, it has no effect other than blocking the DC path and does not alter the admittance spectrum. This makes the measured values of capacitance and conductance [136, 137]:

$$C(\omega) = C_d + \frac{C_{if}}{1 + \omega^2\tau^2}$$

$$G(\omega) = \frac{C_{if}\omega^2\tau}{1 + \omega^2\tau^2} \tag{3.82}$$

where the relation $\tau = C_{if}R_{if}$ is used. Figure 3.21 shows a simulation of a spectrum. Clearly a drop in the capacitance and an accompanying peak in loss is seen at a frequency $f = 1/2\pi\tau$. The loss, $L = G/\omega$, has a maximum

$$L_{\mathrm{max}} = C_{\mathrm{if}}/2 \tag{3.83}$$

for

$$\omega_{\mathrm{max}} = 1/\tau. \tag{3.84}$$

The drop in capacitance, and the amplitude of the peak in loss give information about the density of interface states at the position of the Fermi level. The (dynamic) capacitance is defined as the increment of charge divided by the increment of voltage:

$$C_{\mathrm{if}} = \frac{\delta Q}{\delta V}. \tag{3.85}$$

If the external bias is decreased by an amount δV, the Fermi level at the interface Fermi level is raised by $\delta E_{\mathrm{F}} = q\delta V$, see Figure 3.22. The number of states that have dropped below the Fermi level because of this is

$$\begin{aligned}\delta N &= A\int_{\mathrm{E_F}}^{\mathrm{E_F}+\delta E_{\mathrm{F}}} N_{\mathrm{if}}(E)\mathrm{d}E \\ &\approx AN_{\mathrm{E_F}}\delta E_{\mathrm{F}}.\end{aligned} \tag{3.86}$$

The latter approximation can be made if the movement of the Fermi level is tiny (implying small AC voltage amplitudes). Thus, the charge that has moved in is

$$\delta Q = qAN_{\mathrm{E_F}}\delta E_{\mathrm{F}}. \tag{3.87}$$

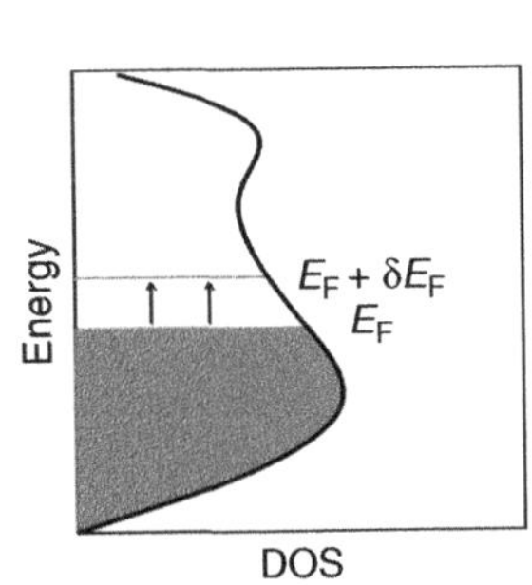

Figure 3.22 Effect of a change of bias on the occupancy of the interface states. All states between E_{F} and $E_{\mathrm{F}} + \delta E_{\mathrm{F}}$ get filled by the increment of voltage

For very thin insulators, the voltage drop across the device is completely taken up by the semiconductor, which means that the band bending is equal to the external bias. Thus,

$$\delta E_F = q\delta V. \tag{3.88}$$

Substituting Equation (3.88) into Equation (3.87) and then into Equation (3.85) yields

$$C_{if} = q^2 A N_{if} \tag{3.89}$$

and then Equation (3.83) becomes

$$L_{max} = q^2 A N_{if}/2. \tag{3.90}$$

The position of the Fermi level can be determined by making a temperature dependence of the loss spectrum, since the position of the maximum in loss, according to Equation (3.84), is equal to the reciprocal relaxation time τ and this relaxation time in its turn depends on the level depth according to Equation (4.13), thus

$$\omega_{max}(T) = \tau_0 T^2 \exp(-E_a/kT) \tag{3.91}$$

with

$$E_a = E_F - E_V - \Delta E_\sigma. \tag{3.92}$$

Thus, ignoring the term ΔE_σ, a plot of the logarithm of the position of the maximum in loss versus the reciprocal temperature is a straight line whose slope reveals the depth of the states under study for a particular bias.

In summary, by varying the bias, a bunch of interface states can be selected. Then, the admittance spectrum (the magnitude of the peak in loss and the drop in capacitance) will yield the density of interface states targeted, and a temperature dependence of the admittance spectrum yields the depth of these states. In this way, the complete DOS profile can be reconstructed.

It was assumed here that changes in occupancy of the states are abrupt, that is to say *all* states below the Fermi level are full and *all* states above it empty, or in other words, the Fermi–Dirac function is a step function. In reality there is a spread of about $3kT$ and this makes the peaks in the C–V plots broader by about $3kT/q$ with a Gaussian line shape. At room temperature this is about 75 mV. Another source for line broadening is the use of noninfinitesimal AC amplitudes. Ideally, the amplitude should be a small as possible, but the trade-off is a reduced signal-to-noise ratio. Reasonable values for the amplitude are 10–100 mV and this causes a proportional broadening of the C–V structures with, in first order, at low frequencies, rectangular line shape. All in all, the resolution that can be obtained is about 0.1 (e)V.

To summarize, the interface states can cause a peak in the capacitance–voltage (C–V) as well as a peak in conductance–voltage (G–V) plots and thus admittance spectroscopy is an excellent way of studying them, as the density and energetic distribution can be determined.

3.6.2.1 *Case Study: Interface State Mapping in a Terrylene Diode*

This subsection describes a study of a Terrylene–insulator–aluminum MIS device [138]. The insulator layer of some tens of ångströms thick, between the semiconductor and the aluminum was unintentionally introduced, but provides a way to study the interface states by reducing the DC conductance, which was about 10^{-5}–10^{-4} S m^{-2} at 1–2 V.

Figure 3.23(a) shows the loss ($L = G/\omega$) as a function of bias. Clearly a peak is visible around 1.5 V which is attributed to the presence of interface states that are resonant with the Fermi level, as schematically drawn in Figure 3.20. For higher frequencies, shallower states are probed and the device has to be driven further in forward bias. In the I–V curves a peak is also visible. Figure 3.23(b) shows the DC conductance corrected for the exponential background conductance $[G_{DC} = (I - I_{exp})/V]$ where the structure can be attributed to the impurity bands [139].

Figure 3.24(a) shows the spectra of the loss-tangent ($\tan\delta = G/\omega C$) for various temperatures. For increasing temperatures, the peak moves to higher frequencies because the thermalization times decrease [Equation (3.91)]. Figure 3.24(b) shows an Arrhenius plot of the position of the maximum which reveals an activation energy of 0.45 eV.

To check the fact that indeed the states responsible for the peak are located at the interface, a bias dependence of the spectrum was made. The peak in loss-tangent moves to higher frequencies upon forward bias. This is in-line with what is expected for interface states. When the bias is increased, the band bending is reduced and shallower levels cross the Fermi level close to the interface. Shallow levels are faster responding at higher frequencies. Were the states distributed in space, but not in energy, i.e. a discrete homogeneous level, the reduced band bending would cause the Fermi level to cross these levels ever closer to

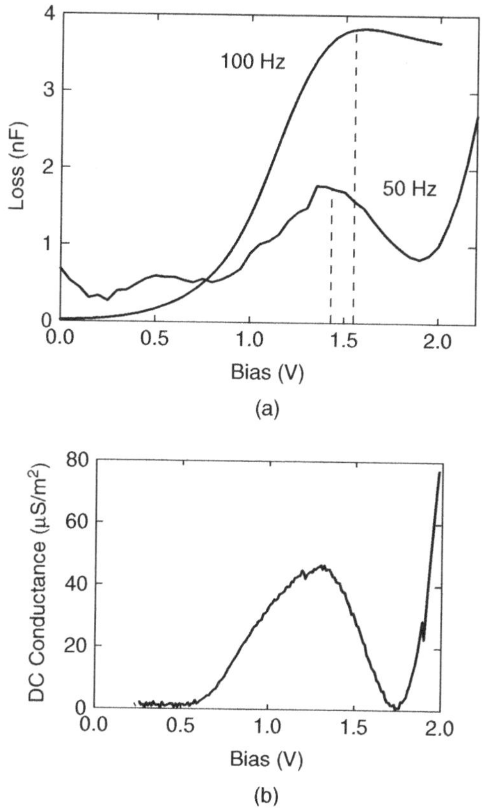

Figure 3.23 (a) Loss as a function of bias for two probing frequencies as indicated. The peak around 1.5 V is attributed to interface states. (b) DC conductance with a peak at 1.3 V. Reproduced with permission from Organic Electronics, Interface state mapping in a Shottky barrier of the organic semiconductor terrylene by P. Stallinga, H. L. Gomes, M. Murgia and K. Mullen, 3, 1, 43 Copyright (2002) Elsevier

the interface, but the activation energy (and hence the response time and spectrum) would remain the same.

Finally, the interface state energetic distribution could be mapped on the basis of the loss spectra. Figure 3.25 summarizes the distribution thus found. The densities are reasonable for inorganic semiconductors, but perhaps a little too high for organic materials.

Some critical remarks about the correctness of this analysis are in order. Analyzing the DOS plot of Figure 3.26, it can be seen that the energy scale goes below 0, implying that the levels are resonant with the valence band. This clearly does not make sense. Moreover, the speed of movement of the loss-tangent peak is not in-line with

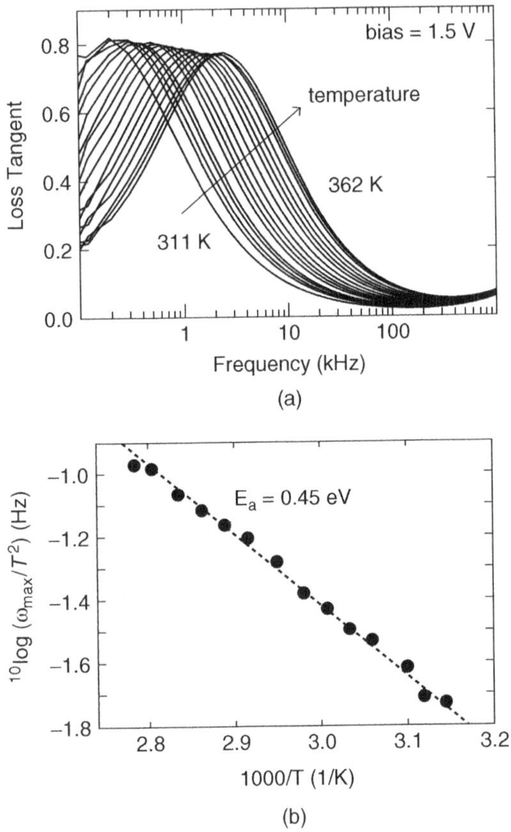

Figure 3.24 (a) Loss tangent ($\tan\delta = G/\omega C$) for various temperatures. (b) Arrhenius plot of temperature-corrected position maximum of loss-tangent ω_{max}/T^2 revealing an activation energy of $E_a = 0.45$ eV (dashed line). Reproduced with permission from Organic Electronics, Interface state mapping in a Shottky barrier of the organic semiconductor terrylene by P. Stallinga, H. L. Gomes, M. Murgia and K. Mullen, 3, 1, 43 Copyright (2002) Elsevier

what is expected of a movement of the Fermi level by 1 eV per V bias change. A 0.2 eV shallower level means a relaxation time that is, at room temperature, $\exp(0.2q/300k) = 2290$ times faster. Yet, a voltage increase of 0.2 V does not even move the peak by a factor of 2, as can be seen in Figure 3.25. Somehow, the Fermi level does not move as fast as expected, or the states are located in a layer further away from the interface. In any case, it seems the scale of the final plot should be changed; instead of the range −0.3 to 0.55 eV, it should be limited to a tiny range around 0.45 eV. This would then also have an effect on the measured density. The same factor 1000 less energy is scanned than

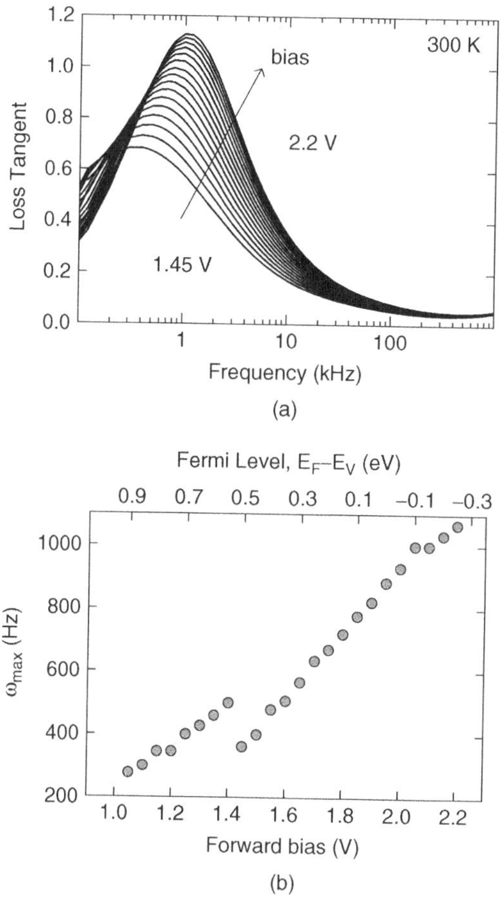

Figure 3.25 (a) Bias dependence of the loss-tangent spectra. (b) Position of the maximum of the loss-tangent as a function of bias. At 1.5 V bias, the activation energy was calibrated by the temperature dependence and found to be 0.45 eV. For other biases, it was assumed that the Fermi level shifts 1 eV V^{-1} with the bias. This is the basis for the scale above the figure. Reproduced with permission from Organic Electronics, Interface state mapping in a Shottky barrier of the organic semiconductor terrylene by P. Stallinga, H. L. Gomes, M. Murgia and K. Mullen, 3, 1, 43 Copyright (2002) Elsevier

predicted by Equation (3.88). This makes N_{if} a factor 1000 larger in Equation (3.90). Indeed, such a high density of interface states would pin the Fermi level and the assumption that the interface states themselves do not contribute to the space charge is no longer valid.

Another questionable assumption made is that the interface states communicate primarily with the bands in the semiconductor. As pointed

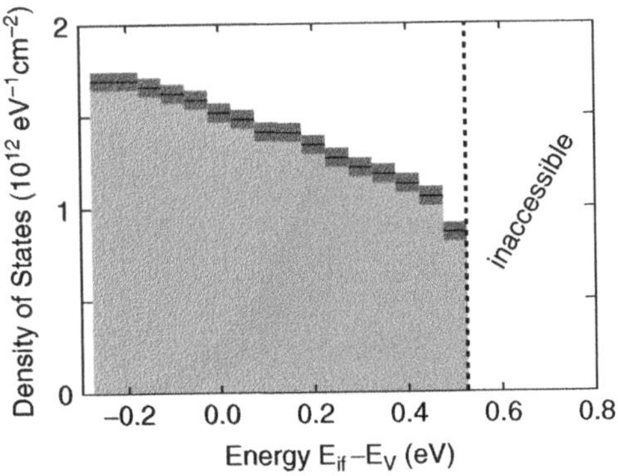

Figure 3.26 Interface states energetic distribution. Inaccessible means that the peak in the loss-tangent falls below the measurement window (50 Hz). Reproduced with permission from Organic Electronics, Interface state mapping in a Shottky barrier of the organic semiconductor terrylene by P. Stallinga, H. L. Gomes, M. Murgia and K. Mullen, 3, 1, 43 Copyright (2002) Elsevier

out by Gray, for thin insulators separating the metal from the semiconductor the tunneling to and from the metal to these interface states can have a much shorter lifetime than the recombination lifetime [139].

3.6.3 MIS of Low-mobility Semiconductors

In standard analysis, it is intrinsically assumed above that the mobility of charges is high enough so that any additional charges needed to maintain the relation $Q = CV$ can be considered to arrive at the place of interest instantaneously. For an MIS diode this is the interface. In case the mobility of the carriers is too small, or the distance the charges have to travel from the injecting electrode to the interface is too large, such that the time it takes for the charges to diffuse to the interface is larger than the time period of the probing signal, the measured capacitance reduces to the geometric capacitance. The cut-off frequency can be found by calculating the diffusion time for carriers from the electrode to the insulator interface. This time is given by

$$\tau = \frac{d_s^2}{D} = \frac{q}{kT}\frac{d_s^2}{\mu} \tag{3.93}$$

where Einstein's Relation was used to convert the diffusion coefficient D into mobility μ. This introduces a cut-off frequency $f_c = 1/2\pi\tau$. Well below this frequency, the capacitance is equal to C_{ox} while well above it, the frequency tends to C_{geo}. As shown in Section 3.3, a process with time-constant τ can be modeled with an equivalent circuit and can be easily measured in admittance spectroscopy. This makes admittance spectroscopy a powerful additional technique for measuring the mobility of materials. It is very similar to the ToF technique described in Chapter 5, but where ToF is in the time domain, admittance measures the transit time in the frequency domain.

Many people have used this technique in various forms to measure the mobility [140–144]. The advantage of using an MIS structure instead of the LED structure used by these authors is twofold. First of all, large DC currents can hinder the observation of (low-frequency) admittance peaks. Moreover, in an LED structure both types of carriers are simultaneously present. Electrons and holes meet somewhere in the device. It is therefore not clear what the exact distance is that the carriers travel [d_s in Equation (3.93)]. It introduces a correction factor to the measurements, which is difficult to determine. In contrast, an MIS diode has zero DC current and the device can either be put into hole accumulation or electron accumulation. In both cases the distance traveled by the charges is the semiconductor layer thickness d_s.

This technique was tested on an organic material nickel bis(dithiolene) (NiDT), which has a comparable hole and electron mobility (and is thus ambipolar) [145]. The work is summarized in the next case study.

3.6.3.1 *Case Study: Mobility Measurement in NiDT via Admittance Spectroscopy*

Figure 3.27 shows the energy diagram of the NiDT material used in this case study. The parameters of the devices were as follows: The area of a device was typically 1.3×10^{-5} m^2. The oxide capacitance thickness was $d_{ox} = 200$ nm, resulting in an oxide capacitance of about 2.2 nF. The thickness of the NiDT layer of $d_s = 50$ nm gives a semiconductor capacitance of $C_s = \varepsilon_s A/d_s = 17.4$ nF, where a value for the dielectric constant of the semiconductor of $\varepsilon_s = 5$ was used. Combining the two gives a geometrical capacitance of $C_{geo} = (C_s^{-1} + C_{ox}^{-1})^{-1} = 2.0$ nF. The energy diagram was based on an ionization potential (difference between the vacuum level and the highest occupied molecular orbital) $I_p = E_{vac} - E_{HOMO} = 5.2$ eV, electron affinity (difference between vacuum level and lowest unoccupied molecular orbital) $\chi = E_{vac} - E_{LUMO} = 4.3$ eV. Here Koopmans' theory was used that links energy differences to levels. Furthermore, gold has a workfunction (difference between vacuum level and top of electron sea in metal) of 5.1 eV [9]. The energy

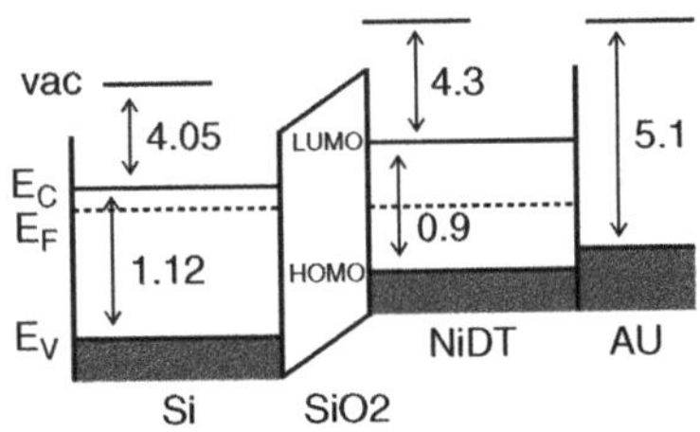

Figure 3.27 Schematic energy diagram of the MIS device based on NiDT. All values in eV. Reproduced with permission from Organic Electronics, Determining carrier mobility with a metal-insulator-semiconductor structure by P. Stallinga, A. R. V. Benvenho, H. L. Gomes *et al.*, 9, 5, 735 Copyright (2008) Elsevier

gap of silicon is 1.12 eV at room temperature and because of the high n-type doping, the Fermi level is expected to be about 100 meV below the conduction band, which is in turn 4.05 eV below the vacuum level. Because the oxide is a good insulator, all voltage drop is absorbed by this material. Combining this information resulted in the band diagram presented in Figure 3.27.

The equivalent circuit to model the admittance data is shown in Figure 3.28. For low frequencies, the semiconductor capacitance is bypassed and the measured capacitance is equal to the oxide capacitance, while for high frequencies the capacitance is the geometric capacitance. The value of the resistance R models the cut-off frequency by defining a relaxation time

$$\tau = R(C_{\text{ox}} + C_{\text{s}}). \tag{3.94}$$

For strong bias, the organic layer is filled with electrons or holes and the resistivity of the organic layer is much lower than that of the insulator. The applied bias is absorbed by the insulating layer and the voltage drop and electric field in the organic material are negligible. Figure 3.29 shows

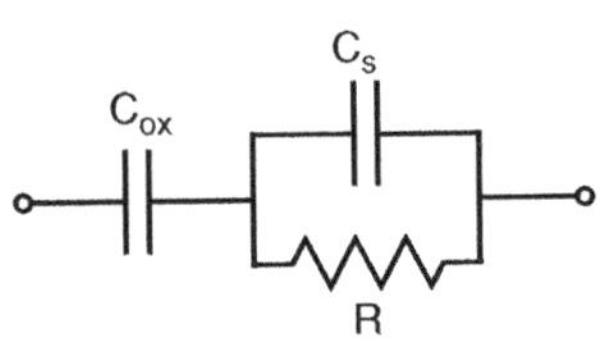

Figure 3.28 Equivalent circuit used to model the admittance data of an MIS diode. Reproduced with permission from Organic Electronics, Determining carrier mobility with a metal-insulator-semiconductor structure by P. Stallinga, A. R. V. Benvenho, H. L. Gomes *et al.*, 9, 5, 735 Copyright (2008) Elsevier

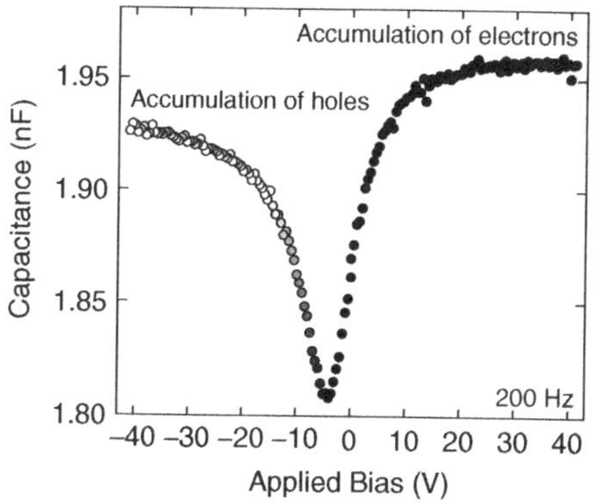

Figure 3.29 C–V plot (measured at a frequency of 200 Hz). At strong bias, either accumulation of electrons or accumulation of holes occurs and the capacitance is equal to the oxide capacitance C_{ox}. In between there is a region where the capacitance reaches a minimum equal to the geometric capacitance C_{geo}. Reproduced with permission from Organic Electronics, Determining carrier mobility with a metal-insulator-semiconductor structure by P. Stallinga, A. R. V. Benvenho, H. L. Gomes *et al.*, 9, 5, 735 Copyright (2008) Elsevier

a characteristic C–V plot which demonstrates how, under strong bias, the capacitance at low frequencies is equal to the oxide capacitance C_{ox}. For positive bias the organic layer is full of electrons and this we call 'accumulation of electrons', whereas for strong negative bias holes are pulled into the organic layer and there exists 'accumulation of holes'. For small biases, the device is depleted, the organic layer neither has holes nor electrons and the capacitance reaches a minimum equal to C_{geo}. It is interesting to compare the device to a standard (doped) MIS device. A standard MIS goes from accumulation to inversion. The transition is abrupt when going to inversion, but stretched when going into accumulation and an asymmetric peak results [9]. Our undoped device, ideally, should go abruptly from accumulation of holes to accumulation of electrons and the dip in the C–V plot should be very narrow. The C–V plot of Figure 3.29 is indeed symmetric, but the switch from accumulation of holes to accumulation of electrons is not instantaneous. This might be due to the presence of amphoteric deep levels. They can be charged either positively or negatively, thereby contributing to space charge and allowing for band bendings and stretching of the transition. More important, most of the applied bias is absorbed by the oxide layer and large voltages must be applied in order to cause any changes in the semiconductor layer. This further broadens the dip.

The minimum capacitance of Figure 3.29 corresponds to the geometric capacitance formed by the silicon and gold electrodes 'filled' with the oxide and organic layers, as described before. It happens for a bias of around −5 V, showing that at 0 V the organic layer is not completely devoid of free electrons.

As described before, the capacitance can be equal to the oxide capacitance C_{ox} *only* if the probing frequency is low enough, otherwise the measured capacitance is the geometric capacitance C_{geo}. Figure 3.30 shows an example of a spectrum of a device in accumulation of holes. Figure 3.30 also shows a simulation of the spectrum based on the equivalent circuit given in Figure 3.28. Apart from the high-frequency dispersion which we attribute to cables and other instrumental artifacts, the fit is quite good and we can see how the capacitance drops from C_{ox} to C_{geo} at a frequency where the loss has a maximum. From this frequency we can determine the (hole) mobility. The spectrum of Figure 3.30 has a maximum at 640 Hz ($\tau = 250\,\mu s$). Equation (3.93) with $d_s = 50$ nm then gives a hole diffusion coefficient $D_p = 1.0 \times 10^{-11}$ $m^2\,s^{-1}$, which translates via Einstein's Relation into a hole mobility of $\mu_p = 3.9 \times 10^{-6}$ $cm^2\,V^{-1}s^{-1}$.

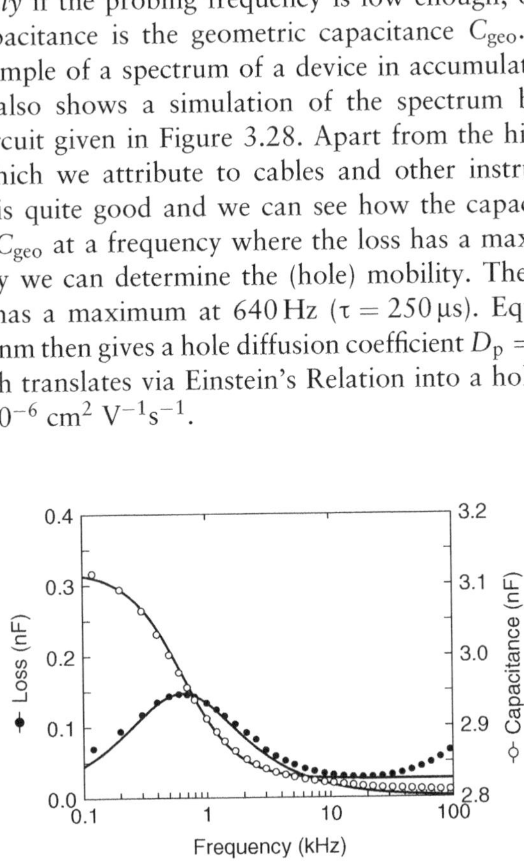

Figure 3.30 Example of a spectrum of loss (•) and capacitance (○) at a bias of $V = -28$ V. For low frequencies the charges have enough time during an AC cycle to reach the interface and the measured capacitance is equal to C_{ox}, while for high frequencies the capacitance is the geometric capacitance C_{geo}, i.e., charges do not move into the organic layer. The turning frequency is given by Equation (3.93) ($f = 1/2\pi\tau$). At this frequency the loss has a maximum. The solid lines are simulations with $C_{ox} = 3.1$ nF, $C_s = 30$ nF and $R = 7.5$ kΩ. Reproduced with permission from Organic Electronics, Determining carrier mobility with a metal-insulator-semiconductor structure by P. Stallinga, A. R. V. Benvenho, H. L. Gomes *et al.*, 9, 5, 735 Copyright (2008) Elsevier

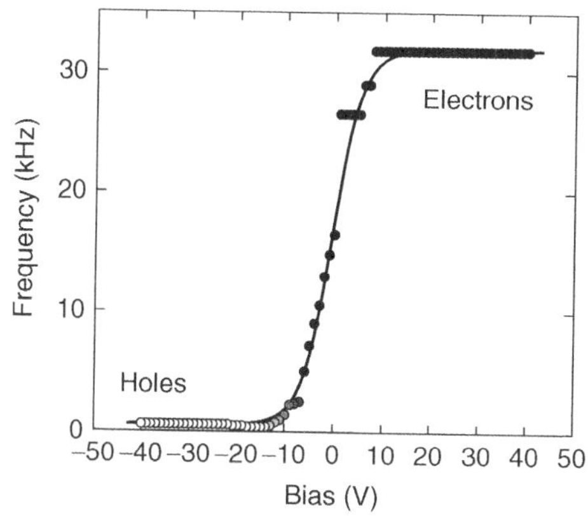

Figure 3.31 Frequency of maximum of loss as a function of bias as shown in Figure 3.30. The solid line is a guide to the eye. At strong positive or negative bias, the frequency is stable and this allows for the determination of the mobility of electrons and holes, respectively, via Equation (3.93) ($f = 1/2\pi\tau$). Reproduced with permission from Organic Electronics, Determining carrier mobility with a metal-insulator-semiconductor structure by P. Stallinga, A. R. V. Benvenho, H. L. Gomes *et al.*, 9, 5, 735 Copyright (2008) Elsevier

Figure 3.31 shows the position of the maximum of the loss as a function of bias. At strong negative bias the frequency is constant and this gives the hole mobility as discussed above. In the same way, for strong positive bias the electron mobility can be determined. The frequency of the maximum in loss is here 31.8 kHz ($\tau = 5\,\mu s$) and this yields an electron mobility of ($D_n = 5 \times 10^{-10}$ m^2 s^{-1}) $\mu_n = 1.9 \times 10^{-4}$ cm^2 V^{-1}s^{-1}.

The mobility is rather low and this hints at a huge abundance of deep localized states, also known as traps. The effective mobility becomes temperature activated, with the activation energy related to the trap depth. Since the frequency of the maximum in loss scales linearly with mobility, this is equal to the activation energy of this frequency.

Figure 3.32 shows an Arrhenius plot of the frequency of the maximum of loss as a function of temperature. From the slope we determine that the mobility of holes has an activation energy of $E_{ap} = 400$ meV and the mobility of electrons an activation energy of $E_{an} = 460$ meV. Considering the fact that the band gap of NiDT is approximately 0.9 eV, this places the responsible traps close to midgap, see inset in Figure 3.32. This is in-line with the observation of an amphoteric trap mentioned earlier.

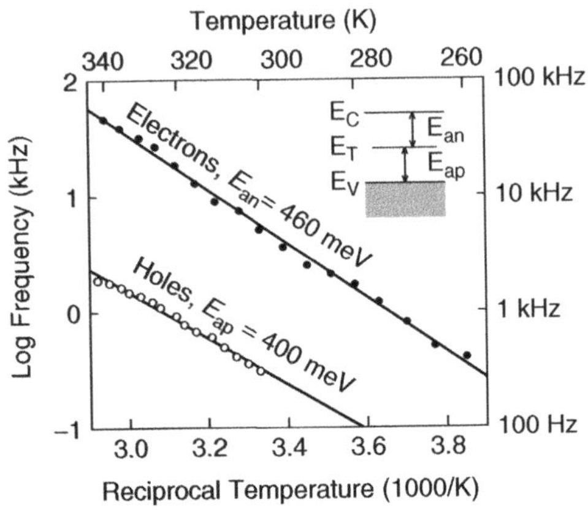

Figure 3.32 Arrhenius plot of the frequency of loss as a function of temperature revealing the activation energies of mobility as indicated. The inset shows the energetic position of the responsible traps. Reproduced with permission from Organic Electronics, Determining carrier mobility with a metal-insulator-semiconductor structure by P. Stallinga, A. R. V. Benvenho, H. L. Gomes *et al.*, 9, 5, 735 Copyright (2008) Elsevier

The mobilities reported in the literature deviate from the values found here. Moreover, we find the electron mobility higher than the hole mobility while it has been reported the reverse [146]. The reason for this is unknown. However, it is very common that mobilities measured with different techniques result in different values.

What is actually measured by this technique is not the mobility, but the diffusion coefficient. They are linked by Einstein's Relation, which has its limitations. As an example, it is valid only when multiple occupancy of the electronic levels is allowed, with uncorrelated particles, like in band conduction. When the particle movements are correlated, the Einstein Relation is no longer valid and diffusion is more rapid than predicted by Einstein's Relation [147]. Another reason can be that Einstein's Relation is violated in nonequilibrium systems [148].

The combination of parameters (film thickness and mobility) puts the interesting frequencies in the range of the experimental set-up used. For materials with mobilities in the order found here admittance spectroscopy is a valuable additional tool in the determination of the carrier mobility. For higher quality materials, such as crystalline silicon, the cut-off frequency falls above the measurement window, while for low-mobility materials or thick devices, the response becomes too slow and the frequency range inconvenient.

3.7 MIS TUNNEL DIODE

Where a Schottky diode is basically a single-carrier-type device, conducting either electrons or holes, but not both, and an MIS diode does not (DC) conduct at all, a peculiar device exists that is between the two. It is an MIS diode that has an insulating layer so thin as to not be able to block the current completely, a current that in this layer is of the tunneling type. For this reason, the device is often called an MIS tunnel diode [149, 150]. Because the insulator can absorb a large part of the external voltage, a large degree of freedom exists between the offset of bands in the metal and the semiconductor. This causes the device to be able to conduct both types of carriers, i.e., both electrons and holes. In some circumstances, these currents can even coexist simultaneously and a minority-carrier injection ratio γ can be defined [151]. This ratio is strongly influenced by the insulator thickness [152, 153]. For LEDs, where both types of carriers are needed in a 1:1 ratio ($\gamma = 0.5$) inside the active layer, such thin interfacial layers can greatly enhance the luminescence efficiency. The same applies to the efficiency of solar cells [154].

The direct visible effects of this dual-carrier-type conductance with minority-carrier injection is a plateau or bend in the I–V curves [149]. In the admittance data, pronounced features can be observed [150]. Instead of the standard depletion capacitance, proportional to $1/\sqrt{V}$, for instance, the C–V plots can have peaks or dips. The amplitude and phase of these structures depend on the frequency and can mimic inductance and negative resistance [155].

This theory is derived for inorganic materials and frequently observed in these materials [156]. Such behavior is also quite common for organic devices. For instance, consider the observation of a frequency-dependent structure in the admittance data of MIS diodes based on polyphenylenevinylene [133], or recently in organic solar cells [157]. Figure 3.33 shows an experimental example of minority-carrier effects of an organic MIS diode, both visible in DC and admittance measurements. It has also been suggested that to observe excess capacitance or inductance caused by minority carriers the insulating layer is not needed and a simple Schottky barrier suffices [158, 159]. However, in MIS structures, a carrier recombination at interface states can also result in frequency-dependent features in the admittance–voltage (Y–V) plots [160].

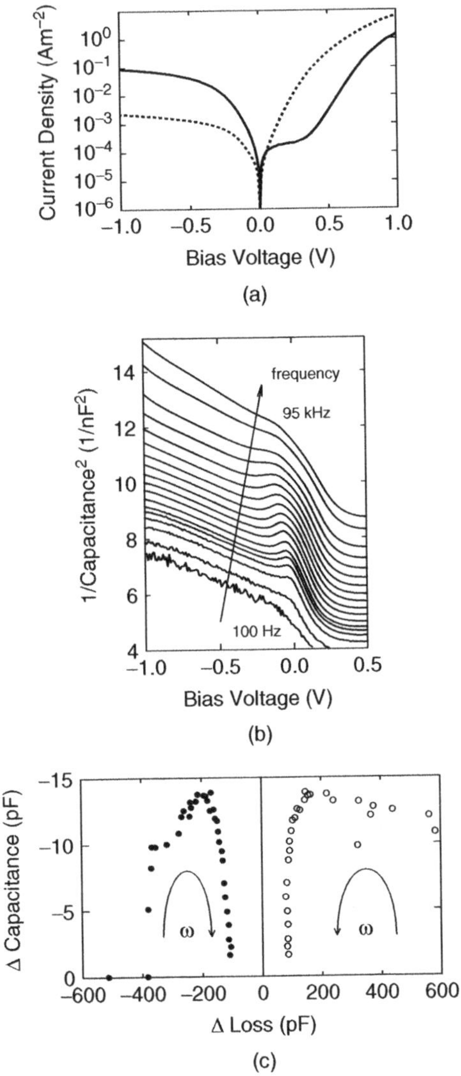

Figure 3.33 Electrical characteristics of an MIS tunnel diode showing minority-carrier effects: (a) I–V curves showing a plateau. The dashed curve shows a normal Schottky barrier for comparison. (b) Mott–Schottky plots of the diode for various frequencies. The dual slope already visible at low frequencies (100 Hz) is caused by a deep level (see Section 3.5.2). At intermediate frequencies the inductive effects of minority-carrier injection are prominent [155]. (c) Amplitude of peak in (b) versus amplitude of peak in loss (not shown) for two different features. From ref. 133. Reproduced with permission from Journal of Applied Physics, Minority-carrier effects in poly-phenylenevinylene as studied by electrical characterization by Peter Stallinga, 89, 3 Copyright (2001) American Institute of Physics

3.8 NOISE MEASUREMENTS

The same information obtained in admittance spectroscopy can also be obtained via noise measurement [161–163]. The difference between the two techniques is that in admittance spectroscopy the measurement is performed in the frequency domain, whereas in noise measurement it is performed in the time domain. The noise consists of random fluctuations of the DC current, and this is a good measurement technique for lower frequencies. A Fourier transform links the two domains.

It can thus study the capture–emission or generation–recombination processes, where the conductance noise is caused by a fluctuation in number of carriers. The study of this noise will yield the associated trap depth. Just as in admittance spectroscopy, a discrete trap will manifest itself in the noise spectrum by a peak at a frequency related to the trap-thermalization time τ. The noise power spectral-density can be written as [163]

$$S(\omega) \propto \frac{\tau N_{\mathrm{T}}}{1 + \omega^2 \tau^2} \quad (3.95)$$

which, when multiplied by the frequency ω, shows a peak at $\omega = \tau$, just like the loss peaks. The characteristic time τ is thermally activated and an Arrhenius plot will reveal the trap depth.

When the number of trapping and detrapping events is limited, the noise can manifest itself as random telegraph noise, i.e. random switching between two (or a limited number of) states [164], as if someone is randomly pressing the key of a telegraph machine.

As shown by Hooge, the omnipresent $1/f$ noise (sometimes also called flicker noise) yields information about the structural imperfections (density of free charge and traps, crystal quality, grains, etc.) [165, 166]. The spectral density is given as

$$S(\omega) = \frac{2\pi V^2 \alpha N}{\omega} \quad (3.96)$$

where α is the Hooge parameter that can be used to compare different materials and devices. Alternatively, $1/f$ noise can be due to mobility fluctuations [167].

This type of noise measurement has also been applied to organic TFTs [168–170]. Jones [171] and Vandamme [172] give good summaries of low-frequency noise and its application in characterizing defects in semiconductor devices.

4

Two-terminal Devices: Transient Techniques

For all the transient techniques of monitoring a system parameter over time we bring the system off-equilibrium in some way and observe how fast it recovers to thermal equilibrium. It can easily be shown that in simple systems the transients are exponential, with a single time constant. However, disordered materials have a distribution of time constants and this makes the transients highly complex. The earliest mention of this nonexponential behavior was reported by Kohlrausch [173] in 1847. Two distinct types seem to be prevalent in the literature, stretched exponential (also sometimes called 'glassy relaxation') [174], i.e.,

$$J(t) = J_0 \exp\left[-\left(\frac{t}{\tau}\right)^{\beta}\right] \tag{4.1}$$

with β between 0 and 1, and power law [175], i.e.,

$$J(t) = J_0 \left(\frac{t}{\tau}\right)^{-\alpha} \tag{4.2}$$

while for the ToF technique often a multi power law is used, as will be discussed in Chapter 5.

The bottleneck process is in most cases the emission and capture of carriers on deep levels, although in some rare cases it is the diffusion of the free carriers out of the depletion zone. First the relaxation process of reaching thermal equilibrium has to be discussed [11]. In general it can be said that a deep level thermalizes slower than a shallow level and the emission from a state depends exponentially on its depth. It is therefore good to start with a description of emission and capture of carriers.

4.1 KINETICS: EMISSION AND CAPTURE OF CARRIERS

Figure 4.1 shows the processes of capture and emission of electrons from an electronic level at E_X with density N_X. The number n_X stands for the number of these levels filled. The rest of the electrons must then be in the conduction band (in this simple model), $n = N_X - n_X$. The terms e_n and c_n stand for the emission and capture rates of electrons (given as fraction per second). The number of electrons actually being emitted and captured is then proportional to these values and the number of available electrons in the states. In equilibrium there is no net transfer of electrons. This requires

$$e_n n_{X0} = c_n n_0 \tag{4.3}$$

with the subscript 0 used to denote the steady-state thermal equilibrium values. This can be rewritten as

$$\frac{e_n}{c_n} = \frac{n_0}{n_{X0}} \tag{4.4}$$

which is the general relation between capture and emission rates. The ratio is given by the ratio of the steady-state occupation of the levels. For

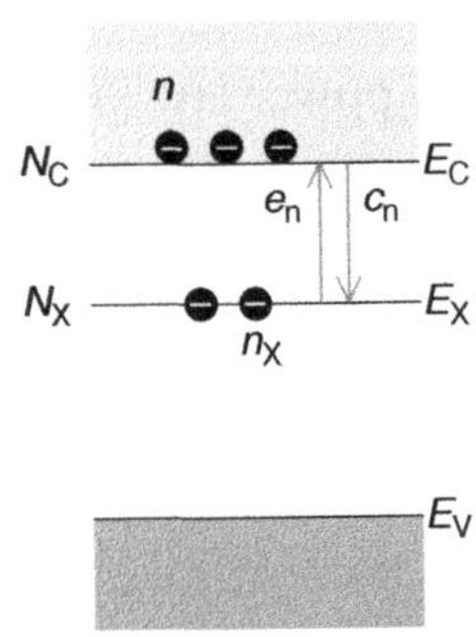

Figure 4.1 Schematic showing the emission and capture processes to and from a level X. N_X, density of X states; E_X, energy of electrons in X states; n_X, density of electrons in X states; e_n, emission rate of electrons from X states; c_n, rate of electron capture to X states

instance, in a system that has few steady-state free electrons, emission is slow and capture is nearly instantaneous.

In equilibrium, the electrons must also obey the Fermi–Dirac distribution:

$$n_{X0} = \frac{N_X}{1 + \exp\left[(E_X - E_F)/kT\right]}$$
$$n_0 = \frac{N_V}{1 + \exp\left[(E_V - E_F)/kT\right]}. \tag{4.5}$$

This can be used to calculate the ratio of emission and capture rates for specific cases. Until this point, the results are general, irrespective of the position of the Fermi level and the DOS. It is interesting to look at some examples.

In a DLTS experiment, a Schottky barrier is used that is first put in forward bias and then the bias is removed (or the device is put in reverse bias) and the capacitance transient measured. The full DLTS experiment will be described later. Here the focus is on the kinetics only. In a filling pulse, the band bending at the interface is removed and the energy diagram is flat (like in the bulk of the material), as in Figure 4.2(a). This shows the case of an n-type semiconductor with a deep level X. The Fermi level is well above the deep level. Consequently, the steady-state occupancy of the deep level is 1 ($n_{X0} = N_X$). Moreover, the Fermi level is still far away from the conduction band so that the

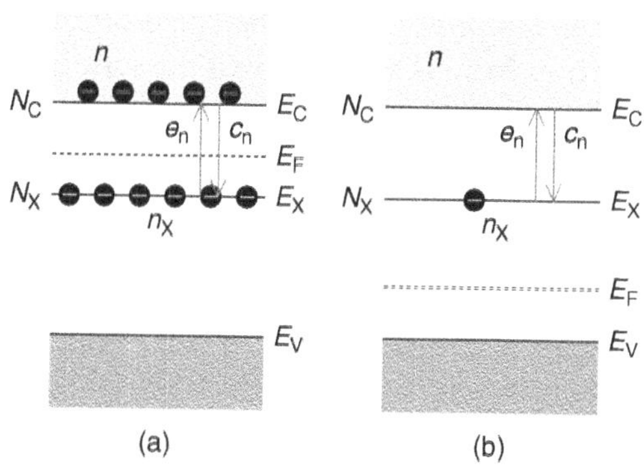

Figure 4.2 Schematic showing the emission and capture processes in a DLTS experiment. (a) Forward bias, with the Fermi level above the deep level causing them to fill up rapidly. (b) Zero bias (depletion) with the Fermi level far below the deep level; electrons are slowly emitted from the deep levels. The electrons shown give an idea of the steady-state occupancies

Boltzmann approximation can be used for the density of free electrons, $n_0 = \exp[(E_F - E_C)/kT]$. The emission/capture ratio of Equation (4.4) thus becomes

$$\frac{e_n}{c_n} = \frac{N_C}{N_X} \exp\left(\frac{E_F - E_C}{kT}\right) \ll 1. \tag{4.6}$$

In other words, the capture rate is very fast. For the experiment this can be considered instantaneous. After this filling pulse, the bias is removed. Part of the device enters into depletion, putting the Fermi level far below both the deep level and the conduction band, as shown in Figure 4.2(b). In this case, for both the occupancy of the conduction band and the deep level the Boltzmann approximation can be used, thus $n_{X0} = N_X \exp[(E_X - E_F)/kT]$, and the emission/capture ratio becomes

$$\frac{e_n}{c_n} = \frac{N_C}{N_X} \exp\left(\frac{E_X - E_C}{kT}\right) \ll 1. \tag{4.7}$$

The difference between the two equations is a factor $\exp[(E_X - E_F)/kT] \ll 1$ (where E_F is the Fermi level in the bulk), or, in other words, the ratio becomes even smaller. The emission is a slow process and governs the transient. Note that this ratio is independent of the position of the Fermi level. This is true when the Fermi level is well below the deep level and the Boltzmann approximation can be used. This is an important observation for DLTS, since it does not matter that there still exist band bendings and associated variations in Fermi level depth, as long as the Fermi level is everywhere well below the deep level.

We now make the assumption that the *emission* of electrons from a state to the conduction band is a property of the state itself. A charge has a certain probability to spontaneously be emitted from the deep level. However, the *capture* of electrons from the conduction band to the deep level is proportional to the following: (1) the capture cross-section, σ_n of the deep levels; (2) the density of capturing sites, N_X; (3) the average thermal velocity of these free electrons, $\langle v_n \rangle$; and (4) the density of free holes in the valence band, n. In other words, the capture rate of electrons is given by

$$c_n = \sigma_n \langle v_n \rangle N_X. \tag{4.8}$$

With this, the emission rate of Equation (4.7) becomes:

$$e_n = \sigma_n \langle v_n \rangle N_C \exp\left(-\frac{E_C - E_X}{kT}\right). \tag{4.9}$$

If we substitute the absolute expressions for the parameters [9]

$$\sigma_n = \sigma_0 \exp(-\Delta E_\sigma / kT) \tag{4.10}$$

$$<v_n> = (3kT/m^*)^{1/2} \tag{4.11}$$

$$N_C = 2M_C[(2\pi m^* kT)/h^2]^{3/2} \tag{4.12}$$

we can calculate the explicit temperature dependence of the emission rate

$$e_n = \gamma T^2 \sigma_{na} \exp(-E_a/kT). \tag{4.13}$$

This defines the prefactors γ and σ_{na}, which are often not interesting for the experiment. However, the above equation demonstrates how a plot of $\ln(e_n/T^2)$ vs. $1/T$ will be a straight line and reveals an activation energy E_a of the deep level, which is related to the level depth according to

$$E_a = (E_C - E_X) - \Delta E_\sigma. \tag{4.14}$$

The relaxation time τ found in a transient experiment, the 'characteristic time', or the 'decay time' of the transient relates directly to the emission or capture parameters e_n and c_n. It depends on the circumstances as to how to describe the transient. For instance, for a system with n_{X0} nearly zero in an 'emission experiment' (at $t = 0$, n_X instantaneoulsy brought off-equilibrium to a nonzero value)

$$\frac{dn_X(t)}{dt} = n(t)c_n - n_X(t)e_n \tag{4.15}$$

with the ratio of e_n and c_n this can be converted to the form

$$\frac{dn_X(t)}{dt} = \frac{n_{X0}}{n_{tot} - n_{X0}} e_n - \frac{n_X(t)}{n_{tot} - n_{X0}} e_n \tag{4.16}$$

where $n_{tot} = n_X + n$ is constant (in this experiment. In others this can be different). This results in an exponential transient with $\tau \approx 1/e_n$ (if $n_{X0} \ll n$). Measuring the relaxation time of the transient will give the emission rate, which, in turn, can give information about the energy levels. In what follows, the link between τ and capture or emission rate, c_n or e_n, is assumed to be reciprocal, for instance $\tau_e = 1/e_n$. The transients are of the form

$$n_X(t) = n_{X0} \exp(-t/\tau_e) \tag{4.17}$$

This gives a possibility of determining the deep level activation energy directly, as we will see later in Section 4.5. With the emission rate of Equation (4.13) in hand we can analyze some measurement techniques. Later, thermally stimulated current (TSC), capacitance transient spectroscopy (CTS), and DLTS will be discussed.

Finally, by combining Equations (4.13) and (4.7) a similar expression can be found for the capture rate

$$c_{\mathrm{n}} \propto N_{\mathrm{X}} T^{3/2} \exp\left(-\frac{\Delta E_{\sigma}}{kT}\right). \tag{4.18}$$

The capture process is only slightly thermally activated. It does not give meaningful information about the deep level and is normally not studied in classical materials. It has to be pointed out that this analysis assumes that the energy diagram consists of an effective discrete conduction and valence bands and that the deep level is not very abundant. In amorphous materials, the situation can be much more symmetric, with the trap levels as abundant as the conductive levels (or in many cases even outnumber the conductive states). In that case it can be expected that capture and emission events are very similar with both processes thermally activated with an activation energy exponentially depending on the energy difference bridged. Moreover, the thermalization times for capture and emission can easily be of the same magnitude, as predicted by Equation (4.4) when the final (steady state) density of trapped and free charge is similar. This is a radical difference compared with the classical (crystalline) semiconductor materials, where capture and emission is highly asymmetric.

An example is a system that mechanically relaxes upon capture and release of carriers, see Figure 4.3. It is not specified what exactly the relaxation means, but it is generalized in a configuration coordinate Q. It might mean the 'breathing' out or in of the atoms surrounding the atom capturing the charge. Q could then represent the distance to the nearest neighbor atoms. Another example might be the dislocation of the atom itself along a certain direction, as in the As anti-site in GaAs. Q then obviously represents the distance of the dislocation from its high-symmetry position. The basic component is that the object X has different stable configurations for the different charge states, represented by a different position of the minimum energy on the Q-axis. Figure 4.3, for instance, has a different atom configuration for the state with the charge trapped on X compared with the charge in the conduction band. It is clear that in this case both capture and emission of the charge

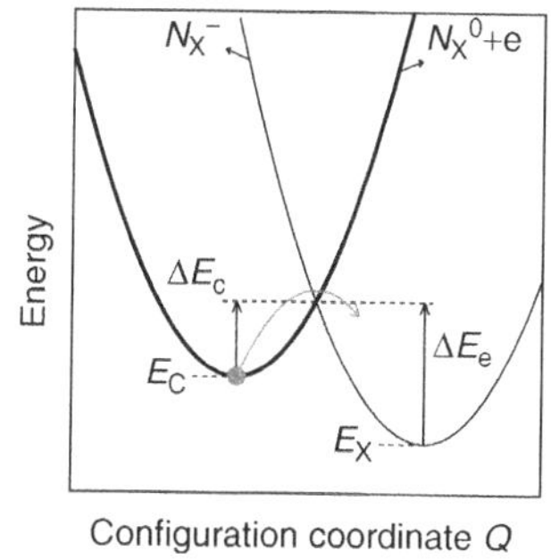

Figure 4.3 Configuration coordinate diagram. A general configuration coordinate (Q) represents an arrangement of the surrounding atoms of the state. The energy of the state depends on this rearrangement (here a parabolic function is assumed). The curved arrow indicates the capture of an electron (e) from the conduction band onto the deep level X which in its minimum energy state has a different configuration. This effect means that both capture and emission of the carriers is thermally activated, with energy barrier of ΔE_c and ΔE_e, respectively

have an effective barrier, ΔE_c and ΔE_e, respectively, that makes both processes thermally activated.

It might well be so that in organic materials every atom can serve as a center for such lattice distortions. Or, in other words, every charge can create its own trap by creating its own distortions. This picture is very similar to polarons, as described in Chapter 1.

4.1.1 Emission and Capture in Organic Materials

For organic materials, or amorphous materials in general, the levels are not discrete but highly dispersed in energy. As discussed before, exponential distributions seem to be adequate for the description of the electrical behavior of the devices. It is therefore interesting to analyze the transient behavior of charge in such systems. Hole conduction may serve as an example. For both the conduction levels ('valence band') and deep levels ('traps') an exponentially decaying function can be used

$$N_V(E) = N_{V0} \exp\left(-\frac{E - E_V}{kT_1}\right) \tag{4.19}$$

$$N_T(E) = N_{T0} \exp\left(-\frac{E - E_V}{kT_2}\right) \tag{4.20}$$

where N_{V0} and N_{T0} are the DOS at the band edge (at E_V) and kT_1 and kT_2 the decay factors, the reciprocal slopes of the DOS in a logarithmic plot. Assuming the emission and capture to be thermally activated,

$$\tau_e(\Delta E) = \tau_{e0} \exp\left(\frac{\Delta E}{kT}\right) \tag{4.21}$$

$$\tau_c(\Delta E) = \tau_{c0} \exp\left(\frac{\Delta E}{kT}\right) \tag{4.22}$$

where ΔE is the energy being bridged by the emission or capture process. It now depends on the starting conditions and the type of measurements as to how this can be used to describe the experiment.

4.2 CURRENT TRANSIENT SPECTROSCOPY

Instead of giving a full set of solutions to all possible conditions and experiments, two examples will be given. In the first example, charge is created on traps in one place of the device. The device is placed in an electrical field. Once the charge is emitted from the deep levels, it momentarily contributes to current, until it is swept out of the device. This is thus a time-derivative experiment. In the second example, more straightforward, the conductance of a homogeneous sample, initially full of free charge is monitored over time. Free carriers are captured by deep states and become unavailable for conduction. As will be shown, both examples will result in power-law transients of the form $J(t) \propto t^{-\alpha}$.

4.2.1 Example of an Emission Experiment

The following is a system similar to the situation for DLTS and related transient techniques, but instead the current is monitored. Charges start on traps in a tiny region of the device, for instance by a light pulse at time zero (this is very similar to a ToF experiment to be discussed in Chapter 5 and because the experimental set-up and procedure is the same, it can easily be confused with it). When the charges are emitted from the traps they move freely and contribute to current until they move out of the device. This can be referred to as a single-trap-and-release (STR) model. It is defined by the following five points:

1. The free-charge (band) mobility is very large.
2. The mobility of trapped charge is zero.
3. Charges begin on traps. If, however, not all charge begins on traps, a spike in current is expected before the regular transient starts.

4. Emission from traps to the band is thermally activated, $\tau \propto \exp(E_A/kT)$.
5. Retrapping does not occur; charges, once emitted from the traps, arrive undisturbed at the counter-electrode. In other words, the trapping time is much longer than the transient time. While transiting they contribute to current, as described before.

For a p-type conduction system with a discrete hole trap level these five points will result in an exponential transient, which can be understood easily if we consider that the current is the density of charge that is moving, and for large free-carrier mobility this is proportional to the rate at which charge is emitted from traps. Thus, for a discrete trap,

$$\rho_t(t) = \rho_{t0} \exp\left(-\frac{t}{\tau}\right) \tag{4.23}$$

$$J(t) \propto \frac{d\rho_t}{dt} \propto \exp\left(-\frac{t}{\tau}\right) \tag{4.24}$$

with the relaxation time given by Equation (4.21),

$$\tau = \tau_{e0} \exp\left(\frac{E_T - E_V}{kT}\right). \tag{4.25}$$

When the traps are not discrete but distributed in energy, things get more complicated. Silver and Cohen made Monte Carlo simulations to determine that for traps distributed exponentially in energy, the emission from the traps will not be exponential, but instead will follow a power law [176]. We can here calculate analytically what happens to the current transient if the traps are distributed in this way. The distribution of the traps is assumed to be as in Equation (4.20) and emission from it is thermally activated and follows Equation (4.21). (See Figure 4.4 for a schematic diagram of this system.) The density of charged traps at energy E and time t is then given as

$$\begin{aligned} N_T^+(E,t) &= H_0(E) N_T(E) \exp\left(-\frac{t}{\tau(E)}\right) \\ &= H_0(E) N_{T0} \exp\left[-\frac{t}{\tau_{e0}} \exp\left(-\frac{E}{kT}\right) - \frac{E}{kT_2}\right] \end{aligned} \tag{4.26}$$

where for convenience the valence band level was used as the reference energy, $E_V = 0$ and with $H_0(E)$ the occupancy fraction at $t = 0$. If we

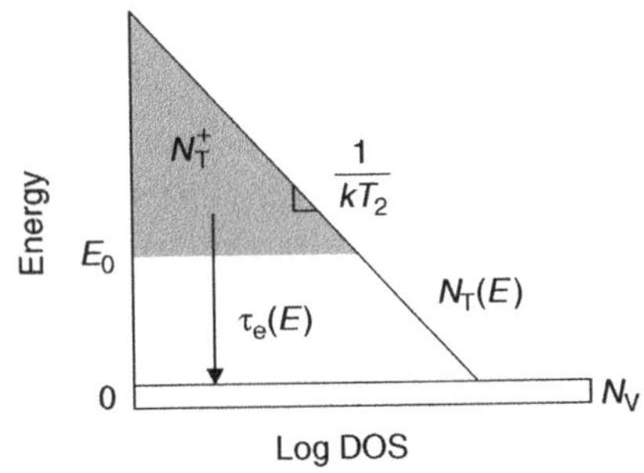

Figure 4.4 Schematic diagram of an STR emission experiment. Holes are trapped on deep states N_T which are distributed in energy. They are slowly emitted with a time constant τ_e that depends on the depth. These released charges then contribute to current

assume that at $t = 0$ all traps are filled from E_0 to infinity (H_0 thus being a heaviside function: $H_0 = 1$ in this range, 0 everywhere else) then the total density of charged trap at any given time t is

$$\begin{aligned} N_T^+(t) &= \int_{-\infty}^{\infty} N_T^+(E,t)\mathrm{d}E \\ &= \int_{E_0}^{\infty} N_{T0} \exp\left[-\frac{t}{\tau_{e0}} \exp\left(-\frac{E}{kT}\right) - \frac{E}{kT_2}\right] \mathrm{d}E \end{aligned} \tag{4.27}$$

and the current is

$$\begin{aligned} J(t) &= -q\frac{\mathrm{d}N_T^+(t)}{\mathrm{d}t} \\ &= \frac{qkTN_{T00}}{\tau_{0T}}\gamma\left(1+\frac{T}{T_2}, \frac{t}{\tau_{0T}}\right)\left(\frac{t}{\tau_{0T}}\right)^{-(1+T/T_2)} \end{aligned} \tag{4.28}$$

with

$$N_{T00} = N_{T0} \exp\left(-\frac{E_0}{kT_2}\right), \tag{4.29}$$

$$\tau_{0T}(T) = \tau_{e0} \exp\left(\frac{E_0}{kT}\right) \tag{4.30}$$

and $\gamma()$ the lower incomplete gamma function (8.350 of Ref. [177], see also 3.331 of Ref. [177]),

$$\gamma(a,x) \equiv \int_0^x e^{-y} y^{a-1} \mathrm{d}y \tag{4.31}$$

with $a > 0$. For small values of x, this function is proportional to x^a, while for large x, the function is constant. In other words, substituting our values of $a = 1 + T/T_2$ and $x = t/\tau_{0T}$ it is clear that the current of Equation (4.28) is constant for small times t and a power law for large times:

$$J(t) \propto \begin{cases} 1 & \text{for } t \ll \tau_{0T} \\ t^{-(1+T/T_2)} & \text{for } t \gg \tau_{0T}. \end{cases} \tag{4.32}$$

For a homogeneous distribution $T_2 = \infty$ and a hyperbolic transient $J(t) \propto t^{-1}$ results. Figure 4.5 shows simulations of this equation. Note that in any case the transient of Equation (4.28) does not depend on the device thickness or mobility. As such, this is not a good technique to determine the mobility, but it will shed some light on the dispersion of energies of trap states.

If also the conduction states are distributed in energy, still a power-law transient is expected. This can easily be reasoned if we imagine the valence band divided into thin slices of energy. Each slice then contributes an emission of the form described above, with an effective E_0 which is the distance of E_0 to the slice of the valence band, instead of the discrete valence band defined at $E_V = 0$. For each slice a power-law contribution to the current is expected. The sum of these contributions is again (the same) power law. Or, in other words, even for distributed final states

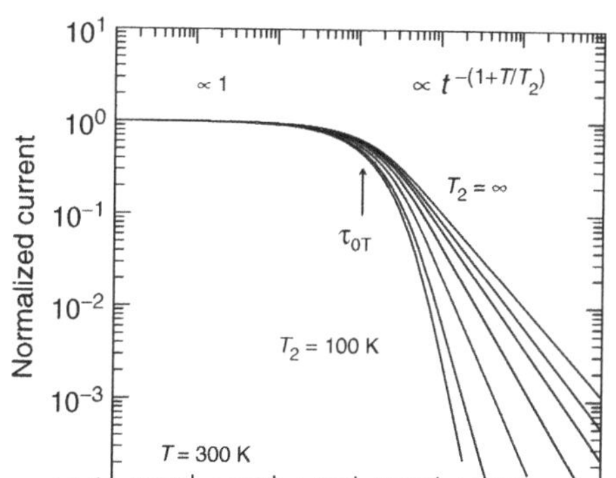

Figure 4.5 Simulations of transients of a STR system in which charge is trapped on one side of the sample from where it is slowly emitted. See Equation (4.28), with $\tau_{0T} = 0.9$ s. Simulations were done with Wolfram Mathematica 5 with output processed with PjotrSoft EPSTool

N_V, the transient is a power law. Another, more formal, way of looking at it is making the observation that the emission rate for each trap level is a convolution over all final states. Assuming each contribution to be exponentially dependent on the energy difference, the total emission rate from a trap level at energy E_T to all the conduction levels E_V is

$$e(E_T) = \int \exp\left(-\frac{E_T - E_V}{kT}\right) F[N_V(E_V)]\mathrm{d}E_V \tag{4.33}$$

$$= \exp\left(-\frac{E_T}{kT}\right) \int \exp\left(\frac{E_V}{kT}\right) F[N_V(E_V)]\mathrm{d}E_V \tag{4.34}$$

$$= e_{00} \exp\left(-\frac{E_T}{kT}\right) \tag{4.35}$$

where the possibility that the emission rate depends on the DOS of the final states was taken into account by the function $F()$. The result is that the emission from each trap level depends exponentially on the trap depth, regardless of the distribution of the destination levels. Since this was the basic assumption of the analysis [Equation (4.21), with $\tau_e = 1/e_p$], the conclusions of a power-law transient remain valid, even for a valence band distributed in energy.

Thus, the conclusion is that for an STR experiment, with the DOS exponentially distributed in energy, a power-law transient is expected, with the exponent in the power law depending on the temperature T and the decay coefficient of the trap distribution T_2. The STR experiment can therefore give valuable insight into the distribution of the traps.

4.2.2 Example of a Capture Experiment

Because the emission and capture processes are very similar in behavior [namely Equations (4.21) and (4.22)], the same type of transient is expected for a system in which all charge starts in conducting states, and a very abundant trap exists that captures the free charges and makes them unavailable for contributing to current. In this section experiments are described with the following components:

- The device is homogeneous: the DOS, free charge and trapped charge, and the electrical field are everywhere the same.
- Since densities are everywhere the same, there is no diffusion current.
- There are no effects of the contacts.

Imagine a system with a discrete valence band and a discrete trap level that is so abundant that, eventually, all charge will wind up being trapped, when given enough time to relax. When the abundant trap density is effectively infinite, re-emission is reduced to zero and the density of free charge is a simple exponential function of time. With the current being proportional to the free charge only ($\mu_T = 0$), the current transient is a simple exponential. Equation (1.1) becomes

$$J(t) = q\mu_p E p(t) = J_0 \exp(-t/\tau). \tag{4.36}$$

Imagine now the system, opposite to the one given before for the STR emission experiment, namely a discrete very abundant trap and a free-charge DOS that is exponentially distributed in energy. The distribution of the valence band states is assumed to be as in Equation (4.19) and the capture by the traps is thermally activated and follows Equation (4.22), see Figure 4.6. The density of free charge at energy E and time t is then given by

$$p(E,t) = H_0(E)N_V(E)\exp\left[-\frac{t}{\tau(E)}\right] \tag{4.37}$$

$$= H_0(E)N_{V0}\exp\left[-\frac{t}{\tau_0}\exp\left(\frac{E}{kT}\right) - \frac{E}{kT_1}\right] \tag{4.38}$$

where for convenience the trap level was used as the reference energy, $E_T = 0$, and where H_0 is the occupancy fraction at $t = 0$. If we assume

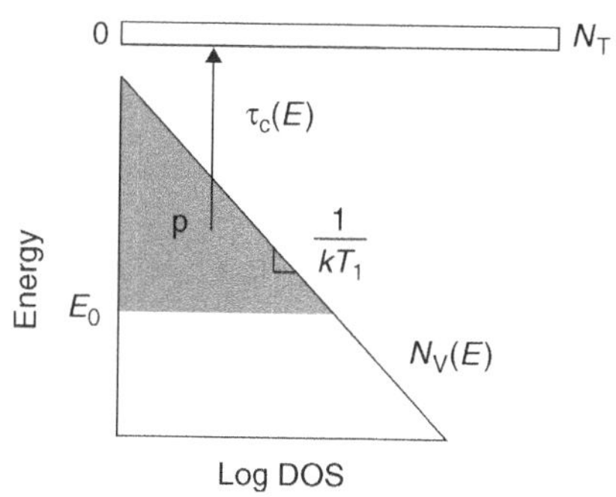

Figure 4.6 Schematic diagram of a current transient trapping experiment. Holes start in the valence band that is exponentially distributed in energy. They are slowly captured by a discrete abundant level N_T with a time constant τ_c that depends on the depth. These trapped charges then become unavailable for conduction and the current drops

that at $t = 0$ all the valence band states are filled from $E = E_0$ to ∞, then the total density of free charge at any given time t is

$$p(t) = \int_{E_0}^{\infty} N_{V0} \exp\left[-\frac{t}{\tau_{c0}} \exp\left(\frac{E}{kT}\right) - \frac{E}{kT_1}\right] dE. \qquad (4.39)$$

The current is proportional to this free-charge density, and thus becomes

$$\begin{aligned} J(t) &= q\mu_p p(t) V/d \\ &= \frac{q\mu_p V N_{V0}}{d} \int_{E_0}^{\infty} \exp\left[-\frac{t}{\tau_{c0}}\left(\frac{E}{kT}\right) - \frac{E}{kT_1}\right] dE \qquad (4.40) \\ &= \frac{q\mu_p VkTN_{V00}}{d} \left(\frac{t}{\tau_{0T}}\right)^{-T/T_1} \gamma\left(\frac{T}{T_1}, \frac{t}{\tau_{0T}}\right) \qquad (4.41) \end{aligned}$$

with

$$N_{V00} = N_{V0} \exp\left(-\frac{E_0}{kT_1}\right) \qquad (4.42)$$

$$\tau_{0T}(T) = \tau_{c0} \exp\left(-\frac{E_0}{kT}\right) \qquad (4.43)$$

and $\gamma(a, x)$ the lower incomplete gamma function, see Equation (4.31) (8.350 of Ref. [177], see also 3.331 of Ref. [177]). For small values of x, this function is proportional to x^a, while for large x, the function is constant. In other words, substituting our values of $a = T/T_1$ and $x = t/\tau_{0T}$ it is clear that the current of Equation (4.40) is constant for small times t and a power law for large times:

$$J(t) \propto \begin{cases} 1 & \text{for } t \ll \tau_{0T} \\ t^{-T/T_1} & \text{for } t \gg \tau_{0T}. \end{cases} \qquad (4.44)$$

Figure 4.7 shows simulations of this equation. The current will thus shed some light on the dispersion of energies of conduction states.

Again, as for the case of the emission experiment described in Section 4.2.1, if also the destination states (traps) are distributed in energy, still a power-law transient is expected. The reasoning that can be followed is the same. The trapping is now a convolution over all possible trap states, instead of just one discrete level. The relaxation time is the result of the convolution of these capture rates. Since at every energy

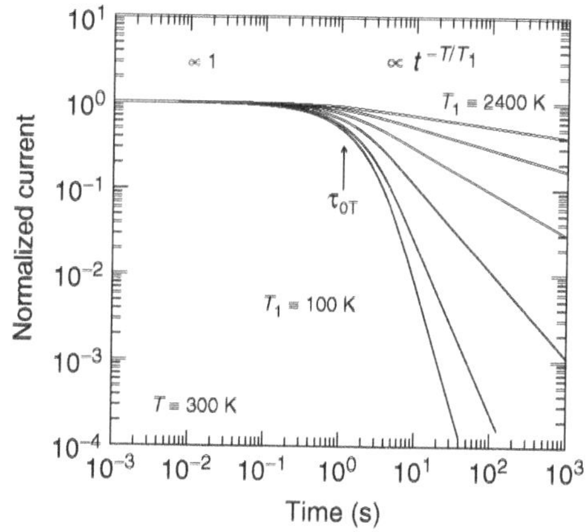

Figure 4.7 Simulations of current transients of a system of capture in which charge is initially free and slowly captured by deep traps. See Equation (4.40). Simulations were done with Wolfram Mathematica 5 with output processed with PjotrSoft EPSTool

the rate depends exponentially on the energy bridged, the total relaxation time also depends exponentially on the average energy bridged. (See Section 4.2.1 for more details.) The conclusions of a power-law transient remain valid, even for the trap states also distributed in energy.

The transients, in the case of exponentially distributed states, thus become of the power-law type. In reality, it is difficult to distinguish this type of behavior from other types, such as stretched exponential predicted by others $\{\propto \exp[-(t/\tau)^{\alpha}]\}$. The differences are minimal and a set-up with low noise and high dynamic range is needed to see them. Often the data are consistent with the chosen model, but cannot be considered proof of it.

4.3 THERMALLY STIMULATED CURRENT

With temperature stimulated current experiments trap level depths and concentrations can be found. For organic materials this is a powerful technique that yields information about the electronic structure [178–180]. The underlying idea is that the current that is needed for restoring thermal equilibrium can be monitored. In practice this means that the sample is cooled down and the thermal equilibrium is disturbed in one way or another. This can be done either by applying a strong bias (forward or reverse, while cooling or afterwards) or by illumination. Figure 4.8 shows an example of a TSC experiment to determine

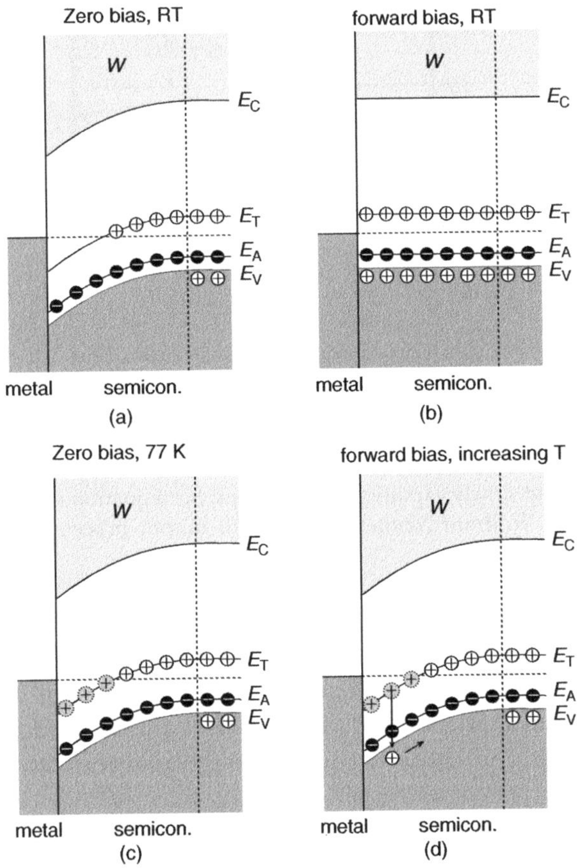

Figure 4.8 Schematic of a TSC experiment: (a) Device at thermal equilibrium at room temperature without bias with a region where the traps are filled and a region where they are empty. (b) Device at thermal equilibrium at room temperature with a bias. All deep traps are filled with holes everywhere. (c) Device off equilibrium at low temperatures with a region where thermalization has not taken place due to low hole emission rates [Equation (4.13)], these off-equilibrium trapped holes are indicated by shading. (d) At higher temperatures the emission of holes becomes significant. They drift towards the electrode and an external current is observed until all holes are emitted and the system returns to situation (a)

a deep acceptor level. At room temperature, the device is at thermal equilibrium; all deep levels below the Fermi level are ionized (negatively charged) (Figure 4.8a). At this point a strong forward bias is switched on. The band bending at the interface disappears and this lifts all deep levels above the Fermi level. Sufficient time is allowed for holes to be captured by the deep traps which at the end are all positively charged

(Figure 4.8b). Then, with the bias on, the device is cooled down. At the desired low temperature, the bias is released; an amperimeter is placed between the two electrodes of the device. The band bending is restored immediately which pulls part of the deep levels below the Fermi level. They now should be neutral, as discussed in Section 1.5.2. However, the emission rate of holes from the deep levels [Equation (4.13)] at these low temperatures is effectively zero. The device is thus persistently off thermal equilibrium (Figure 4.8c). When the temperature is increased, the emission rate gradually increases until a point where the emission of holes from the deep levels becomes substantial (Figure 4.8d). The emitted holes are immediately swept away by the built-in field towards the cathode (right side of the diagrams) where they are collected and contribute to externally observable current. Once all the possible holes are emitted from the deep levels, the device is back to the situation as depicted in Figure 4.8(a) and the current drops back to zero.

From the TSC experiment described above, it is possible to determine the activation energy of the deep level with the help of the master equation for the emission rate [Equation (4.13)]. This can be done by analyzing the current in the heating-up cycle: In the depletion zone, there are no free carriers and the capture of carriers does not take place. Holes are emitted from the traps at a rate e_p and the number of filled deep traps p_T therefore changes with a rate

$$\frac{dp_T}{dt} = -e_p p_T. \tag{4.45}$$

If we assume that all the charges that are emitted immediately drift towards the bulk, without being recaptured, all contribute to the current:

$$J \propto e_p p_T. \tag{4.46}$$

When we make a temperature scan of this current, one important parameter is the temperature at which the maximum of the current occurs. At this place, the derivative of the current is equal to zero, $dJ/dT = 0$. According to Equation (4.46) the derivative of the current is equal to

$$\frac{dJ}{dT} \propto p_T \frac{de_p}{dT} + e_p \frac{dp_T}{dT}. \tag{4.47}$$

Note that

$$\frac{dp_T}{dT} = \frac{dp_T}{dt}\frac{dt}{dT} = -\frac{e_p p_T}{\beta} \tag{4.48}$$

where Equation (4.45) was used and the scanning rate $\beta = dT/dt$ was defined. This can be put into Equation (4.47) and to find the maximum, the result should be set to zero.

$$p_T \left(\frac{de_p}{dT} - \frac{e_p^2}{\beta} \right) = 0. \qquad (4.49)$$

One trivial solution is $p_T = 0$. This occurs at the end of the scan, when thermal equilibrium is restored. The other temperature, T_m, can be found by substituting the emission rate found before see Equation (4.13), resulting in

$$\frac{\sigma\gamma T_m^4}{\beta} \exp\left(-\frac{E_a}{kT_m}\right) = (2T_m + E_a/k) \qquad (4.50)$$

For $kT_m \ll E_a$

$$\ln(T_m^4/\beta) = E_a/kT_m + \ln(E_a/\sigma\gamma k). \qquad (4.51)$$

To determine the deep level activation energy E_a, a set of scans should be made with different scanning speeds β. Each time the temperature T_m at which maximum current occurs is noted and the slope of a plot of $\ln(T_m^4/\beta)$ vs. $1/T_m$ is then proportional to E_a/k.

The activation energy can also be found directly from a single TSC scan, as shown by Cowell and Woods. A peak in a TSC scan follows the equation (no retrapping) [181]

$$J(T) = A \exp\left[-\Theta - B\exp(-\Theta)\Theta^{-2}\right] \qquad (4.52)$$

with $A = qN_X\tau\mu\nu$ and $B = \nu E_a/\beta k$ (τ is the free-carrier lifetime, ν the attempt-to-escape frequency of a trapped charge, $\nu = N_V v\sigma$, assumed to be independent of temperature) parameters depending on the properties of the carriers and the trap and the scanning rate β, and $\Theta = E_a/kT$ the scanning variable. Fitting this equation to the TSC data will yield the activation energy E_a.

For fast retrapping the above equation is substituted by [181]

$$J(T) = C \exp\left[-\Theta - D\exp(-\Theta)\Theta^{-7/2}\right] \qquad (4.53)$$

where C and D are constants similar to A and B above. Karg *et al.* used intermediate retrapping for organic materials and derived an equation [179]

$$J(T) = \frac{A\exp(-\Theta)}{[1 + B\exp(-\Theta)\Theta^{-2}]^2} \tag{4.54}$$

In all these cases, the interdependence of the parameters makes the fitting very difficult.

In either case, the integrated current $J(t)$ over time reveals the number of defects emptying their charge and this can then be related to the defect density if we know the dimensions (enhanced depletion width W_2 and electrode area A) of the active region.

$$N_T = \frac{1}{qW_2A}\int J(t)\mathrm{d}t. \tag{4.55}$$

Figure 4.9 shows an example of a TSC measurement with parameters $E_a = 0.5318$ eV, $A = -0.269$ A, $B = 9.49 \times 10^{12}$ ($\beta = 1\times$), 4.74×10^{12} ($\beta = 2\times$), 2.37×10^{12} ($\beta = 4\times$). Note that the integrated curve (over time!) is equal in all three cases. The biggest curve looks larger, but it is scanned at higher speed.

$$\int J(t)\mathrm{d}t = \int J(t)\mathrm{d}T\frac{\mathrm{d}t}{\mathrm{d}T} = \frac{1}{\beta}\int J(T)\mathrm{d}T. \tag{4.56}$$

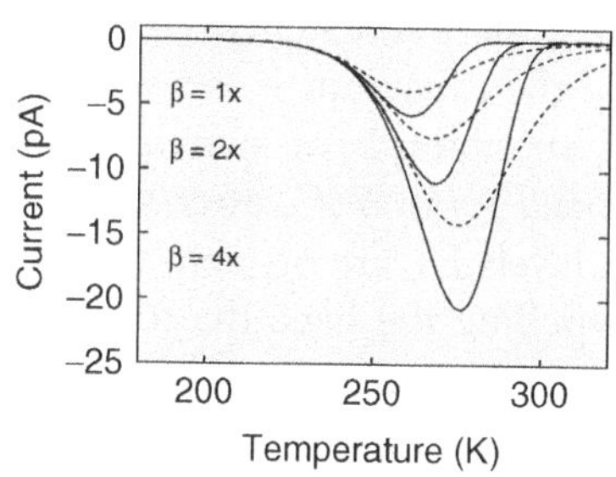

Figure 4.9 Simulation of TSC curves for different relative scanning speeds $\beta = \mathrm{d}T/\mathrm{d}t$, as indicated, for the equation of Cowell and Woods for no retrapping [Equation (4.52), solid lines] compared with the equation of Karg *et al.* for intermediate retrapping [Equation (4.54), dashed lines]. From the fitting of the data to the equation the activation energy of emission from the traps can be found

Finally, the same information can be obtained from measuring the temperature evolution of the capacitance instead of the current, in the so-called thermally stimulated capacitance (TSCap) technique [182], but because of the more difficult instrumental set-up and more complicated analysis this is less used. Scientists prefer to see peaks in their data and whilst the TSC technique directly gives a peak in the scan, the TSCap data have to be presented in a derivative form ($\mathrm{d}C/\mathrm{d}T$ or $\mathrm{d}C/\mathrm{d}t$) to give peaks.

4.4 CAPACITANCE TRANSIENT SPECTROSCOPY

Capacitance transient measurements are a predecessor of DLTS to be discussed later. It consists of putting the system off equilibrium, either by an instantaneous voltage step or by light switching or pulsing. The speed at which the system recovers to equilibrium is determined and when this is done as a function of temperature, information can be obtained about the deep levels. The parameter that is monitored to determine the kinetics is capacitance. As a first observation, the deep levels are slow and thus do not thermalize within a period of the probing signal. This means that the high-frequency approximation can be made for the capacitance, as discussed in Section 3.5.2 (see also Figure 3.12). This means that for all cases

$$C(t) = \frac{\varepsilon_S A}{W(t)}. \tag{4.57}$$

Or, in other words, it suffices to calculate or know the depletion width to estimate a value for the capacitance and vice versa.

Figure 4.10 shows an example of switching a Schottky diode from zero bias to reverse bias. As in TSC, after switching the bias, a region exists where the deep levels E_X are brought off thermal equilibrium this way. In fact, after switching the bias, the following things happen (the numbers in parentheses relate to the numbers in Figure 4.10):

- (1) The free carriers (holes in the valence band) will rapidly move out of the interface region. The space charge increases. A new, larger, depletion width is reached.
- Larger depletion width means smaller capacitance.
- Inside the new depletion zone a region exists (2) where the deep level has dropped below the Fermi level and where its occupancy is off equilibrium. It cannot thermalize so fast and initially remains neutral.

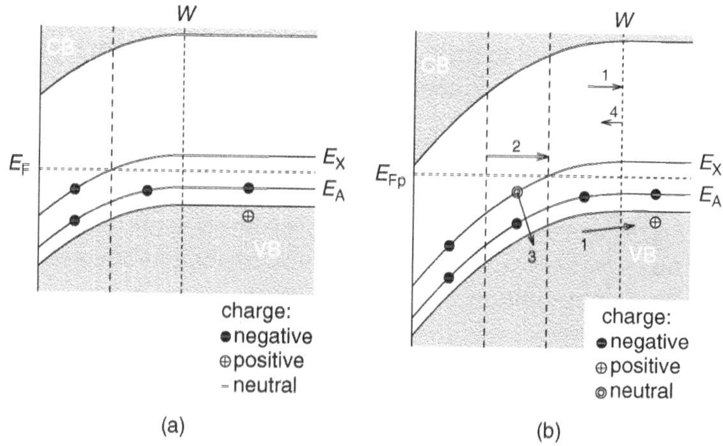

Figure 4.10 Schematic of a Schottky barrier with a deep majority carrier level upon a voltage step. (a) Zero bias system in thermal equilibrium. (b) Processes taking place after a bias step, as discussed in the text

- (3) Charges (holes) are slowly emitted from the deep level to the valence band on a timescale τ. These charges are immediately swept away by the electric field.
- Because the space charge density increases closer to the interface, less depletion width is needed. (Visualize: the integration of Poisson's Equation reaches V_{bb} in a little less space.) W shrinks (4).
- If the depletion width shrinks, capacitance increases. This happens on a timescale of the hole emission rate, $\tau = 1/e_p$, as given by Equation (4.13) (e_n substituted by e_p).

The deep acceptor X captures an electron from the valence band (emits a hole to it, as shown in Figure 4.10) and becomes negatively ionized N_X^-. These negatively ionized deep acceptors then have the same function as normal (shallow) acceptors that are all ionized everywhere ($N_A^- = N_A$). For the depletion width and capacitance the immobile ionized species count. In other words, the acceptor concentration in the calculation of the depletion width has to be replaced by one also including the ionized deep acceptor (and only the ionized ones), $N_A + N_X^-$

$$W = \sqrt{\frac{2\varepsilon(V_{bi} - V)}{q(N_A + N_X^-)}} \tag{4.58}$$

For the calculation it is useful to define the charge p_X that can still be emitted by the deep acceptors. This is equal to the neutral deep acceptors, $p_X = N_X - N_X^-$. Because the emitted holes are immediately swept away by the field in the depletion region, no recapture takes place. Under this condition, the kinetics can be simplified as

$$\frac{dp_X}{dt} = -e_p p_X \tag{4.59}$$

where p_X is the density of charges on the deep levels in the off-equilibrium region, and e_p is the emission rate of holes according to Equation (4.13). This has as simple solution

$$p_X(t) = p_{X0} \exp(-e_p t) \tag{4.60}$$

and, converting back to ionized deep acceptors,

$$N_X^- = N_X [1 - \exp(-t/\tau)] \tag{4.61}$$

where τ is the time constant of the transient equal to the reciprocal emission rate, $\tau = 1/e_p$. Substituting this in Equations (4.58) and (4.57) gives a capacitance transient of the form

$$\begin{aligned} C(t) &= A\sqrt{\frac{q\varepsilon[N_A + N_X^-(t)]}{2(V_{bi} - V)}} \\ &= C_0\sqrt{1 + \frac{N_X}{N_A}\left[1 - \exp\left(-\frac{t}{\tau}\right)\right]} \end{aligned} \tag{4.62}$$

where C_0 is a constant given by

$$C_0 = A\sqrt{\frac{q\varepsilon N_A}{2(V_{bi} - V)}} \tag{4.63}$$

In the absence of deep acceptors, this equation yields the normal depletion capacitance, independent of time, as expected. For small densities of deep levels, $N_X \ll N_A$, Equation (4.62) can be approximated by

$$C(t) = C_0\left\{1 + \frac{1}{2}\frac{N_X}{N_A}\left[1 - \exp\left(-\frac{t}{\tau}\right)\right]\right\}. \tag{4.64}$$

In other words, the capacitance transient of a system with, apart from the shallow acceptor, a low-density deep discrete acceptor is an exponentially decaying function

$$C(t) = C_\infty - \Delta C \exp(-t/\tau) \tag{4.65}$$

with a time constant equal to the reciprocal capture rate of Equation (4.13), $\tau = 1/e_p$ and an amplitude proportional to the deep acceptor density (in the region affected by the bias step), $\Delta C/C_0 = N_X/2N_A$. As described before, determining the capture rate, or relaxation time, as a function of temperature and plotting it in the form $\ln(\tau/T^2)$ vs. $1/T$ will reveal the activation energy E_a of the deep level, which is related to the level depth according to Equation (4.14). Normally, the correction factor E_σ is ignored and it is assumed that directly the level depth is found.

Figure 4.11 shows an example of how a capacitance transient looks for a majority level. The initial step is caused by a rapid moving out of the charges when the bias is changed, the slow, incomplete recovery is caused by an emission of holes from the levels and a shrinking of the depletion width.

The above is true for a majority level, a level that communicates with the majority carrier band, a level that thermalizes following the majority Fermi level. In contrast, there also exist levels that thermalize according to the minority carrier Fermi level because they fill and empty to the minority carrier band. As demonstrated in Section 3.5.3, these levels have the opposite behavior compared with majority carrier levels. When the device is placed in reverse bias for majority carriers, the density of free carriers in the minority carrier band decreases and the minority carrier Fermi level moves away from the band. As an example, when a p-type Schottky barrier is placed in reverse bias, both the hole and electron density decrease; the hole quasi Fermi level moves up and the electron quasi Fermi level moves down. (Note that this only works in the presence of current; without current, the quasi Fermi levels for holes and electrons are always coinciding, regardless of the bias.) Thus, after a voltage step to reverse bias, a level that communicates with the majority band (holes) will slowly get more negatively charged, whereas a level

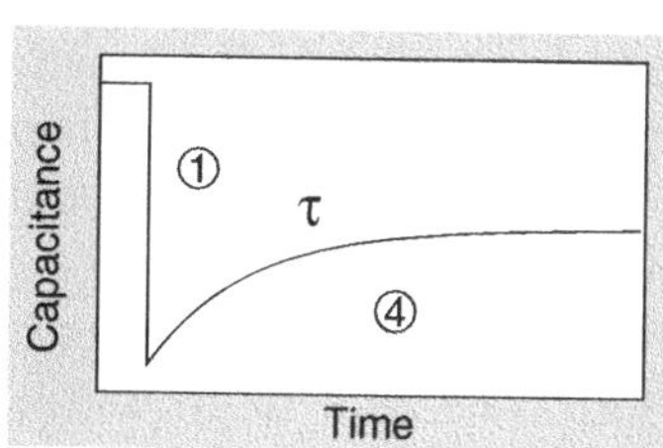

Figure 4.11 Example of a capacitance transient for a majority carrier level (deep acceptor in p-type material). The numbers relate to the processes in Figure 4.10

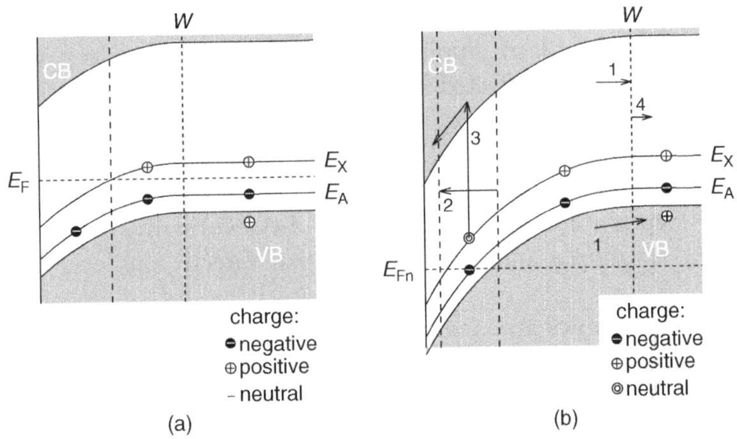

Figure 4.12 Schematic of a Schottky barrier with a deep minority carrier level upon a voltage step. (a) Zero bias system in thermal equilibrium. (b) Processes taking place after a bias step, as discussed in the text

that communicates with the minority band (electrons) will get more positively charged. The effect on the space charge and depletion width is thus opposite and an opposite sign in the transient can be expected. Figure 4.12 shows schematically what happens upon a change in bias to a system with a deep minority carrier (deep donor).

- (1) The free carriers (holes in the valence band) will rapidly move out of the interface region. The space charge increases. A new, larger, depletion width is reached. Larger depletion width means smaller capacitance.
- Due to the reduced presence of minority carriers which have been swept out by the increased field, the minority carrier (electron) quasi Fermi level has moved down. Therefore, inside the depletion zone there now exists a region (2) where the deep level has dropped below the Fermi level and where its occupancy is off equilibrium. It cannot thermalize so fast and initially remains neutral.
- (3) Charges (electrons) are slowly emitted from the deep level to the conduction band on a timescale τ. These charges are immediately swept away by the electric field.
- Because the space charge density decreases closer to the interface, more depletion width is needed; W grows (4).
- If the depletion width grows, capacitance decreases. This thus happens on a timescale of the electron emission rate, $\tau = 1/e_n$, as given by Equation (4.13).

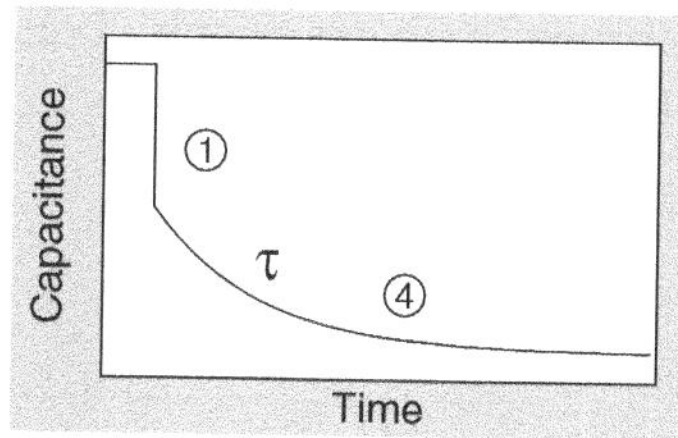

Figure 4.13 Example of a capacitance transient for a minority carrier level. The numbers relate to the processes in Figure 4.12

Figure 4.13 shows an example of how a capacitance transient looks for a minority level. The initial step is caused by a rapid moving out of the charges when the bias is changed, the slow transient is caused by an emission of electrons from the deep donor levels and a consequential growing of the depletion width.

Finally, it is interesting to determine how a capacitance transient will look in the presence of traps. The difference between traps and dopants is their charge states. For example, an acceptor can be either neutral or negatively charged, whereas a hole trap can be either neutral or positively charged. Both acceptors and hole traps communicate to the valence band and thus follow the hole quasi-Fermi level. As can be seen in Figure 4.14, the hole traps cause a capacitance transient similar to a deep acceptor level, i.e., with negative amplitude, as in Figure 4.11. A

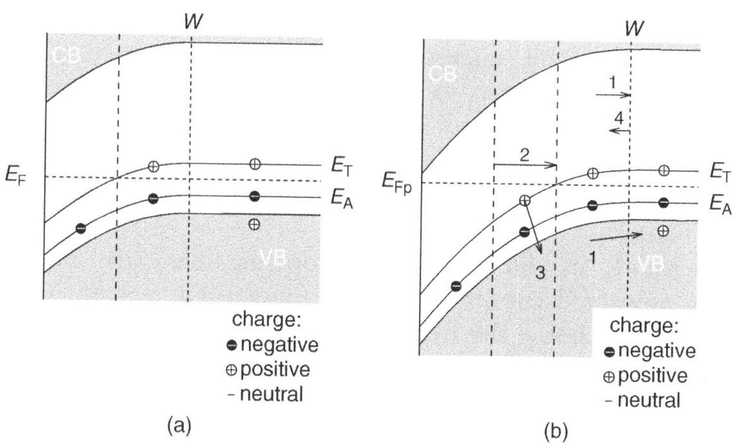

Figure 4.14 Schematic of a Schottky barrier with a deep hole trap level E_T upon a voltage step. (a) Zero bias system in thermal equilibrium. (b) Processes taking place after a bias step, as discussed in the text. The resulting transient is similar to a deep-acceptor-level transient

hole trap, in terms of capacitance transients, behaves like an acceptor. The general equation for the transient becomes

$$C(t) = A\sqrt{\frac{q\varepsilon[N_A - N_T^+(t)]}{2(V_{bi} - V)}}$$
$$= C_0\sqrt{1 + \frac{N_T}{N_A}\left[1 - \exp\left(-\frac{t}{\tau}\right)\right]} \qquad (4.66)$$

where C_0 is a constant equal to the one used before [Equation (4.63)].

Although the above measurement technique was developed for inorganic materials (Si, GaAs, etc.), it works well for organic materials. There are however some important differences between organics and inorganics that have to be taken into account when designing an experiment for inorganics.

First of all, as has already been stated at the beginning, organic materials, due to their wide band gap, can have electronic levels that are much deeper compared with inorganic materials. This makes the emission rate of charges from these levels much smaller and the transients consequently much slower. Assuming that the prefactors γ and σ_{pa} in Equation (4.13) are the same, this equation tells us that at room temperature, for every 70 meV increase in level depth, the transient grows a factor 10 slower. A midgap level in silicon is 0.56 eV, whereas in a typical organic material a midgap is easily 1.3 eV deep. Hence, one can expect the transients to be up to a factor 10^{12} slower. Typical transients for silicon are in the microsecond to millisecond range. Thus, for organics, in the worst case, we can expect transients in the megasecond to gigasecond range (up to 30 years!). These can hardly be called 'transients' anymore, but rather 'meta-stabilities'. These effects are also visible in FETs where they are called 'stressing', as will be discussed in Section 6.10.4.

Furthermore, as discussed before, organic materials suffer from (mostly) unwanted cut-off frequencies for various reasons. Care has to be taken to choose the frequency of the probing signal below any cut-off frequency, in order to make sure that the interface region at the barrier is measured. Commercial equipment developed for the silicon industry typically works at a 1 MHz AC frequency. This is, in most cases, inadequate for organics.

Where in silicon the purity of the material is so high as to make deep levels and traps a small perturbation, in organics they can be substantial. The approximation made for small density of deep levels ($N_X \ll N_A$ and

$N_T \ll N_A$) is then not valid and the full expression for the capacitance transient has to be used, Equations (4.62) and (4.66) for the deep levels and traps, respectively [183, 184]. Because of the square-root nature of the transients, in this case it is useful to work with the square of the capacitance; such techniques are often labeled C^2-DLTS.

Together with the high density of traps comes their nondiscrete distribution in energy. This causes a dispersion in emission rates and a nonexponential character of the transients. In principle, the exponential term in Equation (4.66) has to be replaced by a power law. Such a situation is highly complicated and each case deserves its own analysis. Instead of giving here a solution for all, it is left to the reader to derive an adequate function for the specific cases.

4.4.1 Case Study: Example of a Capacitance Transient Measurement

This section describes a successful capacitance transient measurement in an organic material. The device consisted of a Si/MeH-PPV Schottky barrier with an unintentional thin silicon oxide layer in between. This layer diminished the DC conductance and allowed for the observation of interface effects [133], but played no role in the capacitance transient measurements performed. Figure 4.15 shows the band diagram of the system. Note that the LUMO (lowest unoccupied molecular orbit) plays the role of the conduction band and the HOMO (highest occupied molecular orbit) is equivalent to the valence band. The silicon was highly doped n-type and, because of the immobility of the Fermi level, can be considered a metal for the analysis.

Figure 4.16 shows an example of a transient revealing two minority levels and one majority level. In some cases the type of the transient could be selected by a careful choice of voltages. Figure 4.17 demonstrates how for a certain temperature, the transient could be programmed to probe the minority levels and the majority levels by varying the bias. Figure 4.18 summarizes the capacitance transients. In Figure 4.18, each dot represents a relaxation time measured. A full circle corresponds to a majority type transient (positive sign), and an open circle represents a minority type transient (negative sign). The area of each dot is proportional to the amplitude of the transient (multiplied by 4 in the right panel).

Moreover, the type of the impurity responsible for the deep level and the associated transient could be determined following the reasoning of Omling *et al.* [185]. When the pulse length is varied, point defects, or in

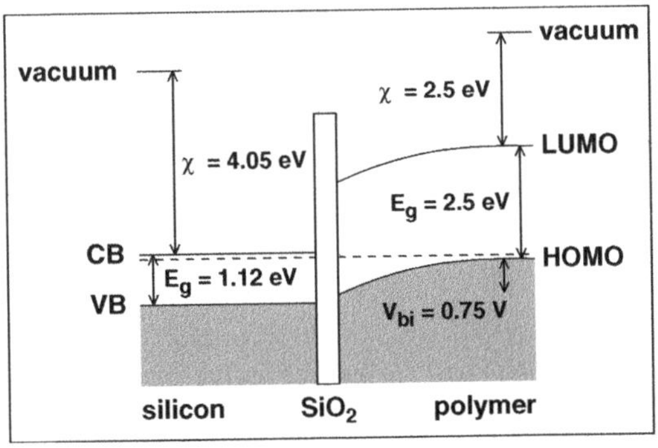

Figure 4.15 Band structure of a Si/MeH-PPV Schottky barrier used in the capacitance transient experiment. Reproduced with permission from Journal of Applied Physics, Minority-carrier effects in poly-phenylenevinylene as studied by electrical characterization by Peter Stallinga, 89, 3 Copyright (2001) American Institute of Physics

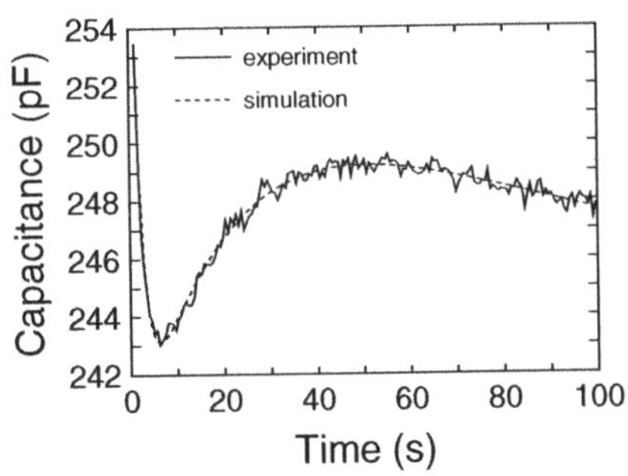

Figure 4.16 Example of a capacitance transient in a Si/MeH-PPV Schottky diode. For this temperature, two minority levels and one majority level were observed. The dashed line is a fit with three relaxation times. Reproduced with permission from Journal of Applied Physics, Minority-carrier effects in poly-phenylenevinylene as studied by electrical characterization by Peter Stallinga, 89, 3 Copyright (2001) American Institute of Physics

general defects that can capture only a single charge, the amplitude of the transient, follow

$$\Delta C(\Delta t) = \Delta C_\infty [1 - \exp(-\lambda \Delta t)] \tag{4.67}$$

where Δt is the pulse length and λ is a constant depending on the trap filling time. ΔC_∞ is the transient amplitude if an infinite time was allowed to fill the levels. Thus, a plot of $\log[\Delta C(\Delta t) - \Delta C_\infty]$ vs. Δt will be a straight line. However, for extended defects, that can trap multiple

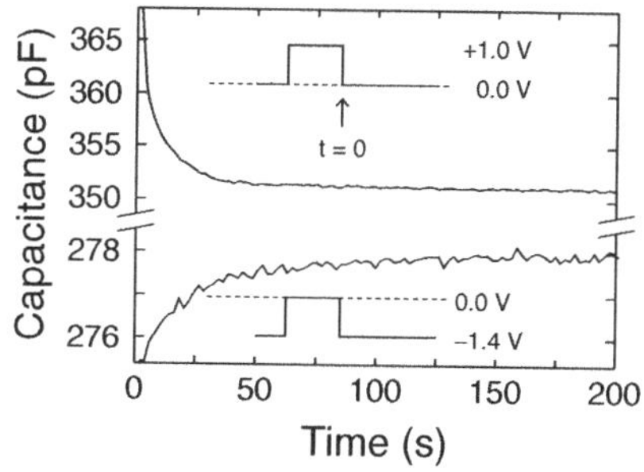

Figure 4.17 By carefully choosing the pulse shape, for some biases the transient can be of minority type (top) or majority type (bottom). Reproduced with permission from Journal of Applied Physics, Minority-carrier effects in poly-phenylenevinylene as studied by electrical characterization by Peter Stallinga, 89, 3 Copyright (2001) American Institute of Physics

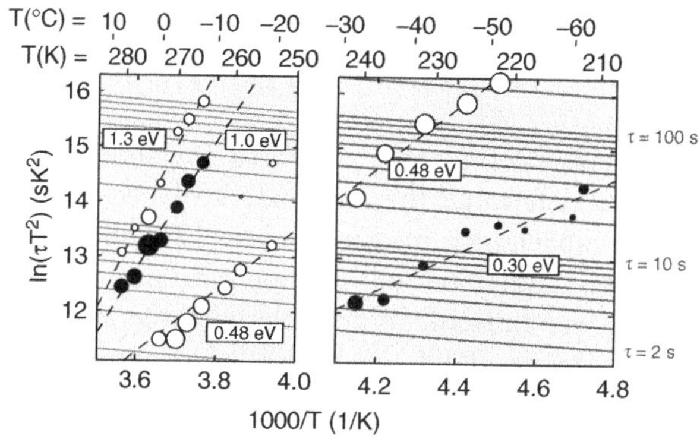

Figure 4.18 Summary of a capacitance transient measurement in a Si/MeH-PPV Schottky diode. Each dot represents a relaxation time τ obtained from a transient. Full circles represent majority carrier (positive) transients as in Figure 4.11 and open circles are minority carrier (negative) transients as in Figure 4.13. The area of a dot represents the amplitude of the transient (multiplied by 4 in the right panel). Reproduced with permission from Journal of Applied Physics, Minority-carrier effects in poly-phenylenevinylene as studied by electrical characterization by Peter Stallinga, 89, 3 Copyright (2001) American Institute of Physics

charges, or for high density defects that feel each other's Coulombic field, the probability of trapping depends on the number of charges already present and thus becomes slower and slower. The above mentioned plot will no longer be a straight line but will bend upwards. Figure 4.19 summarizes the pulse length dependence and prooves that the defect involved indeed is a diluted point defect.

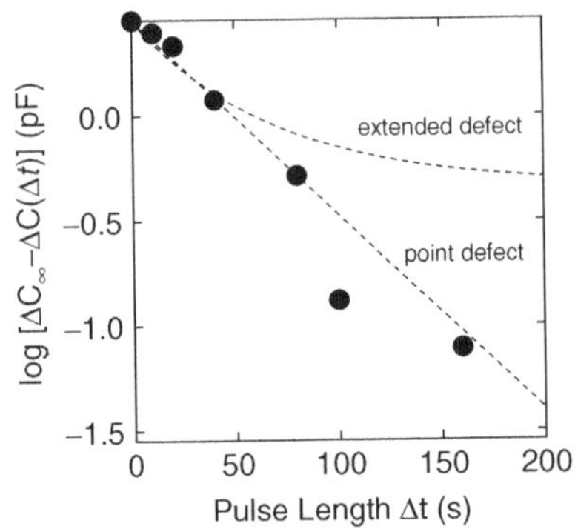

Figure 4.19 Pulse length dependence of a transient amplitude. The straight line indicates point defects rather than extended defects as the source of the transient. Reproduced with permission from Journal of Applied Physics, Minority-carrier effects in poly-phenylenevinylene as studied by electrical characterization by Peter Stallinga, 89, 3 Copyright (2001) American Institute of Physics

4.5 DEEP-LEVEL TRANSIENT SPECTROSCOPY

Deep-level transient-spectroscopy, or DLTS for short, is a very powerful technique to determine the parameters of deep levels by means of time-resolved capacitance measurements [186]. Because it requires computerized data acquisition and off-line automated data analysis as will be shown, it was only made possible with the advent of personal computers in the 1970s. DLTS is basically an automated version of capacitance transient techniques discussed above.

It works in the following way. Capacitance transients are recorded in the same way as described above. The device is placed in the emptying voltage V_E and the system is allowed to reach thermal equilibrium. For a short time the device is placed under the filling voltage V_F. In this time the traps will fill with charges. When the voltage is switched back to V_E a transient is observed as described in the previous sections. From this transient two samples are taken at times t_1 and t_2 after switching the bias. The DLTS spectrum is now the difference in capacitance at these two times as a function of temperature:

$$S(T) = C(t_1) - C(t_2). \tag{4.68}$$

This means (substituting the explicit form of the transient, see Section 4.4)

$$S(T) = [\exp(-t_1/\tau) - \exp(-t_2/\tau)]. \tag{4.69}$$

The maximum of this signal occurs when the relaxation time τ reaches the value τ_{max} which can be found by differentiating $S(T)$ with respect to τ:

$$\tau_{max} = \frac{t_2 - t_1}{\ln(t_1/t_2)} \tag{4.70}$$

The explicit temperature dependence of τ is [see Section 4.1, Equation (4.13)]

$$\tau(T) = \tau_0 T^{-2} \exp(E_a/kT) \tag{4.71}$$

where $E_a = (E_X - E_V) - \Delta E_\sigma$, and E_X is the energy of the deep level under study. From this it is clear that the temperature at which the maximum DLTS signal occurs does not reveal the activation energy directly, unless τ_0 is known. If not, at least two measurements have to be made with different time windows (t_1, t_2) and the two points (T_{max}, τ_{max}) entered into Equation (4.71) will yield the trap activation energy E_a.

Important parameters of DLTS include:

- Repetition rate: how many times per second a transient is recorded. The time between two transients should be long enough to ensure that the system is in thermal equilibrium.
- Time window: The values of the sampling point times t_1 and t_2.
- Filling voltage V_F and emptying voltage V_E.
- Filling time: how long the device is placed in the filling voltage V_F before switching back to reverse voltage V_E.

Figure 4.20 shows a simulation of DLTS. Figure 4.20(a) shows simulations of transients caused by a 0.2 eV deep majority carrier level for various temperatures (44 K to 52 K in 2 K steps). For the lowest temperatures the transients are very slow (bottom trace) and the difference between the capacitance at t_1 and t_2 is small, hence the DLTS signal $S(T)$ is very small. When the temperature is increased the transients become faster and the DLTS signal increases. Note that the signal has negative sign; for majority carrier traps, the transient has an upward trend and therefore $C(t_1)$ is smaller than $C(t_2)$. When the temperature is further increased, the transients become so fast that they have already died out before the time window $(t_1 - t_2)$ and the DLTS signal has vanished again. Figure 4.20(b) summarizes this for the majority carrier trap (lower trace). A comparison is also given with a minority carrier trap (upper trace) whose DLTS signal has the opposite sign because the related

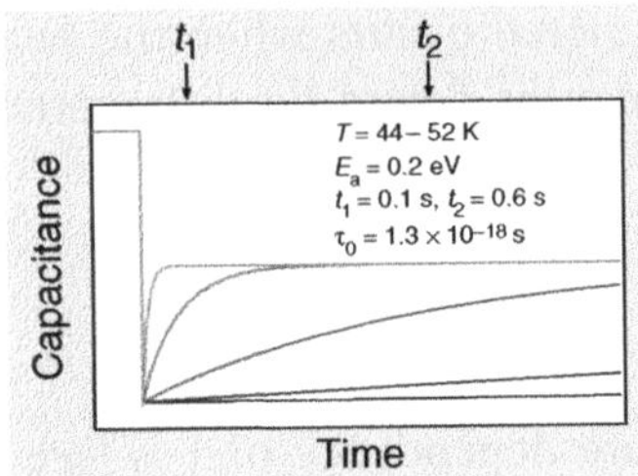

(a)

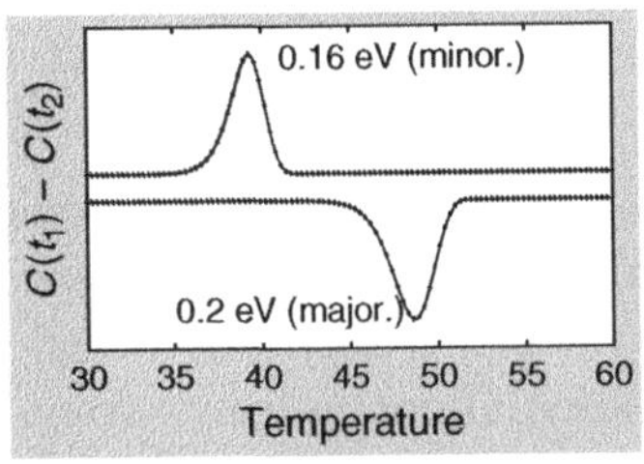

(b)

Figure 4.20 (a) Capacitance transients for various temperatures and (b) resulting DLTS signal for majority levels (bottom) and minority levels (top)

transients have downward trends. In this simulation the minority-carrier trap is chosen to be a little more shallow (0.16 eV). Assuming that the prefactor τ_0 is identical (1.3×10^{-18} s in both cases) it means that the peak has shifted towards lower temperatures.

The advantage of DLTS is that it can be done completely automatically with a minimum of data processing. The disadvantage is that of the entire transient only two points are used; most of the data are thrown away and the duty cycle of the measurement is very low. In Laplace DLTS the duty cycle is much higher because the entire transient is used in a Laplace transformation [187]. The cost is a more elaborate calculation; only modern computers can be used. In reality, DLTS is more often used not to determine the trap activation energies but more to show the presence of certain impurities through their fingerprint spectra and to determine their densities through the intensity of the DLTS spectra and through the underlying amplitude of the transients. This is based on the fact that the amplitude of the transient is linearly proportional to the density of the deep defect relative to the density of the dopant (assuming that this ratio is small), as shown in the previous section,

$$\frac{\Delta C}{C} = \frac{N_{\mathrm{X}}}{2N_{\mathrm{A}}}. \tag{4.72}$$

When the signals are large, a more complicated method has to be used, such as C^2-DLTS [183].

Finally, to be precise, the use of the word 'spectroscopy' in DLTS is misleading, because in a spectroscopy experiment an experimental value is monitored while the frequency is scanned; a spectrum is the visualization of an observable as a function of frequency. In DLTS no frequency is scanned, so the name is slightly inadequate. In DLTS (normally) the temperature is scanned and the DLTS signal intensity is plotted. The end plot resembles a spectrum in that it shows peaks and that it is a 'fingerprint' of the defect in the material. It would have been better if the word 'scanning' was used for the 'S' in the acronym DLTS.

4.6 Q-DLTS

A cross-breed between DLTS and TSC is Q-DLTS. Instead of measuring the capacitance transient, as in conventional DLTS, the charge emitted by the device after a bias step is analyzed [188]. One of the advantages of this technique is that no capacitance is measured and thus no depletion is needed. Where DLTS only works on rectifying contacts (Schottky barriers), Q-DLTS can be performed on any type of device.

Assuming that all charge emitted from the traps is swept by the (tiny) field to one electrode, the current is proportional to the rate of charge emitted by the traps. Since, as shown, this is dependent on the energetic depth of the traps and the temperature, Q-DLTS can reveal the same information as conventional DLTS. Moreover, the similarity between DLTS and Q-DLTS lies in the fact that both take two points of the total transient and thus define a time window for the measurements. Equation (4.69) becomes

$$\begin{aligned} S \equiv \Delta Q &= Q(t_1) - Q(t_2) \\ &= Q_0\left[\exp(-e_p t_1) - \exp(-e_p t_2)\right] \end{aligned} \tag{4.73}$$

where e_p is the emission rate of charge (holes) from the traps [Equation (4.13)]. The Q-DLTS 'spectrum' is defined as this quantity ΔQ as a function of the time-window parameter τ

$$\tau \equiv (t_2 - t_1)\ln(t_2/t_1) \tag{4.74}$$

Similar to the DLTS spectrum based on the capacitance signal, this has a maximum ΔQ_{max} at $\tau_{max} = 1/e_p$. The signal strength at this maximum is a direct measure for the number of traps involved

$$VN_{\mathrm{T}} = 4\Delta Q_{\mathrm{max}}/q \tag{4.75}$$

where V is the volume of the device. The depth of the trap can be found by repeating the transient for various temperatures; an Arrhenius plot of $\ln(1/\tau_{\mathrm{max}}T^2)$ vs. $1/T$ will yield the activation energy of release of charge from the traps.

5

Time-of-flight

5.1 INTRODUCTION

The advantage of organic devices is mainly their possible low cost of production. As a result of the low-technology way of producing the devices, the materials normally wind up in an amorphous state. This makes them notoriously slow (note the comment of Karl *et al.* on the possibility of fast organic materials [189]), but for many applications this is not a problem compared with the large benefit of reduction in cost. Where amorphous silicon is probably the best known and best studied example of amorphous materials, in the last decades organic amorphous materials have shown themselves to be a commercially viable alternative, and, in some cases, even outperform their inorganic counterparts when special applications are needed, such as mechanical flexibility of device, or printability of electronic circuits.

As discussed in earlier chapters, one of the most important parameters in electronic materials in general is the charge carrier mobility (μ), defined as the ratio between the velocity of the carriers (v) and the electrical field (E), $v = \mu E$. Directly measuring the mobility using this relation however is difficult and there exist several indirect techniques, of which I–V curves, Hall effect, field effect in a transistor, and ToF are the most popular (see the work of Karl for a more complete list of measurement techniques and a general discussion of electronic transport in organic semiconductors [190]). Because of the indirect way the mobility is measured, the various techniques can yield widely

Electrical Characterization of Organic Electronic Materials and Devices Peter Stallinga

differing values, and can in the extreme case even give negative values [191]. As will be shown in Chapter 6, in FETs not the mobility but an effective mobility is measured [192]. The current chapter focuses on the ToF technique, a technique that is often used for organic materials [193–201] because it is a fast and relatively cheap method that is adequate for the range of mobilities encountered in organics.

In amorphous materials an anomalous transit-time dispersion is often observed. The anomalous behavior is described as the observation that the mobility, as measured in a ToF experiment, depends on the sample thickness and does not seem to be an intrinsic material property. The work of Scher and Montroll is classic reading material on this subject [202], with others reasoning along similar lines [203, 204]. (Alternatively it can be described as the transit time following a nonquadratic power-law dependence on sample thickness [205].) They have shown that, if one assumes a hopping model of conduction, the anomalous behavior emerges, and tradition in the literature is then to reverse this logic and use the observation of anomalous ToF transients as a proof of the hopping model. This is in spite of work by Silver and Cohen, shortly published thereafter, that has shown that the same results emerge when assuming exponentially distributed traps [176] with a band conduction mechanism. To use their words, 'simple transient measurements cannot by themselves distinguish between them'.

This situation is aggravated by two important facts. First of all, measurements are mostly performed on digital oscilloscopes with a sampling resolution of 8 bits (256 different digitalization levels). Yet, as will be shown here, a lot of information resides in a much larger dynamic range. See for example the power-law behavior discussed later. Secondly, because the transient behavior extends over longer times, and, moreover, they are not simple exponential, it is very difficult to have a device in a well-known, reproducible state at the start of an experiment. Two consecutive ToF measurement can easily give different results. In view of these two points, it is nearly impossible to arbitrate between the various models on the basis of observed data; where the observed transients are often *consistent* with the model of choice, they do not *proove* it. This is a problem often encountered in devices of organic materials and for ToF measurements, as already pointed out before.

In this chapter we will see the range of possible transient types and what causes them. Analytical solutions are given for systems with a single feature highlighted. For instance, the effect is shown of an exponential DOS, without reverting to Monte-Carlo simulations, similar to the work of Tiedje and Rosen [206]. Traps often play an important role in the modeling of ToF behavior found in the literature [207]. It will be shown here how anomalous behavior can result for any low-mobility material, regardless of the conduction mechanism, thus shedding new light on a multitude of reported works. The assumption of traps being responsible for the anomalous behavior makes sense in view of the other devices and measurement techniques presented in this book. Ranging from the SCLC regime in diodes to the gate-bias and temperature dependent mobility in FETs, they all can be explained with the single postulate of traps governing the conduction process.

5.2 DRIFT TRANSIENT

In a ToF experiment the current transient is measured after a light pulse and from this transient the mobility can be determined. In a related technique the determination of the mobility is achieved in the opposite way, namely by detecting the delay of electroluminescence after a bias step [208]. The current chapter only deals with pure ToF experiments.

On one side of a device with a light pulse electron–hole pairs are created, for instance by illumination through an optically (semi) transparent electrode (for example ITO). While alternative structures such as TFTs can also be used [209], the sandwich structure is the most adequate. It is assumed that the penetration of the light into the sample is minimum, so that electron–hole pairs are mainly generated close to the electrode. Moreover, it is assumed that the light pulse is much shorter than the transit time, thus ensuring a delta function for the initial carrier distribution in space. The external electrical field separates the charges and drives one type of carrier across the interface towards the other electrode. This immediately shows one of the advantages of the ToF technique over other techniques such as delayed electroluminescence, namely the fact that mobilities of both electrons and holes can independently be determined by selecting the sign of the bias.

The time it takes for the carriers to arrive at the counter-electrode, labeled τ_{ToF}, is given by the device thickness L divided by the average speed of the carriers, $v = \mu E$,

$$\tau_{\text{ToF}} = \frac{L}{v} = \frac{L}{\mu E} = \frac{L^2}{\mu V} \tag{5.1}$$

where V is the applied bias. This is the so-called quadratic equation telling us that the time depends quadratically on sample thickness, other things being equal. The mobility therefore, can be calculated as

$$\mu_{\mathrm{ToF}} = \frac{L^2}{\tau_{\mathrm{ToF}} V}. \tag{5.2}$$

One essential condition is that the material must be isolating, meaning that the amount of thermal equilibrium charge in the device (qp_0L) is much smaller than the amount of charge on the electrodes ($\varepsilon_s\varepsilon_0 V/L$) because the material dielectric relaxation time has to be larger than the transit time [210]. This gives an underlimit to the bias used in the experiment,

$$V_{\mathrm{min}} = \frac{qp_0L^2}{2\varepsilon_s\varepsilon_0} \tag{5.3}$$

However, the bias cannot be too large either for reasons explained later. Some of the anomalous behavior of the transients can be ascribed to not meeting these conditions.

The time of flight τ_{ToF} can be determined from the current transients. In the ideal case, there is a current from 0 to τ_{ToF} and zero current after that, with the integrated current equal to the charge induced by the light pulse. This can easily be understood when we analyze the system. For this simple analysis we assume the delta distribution of induced holes to move linearly across the device from one electrode to the other (the electrons are assumed to have zero mobility and remain at one side of the device until they are neutralized). Initially, there is a thin layer of free holes p close to the left electrode, compensated by the same amount of free electrons n keeping charge neutrality (units per square meter). When the holes move towards the right electrode, they induce a density of electrons in this electrode. Because charge neutrality is maintained, the density of electrodes in the left electrode is diminishing. Actually, electrons move from the left electrode via the external leads and amperimeter towards the right electrode and a current is observed. The magnitude of this current can be calculated if we imagine the device as consisting of two serial capacitors. The value of the first capacitance is given by a metal plates system, with one plate at 0 and the other positioned at x, filled with a dielectric ε. The capacitance density is thus

$$C_1 = \varepsilon/x \tag{5.4}$$

where x is the actual position of the sheet of holes moving across the sample. The other capacitance is in the same way given as

$$C_2 = \varepsilon/(L - x). \tag{5.5}$$

The holes p are compensated partially on the left side by electrons n_1 at the left electrode and partially on the right side by n_2 electrons at the right electrode. It can be imagined that these charges reside in capacitors 1 and 2, as shown in Figure 5.1(b). Thus, $n_1 = p_1$, $n_2 = p_2$ and $p_1 + p_2 = p$. The voltage drop induced by this charge in each capacitor can be calculated as ($V = Q/C$)

$$\Delta V_1 = \frac{qp_1}{C_1} = \frac{qp_1}{\varepsilon}x \tag{5.6}$$

$$\Delta V_2 = \frac{-qp_2}{C_2} = -\frac{qp_2}{\varepsilon}(L - x). \tag{5.7}$$

Moreover, because of Kirchhoff's Law, the two voltage drops have to be equal and of opposite sign, $\Delta V_1 = -\Delta V_2$. (The external bias is not considered and is only used to determine the speed of movement. Including this bias here would not change the analysis, as it would only offset the charge distributions n_1 and n_2 and not change their time dependence.) Combination of the above ideas gives

$$n_2 = p_2 = p\frac{x}{L} \tag{5.8}$$

$$n_1 = p_1 = p\left(1 - \frac{x}{L}\right). \tag{5.9}$$

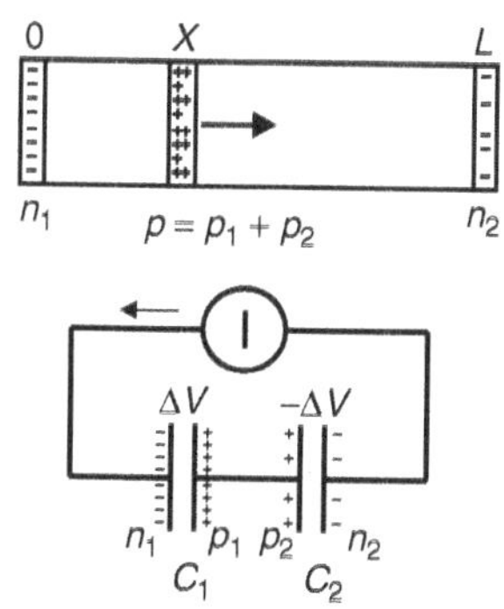

Figure 5.1 Modeling of ToF signal. A layer of charges with a density per area equal to p is located somewhere in the device at position x. This can be thought of as consisting of two capacitors, C_1 and C_2, each with part of the charges, p_1 and p_2, respectively. The value of the capacitance follows metal plates capacitors $C = \varepsilon A/d$, where d is equal to x and $L - x$ for capacitors 1 and 2, respectively. The positive charge on each is compensated by an equal negative charge on the counter-electrode. The positive charges on the capacitors add up to p

The current going through the amperimeter is proportional to the time derivative of n_2 (or $-n_1$), since these charges have to pass through the amperimeter (at least until the holes reach the electrode, whence electron–hole recombination can change n_2):

$$J(t) = q\frac{dn_2}{dt} = \frac{qp}{L} \cdot \frac{dx(t)}{dt}. \tag{5.10}$$

In other words, the current density is proportional to the movement of the charge and is constant until the layer of charge reaches the other electrode. The charges then rapidly recombine with the electrons present there and the current immediately drops to zero. The constant current can be calculated when we realize that the velocity of charge is equal to the mobility μ and the electric field $E_x = V/L$. Thus, the current density is given by

$$J(t) = J_0 = \frac{qp}{L} \cdot \mu \cdot \frac{V}{L} \tag{5.11}$$

until the charge reaches the other side at a time τ given by Equation (5.1). The integrated current is thus equal to qAp, as expected. Figure 5.2(a) shows a simulation of a transient for this simple model with parameters as in Table 5.1.

In order to have undistorted signals and to be able to reliably extract the mobility, requisites for ToF measurements are:

- The light pulse only generates electron–hole pairs. Especially when using sub band-gap light, charges can be placed on (deep) traps and the experiment becomes a detrapping experiment instead. (For comparison, such an experiment is described later on, see Section 5.6).

Table 5.1 Parameters used in simulation (unless otherwise specified)

Parameter	Value	Unit
L	1	μm
μ	10^{-6}	$cm^2\ V^{-1}s^{-1}$
V	1	V
T	300	K
τ_{ToF}	10	ms [Equation (5.1)]

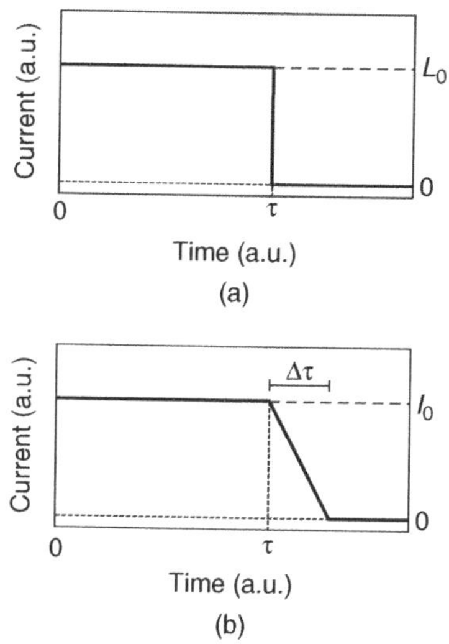

Figure 5.2 Simulation of simple ToF experiment. (a) Simple pulse: The current is equal to I_0 of Equation (5.11) until time τ given in Equation (5.1). (b) Square pulse: Until τ the current is equal to I_0, after that the current drops to half immediately and then linearly decays in a time $\Delta\tau$

- The light pulse only generates electron–hole pairs in a very thin region of the device. When the absorption coefficient is too low, electron–hole pairs can be generated all throughout the layer and the transient becomes distorted as will be shown shortly.
- The pulse length is much shorter than the relevant measurement time (τ). This is in order to make sure that the initial distribution is a delta function.
- No electron–hole recombination takes place during a transit. All the charges (of one polarity) contribute to the current, until they reach the other side of the sample where they are absorbed by the electrode.
- The contacts are ohmic. This to ensure that the external field is the only electric field and it is homogeneous throughout the device.

The first criterion is easily met by selecting a proper wavelength of excitation, namely with energies larger than the band gap. In this case all excited carriers are free carriers. This immediately also ensures the second condition, since for such short wavelengths, across band gap excitation of carriers is efficient and the absorption coefficient normally large enough to ensure that absorption only takes place at the surface.

The third condition dictates that the pulse length be much shorter than the transient time. As an example, Campbell *et al.* measure the mobility of a PPV derivate (MeH-PPV) in a ToF experiment [211]. The absorption coefficient in this material is 3×10^5 cm^{-1}. The penetration depth is therefore about 30 nm, much shorter than the film thickness of 2 μm. The electron–hole pairs can therefore be considered to be generated at the electrode. The found mobilities are in the order of 10^{-6} cm^2 V^{-1}s^{-1}, and with fields in the order of 10^5 V cm^{-1}, the transit time is in the order of 1 ms, which is much longer than the optical pulse length used of 500 ps.

It is easy to understand what will happen when the initial distribution is not a delta function. As an example, consider a packet of holes of density p_0 and finite width W [$P(0) = p_0 W$] moving monotonously from one electrode to the other (no diffusion). No holes disappear before the packet reaches the right electrode, the current is given by Equation (5.11), with $p = P(0)$. When the packet reaches the electrode, at time $t = \tau$, the total charge $P(t)$ is linearly reduced to zero in the time it takes the left edge to reach the electrode, $\Delta\tau = W/\mu E_x$. The current thus drops linearly to zero in this time, see Figure 5.2(b). The same result would have been found if we had divided the charge packet into layers and used the simple approximation given before for each layer. The total transient as a superposition of the individual contributions would be the same.

A special case is one in which the absorption coefficient is not infinite and light is absorbed in a substantial depth of the sample. In the case of normal linear absorption using the Beer–Lambert Law,

$$P(x, t = 0) = P_0 \exp(-\alpha x) \tag{5.12}$$

where α is the absorption coefficient. This distribution moves monotonously towards the right electrode and the current that is still in the device contributes to the current according to Equation (5.11),

$$\begin{aligned} J(t) &= \frac{q\mu V}{L^2} \int_0^{L-\mu Et} P(x, t = 0)\mathrm{d}x \\ &= \frac{q\mu V}{L^2} \int_0^{L-\mu Et} P_0 \exp(-\alpha x)\mathrm{d}x \\ &= \frac{q\mu V P_0}{\alpha L^2} \{1 - \exp[-\alpha(L - \mu Vt/L)]\} \end{aligned} \tag{5.13}$$

until $t = L^2/\mu V$ and zero afterwards. For large α, the light is absorbed in a tiny region and the transients resemble the one given before. For

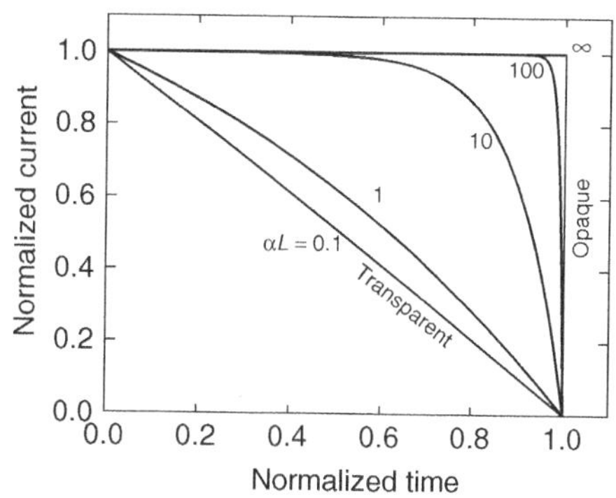

Figure 5.3 Normalized ToF current transients in the case of a finite absorption coefficient α. For strong absorption the transients are rectangular because all charge starts close to the surface, while for small α, the charge is created homogeneously throughout the sample and the transient is triangular

small α, the light is absorbed homogeneously throughout the sample and a triangular ToF current transient results, see Figure 5.3. Note that this assumes that only one type of charge contributes to current, with the other type either being very fast–so that they have moved out rapidly and caused only a current spike in the beginning – or very slow – so that they do not move in the timescale of the transient.

The result for the small absorption coefficient, namely the triangular transient, we will also find in a ToF experiment using a TFT structure when using full-channel illumination. To obtain a rectangular transient as in conventional ToF, in these devices we can focus the laser beam and confine it to a tiny region close to the electrode that repels the charges under study.

If electron–hole pair recombination takes place during a transit of the carriers, the transient signal becomes distorted. In the extreme case, when no charge makes it to the counter-electrode, the transient no longer represents a ToF experiment, but becomes a recombination experiment instead (see the discussion of decay of photoexcited carriers in the book by Sze [9]). The current is caused by the charge still available in the device. Assuming a constant recombination rate, yielding a constant relaxation time of τ_r, the current is

$$J(t) = qp(t)\mu\frac{V}{L} = J_0 \exp(-t/\tau_r). \tag{5.14}$$

The best way to prevent this is to have transients shorter than the relaxation time τ_r and having no free carriers of opposite polarity available in the sample. For instance, use p-type (or intrinsic) material when measuring hole mobility, since there will nearly be no electrons available for recombination. Also, make sure the bias is small, so as not to inject these minority carriers at the counter-electrode.

Finally, the contacts (especially the source electrode) cannot be rectifying. If electric fields are present in the device in the absence of bias, the current transient can be very distorted. In the extreme case, the device can work like a photovoltaic cell and can, for instance have a negative current for zero (or small positive) bias, the so-called short-circuit current I_{SC} and the current only becomes positive for a bias beyond the open-circuit voltage V_{OC}. It is obvious that such a transient becomes difficult to analyze in the ToF framework.

Some of the effects described above can mimic ToF signals, showing that ToF is not a plug-and-play experiment. In what follows, it is assumed that all these conditions are met. We will now look at some intrinsic effects on the transients.

5.3 DIFFUSIVE TRANSIENT

In the basic drift transient of the previous section it was assumed that the charge moves as a sheet linearly from one side to the other. This represents the extreme case when drift ('field driven') is much larger than diffusion ('density gradient driven'). An approximation is often made to ignore diffusion altogether, since 'the spreading of the packet (in crystalline materials) seems to be unimportant' [203]. However, Einstein's Relation [212] links mobility and diffusion in a linear way, and in *any* material (even noncrystalline), the transients have the same shape, and diffusion cannot be ignored, regardless of the value of the mobility. (Later we will see diversions from Einstein's Relation.)

When diffusion is included, apart from a movement of the carrier packet along the field direction as caused by drift, the packet itself broadens over time by diffusion. The diffusion itself causes no current, since it spreads the charges equally in both directions. However, the current is no longer simply proportional to the movement of the maximum of the peak (contrary to what is stated by Scher and Montroll [202, 213]) because holes arrive at the electrodes (on both sides) by diffusion and disappear from the system. The current is only determined by movement of the charges that remain in the device. These remaining charges can be divided into layers, each contributing to the current as described above. Since this is a linear function, the current is proportional to the movement of the center of mass of this distribution and the total charge contained within it:

$$n_2(t) = \int_0^L p(x,t)\frac{x}{L}\mathrm{d}x = \frac{\overline{x}(t)}{L}P(t) \tag{5.15}$$

where $\overline{x}(t)$ is the mean position of the charge distribution and $P(t)$ is the total amount of positive charges in the device at time t. The current density is proportional to the derivative of this

$$\begin{aligned} J(t) &= q\frac{\mathrm{d}n_2(t)}{\mathrm{d}t} \\ &= \frac{q}{L}\left[\frac{\mathrm{d}\overline{x}(t)}{\mathrm{d}t}P(t) + \overline{x}(t)\frac{\mathrm{d}P(t)}{\mathrm{d}t}\right] \\ &= \frac{q}{L}\frac{\mathrm{d}\overline{x}(t)}{\mathrm{d}t}P(t) \\ &= \frac{q\mu E_x}{L}P(t) \end{aligned} \tag{5.16}$$

where the third step, the cancelation of the second term inside the square brackets, is justified by the fact that disappearance of holes from the device when occurring at the electrodes through an electron–hole recombination process causes no external current. It is thus obvious that the current is not simply proportional to the time derivative of the center of the distribution. The current can already start dropping before the center of the package reaches the other side because $P(t)$ is diminishing. This can be because the leading front of the distribution reaches the other side, or the trailing tail recombines at the electrode of origin.

On basis of this, we can determine what happens when diffusion of charge starts playing a role on the timescale of the experiment (when the electric field is small). Imagine we start with a thin layer of P_0 holes at $x = 0$. This layer, apart from moving towards the right electrode $[\overline{x}(t) = \mu E_x t]$, also broadens with a speed determined by the diffusion coefficient $[\Delta x(t) = \sqrt{Dt}]$. The distribution of holes can thus be described as

$$p(x,t) = \frac{P_0}{\sqrt{\pi D t}}\exp\left[-\frac{(x-\mu E_x t)^2}{Dt}\right] \tag{5.17}$$

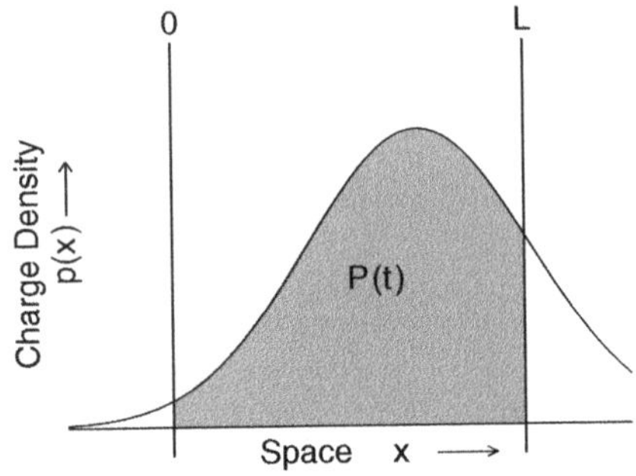

Figure 5.4 Distribution of charge spread by diffusion and moved by an external electrical field. The total charge between the two electrodes $P(t)$ is given by Equation (5.18). Reproduced with permission from Organic Electronics, Thin-film field-effect transistors: The effects of traps in the bias and temperature dependance of field-effect mobility, including the Meyer-Neldel rule by P. Stallinga and H. L. Gomes, 7, 6, 592–599 Copyright (2006) Elsevier

(see Figure 5.4 for a graphical representation of this function). The total amount of charge in the device at given time t is equal to the integral between the electrodes

$$P(t) = \int_0^L p(x,t)\mathrm{d}x \tag{5.18}$$

where it is intrinsically assumed that the electrodes themselves do not influence the distribution (image charge effects, etc.).

Substituting Equation (5.17) into Equation (5.18) will yield $P(t)$ and via Equation (5.16) the current can be found as

$$J(t) = \frac{qP_0\mu E_x}{2L}\left[\mathrm{erf}\left(\frac{L-\mu E_x t}{\sqrt{Dt}}\right) + \mathrm{erf}\left(\frac{\mu E_x t}{\sqrt{Dt}}\right)\right] \tag{5.19}$$

where erf() is the error function [177]. When this current transient is plotted in a log–log format (see Figure 5.5) some important conclusions can be drawn.

It is immediately clear from Figure 5.5(a) and 5.5(c) that for large temperatures or small biases, where diffusion is relatively more important, the transients behave in a different way. In fact, for these conditions, three distinct regions can be distinguished. For the shortest time, and with the knowledge that $\mathrm{erf}(0) = 0$ and $\mathrm{erf}(\infty) = 1$, it can be shown that the current is constant at $q\mu E_x P_0/2L$, which can be understood as: half of the charge is immediately absorbed by the source electrode and the

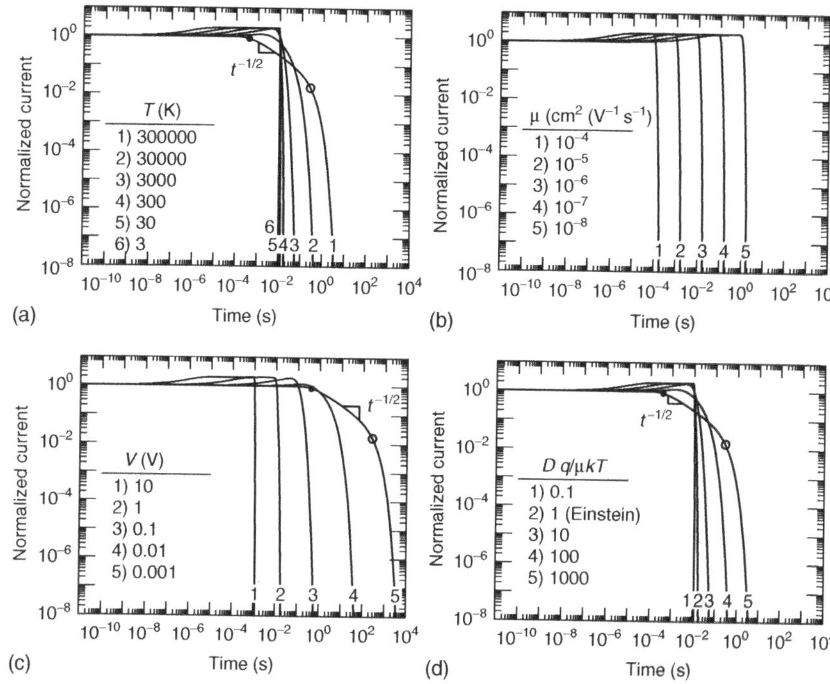

Figure 5.5 Simulation of transients as a function of temperature (a), mobility (b), bias (c) and diffusion coefficient (d). The open circles indicate $\tau_{1\rightarrow 2}$ and the full circles $\tau_{2\rightarrow 3}$. Bold values in a specific panel, corresponding to the thick trace, are the values of Table 5.1 and are also used for the other panels. Simulations were done with MatLab 7 with output processed with PjotrSoft EPSTool

other half moves toward the counter-electrode. It defines the first region in the current transient. For intermediate times, we can Taylor expand the error function,

$$\operatorname{erf}(x) = \frac{2}{\sqrt{\pi}} \sum_{n=0}^{\infty} \frac{(-1)^n x^{2n+1}}{(2n+1)n!} = \frac{2}{\sqrt{\pi}}\left(x - \frac{x^3}{3} + \ldots\right). \tag{5.20}$$

For a small time (but large compared with $L/\sqrt{D}$) the current is equal to

$$J(t) = \frac{qP_0\mu E_x}{\sqrt{\pi D t}} \tag{5.21}$$

or a slope -0.5 in the power-law plot shown in Figure 5.5. This characterizes the second region. The crossover from the first to the second region occurs when $\sqrt{Dt}$ becomes larger than L, thus

$$\tau_{1\to 2} = \frac{L^2}{D}. \tag{5.22}$$

For very long times, $t \to \infty$, Equation (5.19) can be approximated by

$$J(t) = \frac{qP_0\mu E_x}{2}\frac{1}{\sqrt{Dt}}\frac{\mathrm{d}}{\mathrm{d}x}\mathrm{erf}(x) \tag{5.23}$$

evaluated at $x = \mu E_x t/\sqrt{Dt}$,

$$J(t) = \frac{qP_0\mu E_x}{L\sqrt{\pi Dt}}\exp\left(-\frac{\mu^2 E_x^2 t}{D}\right) \tag{5.24}$$

The transition from the second to the third region occurs when the exponent becomes important and starts deviating from unity, roughly at

$$\tau_{2\to 3} = \frac{DL^2}{\mu^2 V^2} \tag{5.25}$$

In the limit for $D \to 0$, in other words when diffusion plays no role (for $V \to \infty$ or $T \to 0$), $\tau_{2\to 3}$ goes to zero and $\tau_{1\to 2}$ goes to infinity, i.e. zones 1 and 3 are reversed. Note that in these cases the transients can have a region where the current is rising, as is visible in the plots, but even more visible when the transients are represented on a linear scale, see Figure 5.6. Observations such as those done by Gill [214] or Lin *et al.* [216] might be caused by these effects.

In any case, the original ToF of Equation (5.1) can be found as

$$\tau_{\mathrm{ToF}} = \sqrt{\tau_{1\to 2} \times \tau_{2\to 3}} \tag{5.26}$$

(on a logarithmic scale this means half way between the two times).

With the help of the Einstein Relation [212] that couples the mobility to the diffusion coefficient, $D = kT\mu/q$, the diffusion coefficient can be eliminated from the above equations,

$$\tau_{1\to 2} = \frac{qL^2}{kT\mu} \tag{5.27}$$

$$\tau_{2\to 3} = \frac{kTL^2}{q\mu V^2}. \tag{5.28}$$

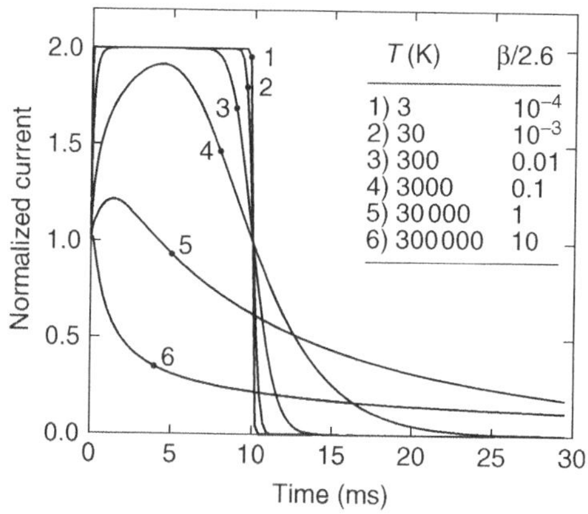

Figure 5.6 Curves of Figure 5.5(a) in a linear scale. When β [$= kT/qV$; Equation (5.30)] is large, either by small voltages or high temperatures, the transients are no longer rectangular and an initial rise is seen [214, 215] and can have a part that follows $t^{-0.5}$. Simulations were done with Mathematica 5 and data processing with PjotrSoft EPSTool

The ratio of these two times is given by

$$\frac{\tau_{2\to3}}{\tau_{1\to2}} = \left(\frac{D}{\mu V}\right)^2 = \left(\frac{kT}{qV}\right)^2 \tag{5.29}$$

with the second step valid for Einsteinian materials. This enables the definition of a parameter, namely

$$\beta = \frac{D}{\mu V} = \frac{kT}{qV}. \tag{5.30}$$

If this parameter has a value below unity the transients are governed by drift and are 'square', while when β is larger than unity, the transients are 'diffusive' and a region exists where they follow a square-root behavior, $J(t) \propto t^{-1/2}$. The factor kT/q is called the thermal voltage and at room temperature it is equal to 26 mV. Thus, for materials obeying the Einstein Relation, only for very tiny biases ($V < 26$ mV at room temperature) or unrealistic high temperatures ($T > 11\,000$ K, at $V = 1$ V) do we observe diffusive transients. Figure 5.6 gives some examples of the transient as a function of temperature (and β). As we will show, for non-Einsteinian materials this can be different and diffusive transients can also be observed for large biases and low temperatures.

Figure 5.5 gives a total summary of the results from which additionally the following observations can be made:

The shape of the transient is independent of the mobility, since it enters in the denominator of both $\tau_{1\to 2}$ and $\tau_{2\to 3}$ (Figure 5.5b); changes in mobility result only in a shift of the transients along the (logarithmic) time axis. The transients can be changed by temperature or the bias (Figure 5.5a and 5.5c, respectively).

The transient shape is independent of the sample thickness L since it appears quadratically in the nominator in both $\tau_{1\to 2}$ and $\tau_{2\to 3}$ [Equations (5.22) and (5.25), respectively]. It can also easily be understood once we realize that both the drift and diffusion times depend quadratically on L. This, for instance, contradicts the observation of Hirao *et al.* (for any system), that 'the contribution of the diffusion length increases with decreasing sample thickness' [217].

The conclusion is that diffusion is only important for situations with extremely low bias (Figure 5.5c), extremely high temperature (Figure 5.5a), or non-Einsteinian systems (Figure 5.5d). In these cases, part of the transient has a shape of $J(t) \propto t^{-0.5}$. The non-Einsteinian systems will be analyzed in more detail in the following sections.

5.4 VIOLATING EINSTEIN'S RELATION

Einstein's Relation couples the diffusion process to the drift process. It is interesting to analyze what happens to the ToF transients if the relation no longer holds. This can occur for several reasons. When only a single occupancy of the electronic level is allowed, Einstein's Relation can be violated. An example is the one-dimensional model of random walk with correlated jumps [147]. In this model, the diffusion and mobility are related as

$$D = \frac{kT}{q}\mu\frac{y_\mathrm{f} + y_\mathrm{b}}{2y_\mathrm{b}} \tag{5.31}$$

where y_f and y_b are the forward and backward jump probability of charges between sites, respectively. Because $y_\mathrm{f} \geq y_\mathrm{b}$, diffusion is always equal or faster than predicted by Einstein's Relation. The effect of this, as can be seen in Equation (5.31), is similar to increasing the temperature and thus an effective temperature can be defined [147]. Note, however, that this hopping conduction model still does not reproduce the observation of Scher and Montroll [202], amongst others. Sometimes, violation of Einstein's Relation is considered enough for producing a long tail [217]. Indeed, diffusion coefficients large compared with $\mu kT/q$ will cause transients of the form $J(t) \propto t^{-0.5}$ (Figure 5.5d), however it will not produce the results of Scher and Montroll, with two distinct time regions, $J(t) \propto t^{-\alpha}$ and $J(t) \propto t^{-1+\alpha}$, respectively.

Often the mobility is field dependent. The Poole–Frenkel mode of conduction incorporates the effect that charges are more easily emitted from a trap into the conduction (or valence) band if they are subjected to a field. The mobility then becomes [9]

$$\mu = \mu_0 \exp\left[\frac{-q(\phi_B - \sqrt{qE_x/\pi\varepsilon})}{kT}\right] \tag{5.32}$$

where ϕ_B is the trap depth (V) and ε is the permittivity of the material. Then the link between mobility and diffusion is lost, since the latter is obviously field independent. If at low fields Einstein's Relation holds, then Equation (5.32) tells us that mobility can be equal or larger than expected on the basis of diffusion and Einstein's Relation.

Figure 5.5(d) shows transients where Einstein's Relation is relaxed and mobility and diffusion are decoupled. In no case is the dual-slope transient observed for violations of Einstein's Relation.

5.5 MULTI-TRAP-AND-RELEASE

A stripped version of the Poole–Frenkel type of conduction is the multi-trap-and-release (MTR). It has the behavior of Poole–Frenkel with the field-dependent part removed. This, as will be shown, results in a standard transient described before, with an effective, reduced mobility.

If per second a carrier has a chance of r_t of being trapped and a trapped carrier a chance r_r of being released, the following chemical reaction can be written

$$p \underset{r_r}{\overset{r_t}{\rightleftarrows}} p_t \tag{5.33}$$

$$\frac{dp_t}{dt} = r_t p - r_r p_t. \tag{5.34}$$

At steady state, $dp_t/dt = 0$,

$$p_t = \frac{r_t}{r_r} p. \tag{5.35}$$

Further, assuming that the capture rate is constant and that the emission rate is thermally activated, the density of free charge follows

$$p = \frac{r_r}{r_t} p_t = p_0 \exp\left(-\frac{E_A}{kT}\right) \tag{5.36}$$

or, in other words, the density of free charge is thermally activated. The current is caused only by the free charge. This can be described by assigning a zero mobility to the trapped charge,

$$J = 0 \times q p_t E + \mu_0 \times q p E. \tag{5.37}$$

If the current is defined with an effective mobility μ_{mtr} for all charge,

$$J = \mu_{\mathrm{mtr}} \times q(p + p_t)E \tag{5.38}$$

then, if the ratio of trapped-to-free charge is large, this mobility can be approximated

$$\mu_{\mathrm{mtr}} = \mu_0 \frac{p}{p + p_t} \approx \mu_0 \exp\left(-\frac{E_A}{kT}\right) \tag{5.39}$$

or, in other words, the effective mobility is thermally activated. This is the basis for the mobility in the frameworks of MTR and Poole–Frenkel in low electric fields.

In fact, this averaged mobility is also the one found in a ToF experiment. This is based on the Law of Large Numbers. As long as the average number of trapping events M for a charge transiting the device is large, the average transit time is given by the average mobility with a small relative spread that is of the order $1/\sqrt{M}$. To explain this and show graphical results, it is easiest to analyze a slightly modified system in which the time a charge stays at a trap is constant, but a charge can be any number of times trapped during the transit. Charges that are never trapped arrive first at the other side of the device, in a time given by

$$\tau_0 = \frac{L}{\mu_0 E_x}. \tag{5.40}$$

Any time a charge is trapped, a constant time τ_t is added to this flight time, thus, a charge that is M times trapped arrives at the counter electrode after a time given by

$$\tau = \tau_0 + M\tau_t. \tag{5.41}$$

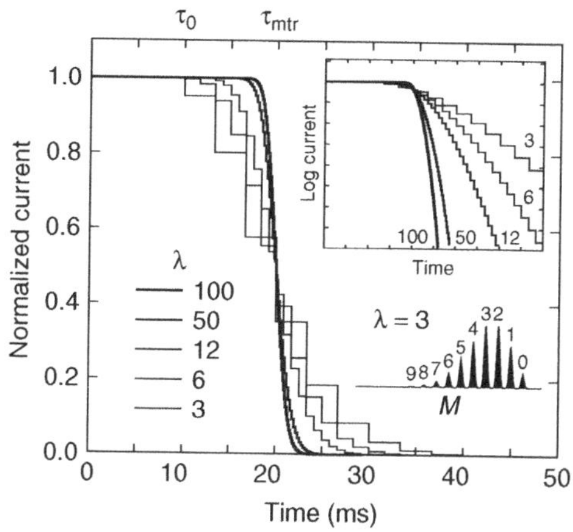

Figure 5.7 Simulations of MTR transients. The charge is divided into packets that have been trapped a certain number of times (M). Because the resident time at a trap is assumed to be constant (τ_t), the packets arrive at the end of the device in periodic intervals, with the amount of charge in each packet governed by a Poisson distribution. Each time a packet arrives, the ToF current drops because it is caused by the remaining charge and an average speed given by the MTR mobility. As can be seen, when the average number of trapping events per carrier (λ) increases, while maintaining the total average time $\lambda\tau_t$ spent on traps (thus reducing the trap resident time τ_t), the ToF transient starts resembling standard ToF transients, with a ToF time τ_{mtr} representing the MTR mobility μ_{mtr}. These simulations were made with $\tau_0 = 10$ ms and $\lambda\tau_t = 10$ ms. The ToF is given by $\tau_{mtr} = \tau_0 + \lambda\tau_t$, with $\tau_0 = L/\mu_0 E_x$ the trap-free transit time. In this case $\tau_{mtr} = 20$ ms

The inset in Figure 5.7 gives a graphical representation of this system, where the horizontal axis either represents the time of arrival of charge at the other side, or a freeze-frame of the charge in the system that is *moving* at a certain time. For this system of fixed time spent at the traps, the distribution is in small packets, corresponding to right-to-left to charges that have not (yet) been trapped, charges that have once been trapped, etc., as indicated by the M number in the figure. The amount of charge in each packet (when talking about arrival time) is given by a Poisson distribution; the fraction of charges that have been trapped M times is

$$f_M = \frac{\lambda^M e^{-\lambda}}{M!}, \tag{5.42}$$

where λ is the expectation value ('average') for the number of times a charge is trapped during the flight. Thus, the number of carriers that are still in the device with packets M still in the device is

$$P_{\mathrm{M}} = P_0\left(1 - \sum_{\mathrm{m}=0}^{M-1} f_{\mathrm{m}}\right) = P_0\left(1 - \sum_{\mathrm{m}=0}^{M-1} \frac{\lambda^m e^{-\lambda}}{m!}\right). \tag{5.43}$$

By definition, those charges move with an average speed v determined by the effective mobility of MTR. The current is thus proportional to the charge still in the device and the MTR mobility

$$J_{\mathrm{M}} = \frac{qv}{L} P_{\mathrm{M}} = \frac{q\mu_{\mathrm{mtr}} E_{\mathrm{x}}}{L} P_{\mathrm{M}}. \tag{5.44}$$

The integral of this current is

$$\begin{aligned}\int_0^\infty J(t)\mathrm{d}t &= \tau_0 J_0 + \sum_{\mathrm{M}=1}^{\infty} \tau_{\mathrm{t}} J_{\mathrm{M}} \\ &= \frac{\tau_0 q\mu_{\mathrm{mtr}} P_0 E_{\mathrm{x}}}{L} + \sum_{\mathrm{M}=1}^{\infty} \frac{\tau_{\mathrm{t}} q\mu_{\mathrm{mtr}} E_{\mathrm{x}} P_0}{L}\left(1 - \sum_{\mathrm{m}=0}^{M-1} \frac{\lambda^m e^{-\lambda}}{m!}\right) \\ &= \frac{q\mu P_0 E_{\mathrm{x}}}{L}(\tau_0 + \lambda\tau_{\mathrm{t}}).\end{aligned} \tag{5.45}$$

Every time a charge packet leaves the device, at periodic intervals, the current drops one level, from J_{M} to $J_{\mathrm{M}-1}$. Figure 5.7 gives an example of a system with $\tau_0 = 10$ ms. The first trace (the most jagged) shows the transient for $\lambda = 3$ and $\tau_{\mathrm{t}} = 3.33$ ms. The lower inset shows the Poisson distribution of the packets with charge that has been trapped M times. (Note that the average values lies indeed at $M = \lambda = 3$.) The distance between the packets is τ_{t} when talking about arrival time (or $\mu_0\tau_{\mathrm{t}}E_{\mathrm{x}}$ when talking about space). From the main curve, it can be seen that the current drops every time a packet arrives. It begins at the arrival of the packets that have never been trapped during their transit ($M = 0$), at $t = L/\mu_0 E_{\mathrm{x}}$. The second packet ($M = 1$) arrives $\tau_{\mathrm{t}} = 3.3$ ms later, dropping the current to the value given in Equation (5.44). As can be seen in the linear–log plot of the same data (see inset), the current transient has a long tail. When the average number of trapping events (λ) is increased, while maintaining the average total time spent on traps by keeping $\lambda\tau_{\mathrm{t}}$ constant, it can be seen that the curve becomes more square, i.e., resembling a regular transient. The current steps become smaller, because the fraction in each packet [Equation (5.42)] decreases, but also the interval between steps τ_{t} decreases. For an infinite average number of trapping events per transit, $\lambda \to \infty$, the transient becomes rectangular and the ToF found is

$$\tau_{\mathrm{mtr}} = \tau_0 + \lambda\tau_{\mathrm{t}} \tag{5.46}$$

(in this example 20 ms). Because the current is constant from $t = 0$ to $t = \tau_{\mathrm{mtr}}$ at a level of $J = J_0 = q\mu_{\mathrm{mtr}}E_{\mathrm{x}}P_0/L$, the total integral of the current is

$$\int_0^{\infty} J(t)\mathrm{d}t = \tau_{\mathrm{mtr}}J_0 \tag{5.47}$$

$$= \frac{\tau q\mu_{\mathrm{mtr}}P_0E_{\mathrm{x}}}{L}. \tag{5.48}$$

This integral must be equal to the total charge qP_0, thus, the ToF found experimentally is

$$\tau_{\mathrm{mtr}} = \frac{L}{\mu_{\mathrm{mtr}}E_{\mathrm{x}}} \tag{5.49}$$

Or, in other words, the ToF found represents the MTR mobility of the carriers. We can take this a step further, by using the definition of the MTR mobility [Equation (5.39)] and the MTR ToF [Equation (5.46)],

$$\tau_0 + \lambda\tau_{\mathrm{t}} = \left(\frac{p + p_{\mathrm{t}}}{p}\right)\frac{L}{\mu_0E_{\mathrm{x}}} \tag{5.50}$$

or

$$\frac{\tau_0}{\tau_0 + \lambda\tau_{\mathrm{t}}} = \frac{p}{p + p_{\mathrm{t}}}. \tag{5.51}$$

This tells us that, on average, the fraction of time the charges are moving freely is equal to the ratio of number of charges that are free to the total number of charges. This makes sense, and, in fact, we could have used this as an axiom and reason the other way around.

The same reasoning applies for systems in which the time spent at a trap is not constant, although it is more difficult to calculate and simulate. As long as the Law of Large Numbers applies, the mobility found in the ToF experiment represents the MTR (Poole–Frenkel) mobility. For the MTR system, the transients behave normally, with an effective mobility that is the same as the one found in conductivity experiments. This is valid when the average number of trapping events is high. However, there might be systems that do not obey the Law of Large Numbers. These can be for instance systems in which the traps are so deep that the time spent at a trap can be much larger than the average transit time. In these cases the analysis is much more complicated.

5.6 ANOMALOUS TRANSIENTS

The treatment until now has been based on a simple assumption, namely that all charges are the same and they all have the same mobility and are governed by the same diffusion coefficient, caused by the Law of Large Numbers that tells us that any process (such as trapping), when occurring on average in many events, can be treated by its average value. A charge can be trapped at a deep charge and can there be stuck for a long time, but, on average, the charges will all spend the same time in such deep traps. The charges have no 'history'. There are no 'slow' charges and 'fast' charges. Because diffusion and drift are stochastic processes, they only have meaning when there is a large number of interaction 'events' during a transit across the sample, and the parameters are the result of averaging over the time of interest. Note, for example, that the mobility is defined as $v = \mu E$, while for free electrons, we would have to use Newton's $dv/dt = qE/m_e$, where m_e is the mass of the electron.

The other extreme is when each charge has its own parameters. Imagine a situation in which the charges travel in different bands, each band with its own mobility (and diffusion coefficient), similar to the case presented by Scharfe with a distribution of mobilities [218]. A device consisting of many independent parallel channels can also be imagined. The ToF transient is then obviously no longer of the form described above, but, instead, is a convolution of transients of each energy band or each channel and the final shape can become very complicated. We will discuss here some examples.

As a first example, consider the (maybe somewhat academic) system as depicted in Figure 5.8, with a constant density of charge in energy from $-\infty$ to 0, $P(E, t = 0) = P_0$ (unit: $J^{-1}m^{-2}$). Without going into detail about what the origins are, the charges are assumed to have a mobility that is thermally activated,

$$\mu(E) = \mu_0 \exp\left(\frac{E}{kT}\right). \tag{5.52}$$

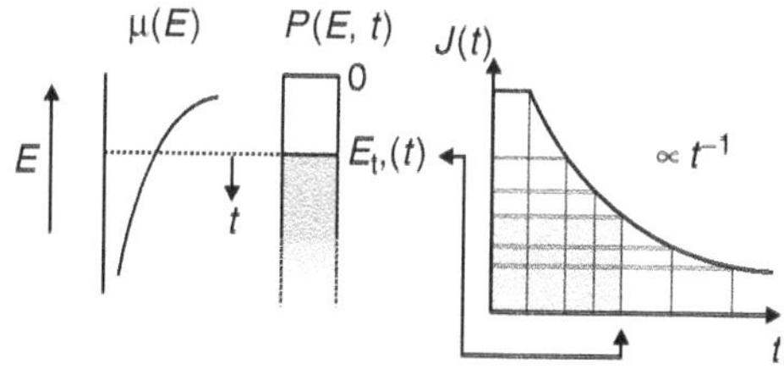

Figure 5.8 A system in which each charge has its own parameters (μ and D), here with constant DOS, $P(E,t) = P_0$. The resulting ToF transient is then a convolution of contributions of each DOS slice. Each slice contributes to the current in a rectangular way, as shown on the right. The charges with high mobility move out first and the remaining charges have ever-lower mobility, making the transient progressively slower, $J(t) \propto t^{-1}$

Each slice in energy contributes to the current in a rectangular way, similar to Equation (5.11),

$$J(E,t) = \frac{q\mu(E)V}{L^2}P(E,t) \tag{5.53}$$

where $P(E,t)$ is the amount of charges in the slice with energies from E to $E + \mathrm{d}E$ at time t. The charges contribute to the current until they reach the opposite side given by

$$t = \frac{L^2}{\mu(E)V} = \frac{L^2}{\mu_0 V}\exp\left(-\frac{E}{kT}\right) \tag{5.54}$$

or, reasoning the other way around, at time t, in the device, there are charges with energies ranging from $-\infty$ to a certain energy given by

$$E_t(t) = kT\ln\left(\frac{L^2}{\mu_0 Vt}\right). \tag{5.55}$$

The current can then be calculated as

$$\begin{aligned} J(t) &= \int_{-\infty}^{E_t(t)} \frac{q\mu_0 V}{L^2}\exp\left(\frac{E}{kT}\right)P_0\mathrm{d}E \\ &= \frac{qP_0kT}{t} \end{aligned} \tag{5.56}$$

for $t > L^2/\mu_0 V$ (and constant, $qP_0kT\mu_0 V/L^2$, before that time). In other words, the ToF transient is a hyperbola of the form $J(t) \propto t^{-1}$. Note that the integral of the current of Equation (5.56) does not converge because of the assumption of initial infinite total charge (P_0 over all energies from 0 to $-\infty$). If we take the time at which the current starts to drop as the ToF, τ_{ToF}, our measured mobility is μ_0, namely that mobility belonging to the fastest charges of the top slice at $E = 0$.

Next we can try an exponentially distributed DOS,

$$P(E,t) = P_0 \exp\left(-\frac{E}{kT_0}\right) H(E,t) \tag{5.57}$$

as shown in Figure 5.9, where P_0 and T_0 are parameters describing the distribution, and $H(E,t)$ is a binary function (0 or 1) representing if the charges with energy E at time t are still inside the sample and still contribute to current. Equal to the case of the constant DOS, it means

$$H(E,t) = \begin{cases} 1 & \text{for } E < E_t(t) \\ 0 & \text{for } E > E_t(t) \end{cases} \tag{5.58}$$

where E_t is given by Equation (5.55). Substituting this into Equation (5.53) and integrating over all energies yields

$$J(t) = \frac{kTT_0}{T_0 - T} qP_0 \left(\frac{L^2}{\mu_0 V}\right)^{-\alpha} t^{-1+\alpha} \tag{5.59}$$

with

$$\alpha = T/T_0 \tag{5.60}$$

for $t > L^2/\mu_0 V$ [and constant, $qP_0kTT_0\mu_0 V/(T_0 - T)L^2$, before that time]. See Figure 5.10 for simulations of these currents on a log–log

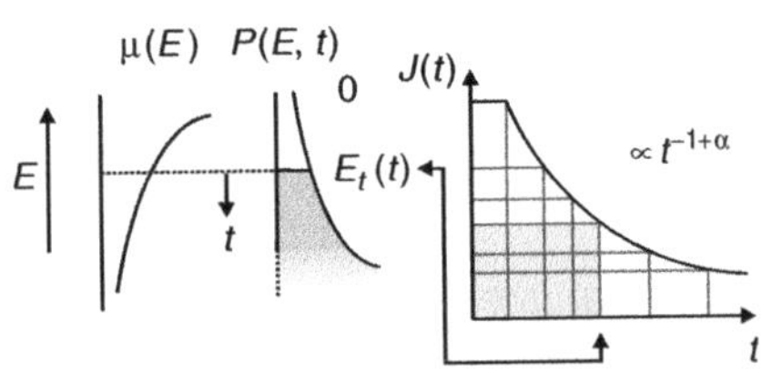

Figure 5.9 A system similar to the one depicted in Figure 5.8, but with an exponential DOS. The resulting ToF transient is of the form $J(t) \propto t^{-1+\alpha}$

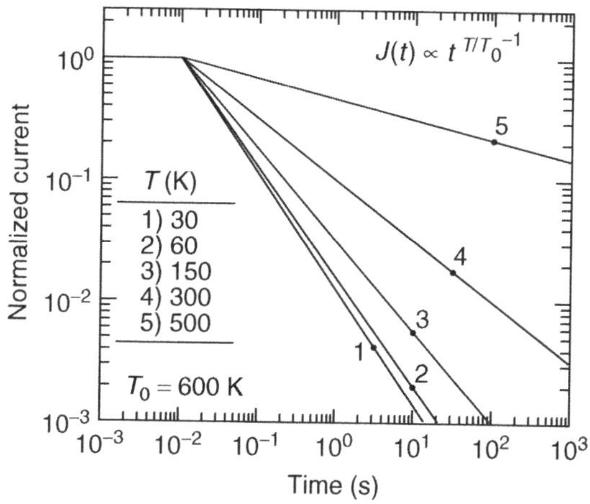

Figure 5.10 Simulations of transients for systems of parallel conduction channels (either in space or energy), each channel with its own conduction parameters. Channels are distributed exponentially and mobility has thermal activation (see Figure 5.9)

scale. Interestingly, the experimentally found ToF (where the current starts dropping) gives the mobility of the fastest channel, μ_0, for instance the top of the charge sea, or the filament with highest mobility.

For $T_0 \to \infty$ ($\alpha \to 0$) we return to the equation given for the constant DOS [Equation (5.56)], as expected. Equation (5.59) shows that this (peculiar) system can indeed predict part of the behavior observed by Scher and Montroll [202], namely current transients of the form $\propto t^{-1+\alpha}$. The reasonability of the system remains to be shown. However, many authors claim to have observed 'filamentary' currents in organic devices. In view of the above, this can be translated into spatial channels, each with its own mobility and each contributing a rectangular drift current profile to the overall ToF transient. See Figure 5.11 for a graphical sketch of such a system. However, for such a situation maybe a better distribution of the mobilities is a Gauss function. Assuming all states to be full from $E = -\infty$ to $E = +\infty$, this implies substituting $P(E, t)$ of Equation (5.57) with

$$P(E,t) = P_0 \exp\left[-\left(\frac{E - E_0}{kT_0}\right)^2\right] H(E,t) \qquad (5.61)$$

with E_0 and T_0 parameters describing the distribution of channels (filaments, see Figure 5.12) and $H(E, t)$ the same as before [Equation

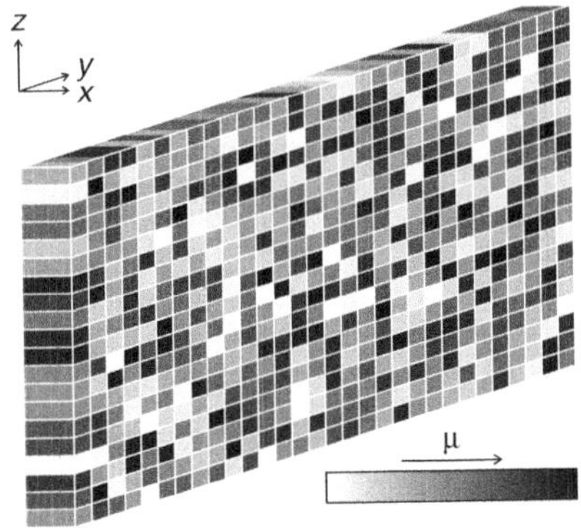

Figure 5.11 A graphical sketch illustrating a system with filamentary currents. The device is divided into a multitude of parallel devices, each with its own parameters (mobility and diffusion). High mobility is indicated with a darker color. The resulting current transient follows the power-law behavior (i.e. $\propto t^{-1+\alpha}$) of Equation (5.59) if the distributions are exponential, or is even more complex [see Equation (5.62)] if the DOS is Gaussian

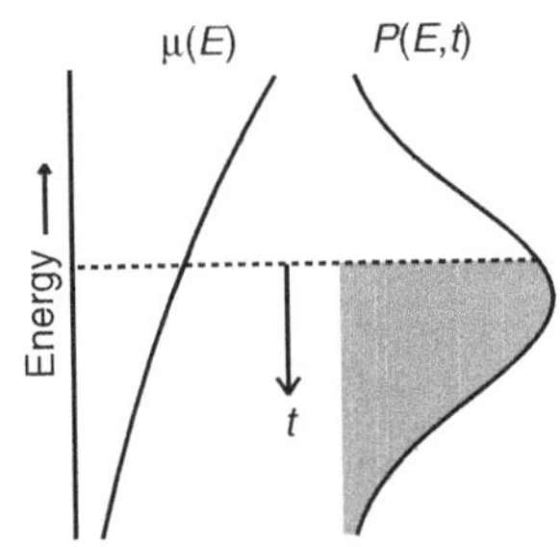

Figure 5.12 A system similar to the one depicted in Figure 5.9, but with a Gauss distribution for the channels (for example filaments of Figure 5.11). The resulting ToF transient is of the form of Equation (5.62)

(5.58)]. This results in a current transient of the form

$$J(t) = J_0 \left\{ 1 - \operatorname{erf}\left[\frac{E_0}{kT_0} + \frac{T}{T_0} \ln\left(\frac{\mu_0 V t}{L^2} \right) + \frac{T_0}{2T} \right] \right\} \tag{5.62}$$

$$J_0 = \frac{q V P_0 \mu_0 \sqrt{\pi} k T_0}{2L} \exp\left[\left(\frac{E_0}{kT_0} + \frac{T_0}{2T} \right)^2 - \left(\frac{E_0}{kT_0} \right)^2 \right] \tag{5.63}$$

for all times. Simulations of this equation are shown in Figure 5.13. Because of the complex form of the transient, the plots are not easy linearizable (log–log and linear–log representations are shown) and a

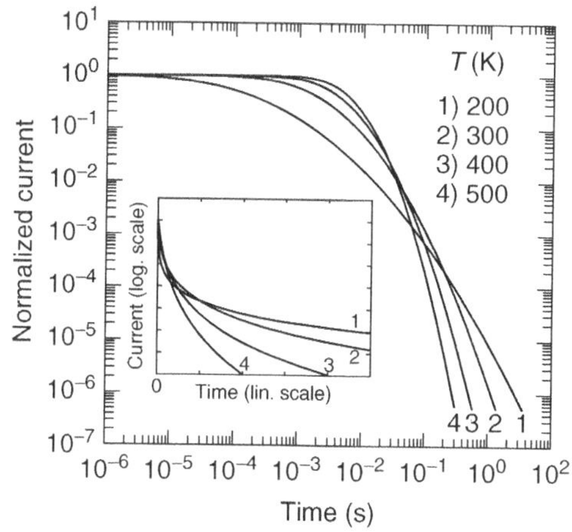

Figure 5.13 Simulations of transients for systems of parallel conduction channels (either in space or energy), each channel with its own conduction parameters. Channels have Gauss distribution and mobility has thermal activation (see Figure 5.12). Because of the complexity of the transient [see Equation (5.62)] the plots are not easily linearizable. Here two trials are shown, log–log and linear–log (inset). An example of this system might be filaments as depicted in Figure 5.11. See also Figure 5.10 for a different distribution

clear ToF cannot be defined. If we use the time at which the current drops to a certain factor of the starting current for a measure of the ToF, our experiment would tell us that the ToF mobility depends on temperature and that it can either increase or decrease with temperature, depending on the current level used for the determination of τ_{ToF}. This shows how one has to be very careful when trying to interpret ToF data.

When doing a ToF experiment, one has to take care to be sure that indeed the experiment is of the ToF type. In Chapter 4 a STR experiment was described that is very similar to ToF, up to the point that the experimental procedure is identical. It is a system where charges are created on traps on one side of the device and from there are slowly emitted. When emitted, they momentarily contribute to current. As it was shown there, the resulting transient looks very much like a ToF transient, having two regions with different slopes in the log–log plot. With the experimental set-up and mode of operation indistinguishable, a wrong interpretation of the results is easily made. Note that in any case the transient of Equation (4.28) does not depend on the device thickness. As such, changing the device thickness is a good way to check if everything is under control.

There now exists a wide range of possibilities. On the one extreme there is the STR system described above where charge is initially trapped on one side of the sample. Once released the charge moves freely to the other side of the sample. This gives currents of the form shown in Figure 4.5, with two time domains, with currents $J(t) \propto 1$ and $J(t) \propto t^{\alpha-1}$, respectively. On the other extreme is the MTR model. Charge, moving from one electrode to the other, is trapped many times during its transition. In fact, it is trapped so many times that the Law of Large Numbers applies and statistics can be performed to give an average behavior of the carriers. Another name for this is the Poole–Frenkel conduction mechanism. Because mobility is field dependent (and diffusion is not), the transients can become 'diffusive', as described earlier. Much more complicated is the situation where only a few trapping events take place in a transit of a carrier across the device. In this case, Monte-Carlo simulation or numerically solving are the best ways to analyze the system. The most famous of these studies is the one by Scher and Montroll [176, 202]. Noolandi [219] takes it a step further and defines a parameter M that quantizes the average number of trapping events for a carrier traversing the device. They find slopes in a log–log plot of $-1+\alpha$ and $-1-\alpha$. It is worth noting that Noolandi manages to approximate these curves with only three discrete trap levels. In view of this, the above described STR transient has $M = 1$ (never again trapped) and the MTR transient $M \rightarrow \infty$. A helpful conclusion is that, to avoid difficult transient shapes, one could make the sample much thinner and perform an STR/DLTS experiment, or make the device much thicker and perform an MTR experiment.

5.7 HIGH CURRENT (SPACE CHARGE) TRANSIENTS

Until here it was assumed that the charges themselves do not influence each other. In other words, the charges feel the external electric field, but do not perturb it or create their own field. For small charge densities this picture is correct. However, for larger densities, the fields created by the space charge can seriously change the situation. Figure 5.14 shows a schematic of a situation in which a high-density sheet of charge has moved a distance x_0 from the originating electrode. To find the electric field and potential, Poisson's Equation can be used, with as boundary conditions $V(0) = V(L) = 0, p = n_1 + n_2$, as described earlier. The potential at maximum is then

$$V(X) = \frac{1}{\varepsilon} q p x_0 \left(1 - \frac{x_0}{L}\right). \tag{5.64}$$

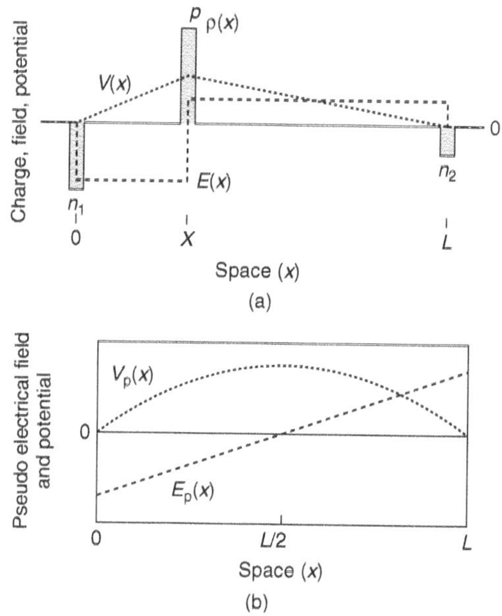

Figure 5.14 (a) Schematic of the charge distribution, electric field and potential in a situation with a sheet of high density (space) charge somewhere on its way in the device. (b) Effective electric field of the system

To find out how the system evolves, it is not allowed to use the static values of the field or potential shown [like a classical mechanics force $F_x = qE(x)$]. This would be valid for a single (infinitely small) charge in the (otherwise static) sheet p. Instead, a more complex energy equation has to be used, $F_{x0} = -\mathrm{d}U(x_0)/\mathrm{d}x_0$, where U is the energy of the system. This energy for the system shown in Figure 5.14 is equal to $U(x_0) = qpAV(x_0)$. (Alternatively, the same result can be obtained by calculating the total electric field energy, $U = \int \varepsilon |E(x)|^2 \mathrm{d}x$, or by using the electronics equation for the energy of a charged capacitor $U = Q^2/2C$, for the two capacitors as described in Section 5.2.) The force on the charge sheet evolving in the direction x is then

$$F_{x0} = -\frac{\mathrm{d}U(x_0)}{\mathrm{d}x_0} = \frac{A}{\varepsilon}\left(\frac{2x_0}{L} - 1\right) q^2 p^2. \qquad (5.65)$$

Knowing that the total number of particles is Ap, the force per particle is

$$F_\mathrm{p} = \frac{F_{x0}}{Ap} = \frac{1}{\varepsilon}\left(\frac{2x_0}{L} - 1\right) q^2 p. \qquad (5.66)$$

Comparing this with the static field force $F = qE$, an effective electric field works on the particles that is

$$E_{\mathrm{p}}(x) = \frac{F_{\mathrm{p}}}{q} = \frac{1}{\varepsilon}\left(\frac{2x}{L} - 1\right) qp. \tag{5.67}$$

Figure 5.14 shows the effective electrical field and potential in such a system. As can be seen, the force is driving the positively charged sheet towards the nearest electrode. Moreover, a meta-stable zero-force position exists exactly half way in the device. The maximum strength of the field close to the electrodes ($x = 0$ and $x = L$) depends on the density of initial charge, p, as can be seen in Equation (5.67), $|E_{\mathrm{p,max}}| = qp/\varepsilon$. To give an idea of the importance of this field, for a silicon device with parameters of Table 5.1, the external field is $E = V/L = 1$ MV m^{-1}. For a laser operating at 700 nm with a 1 W pulse of 100 ps and a beam diameter of 400 μm, the dose is 2.8×10^{15} photons m^{-2} and the pseudo field strength $|E_{\mathrm{p,max}}| = 4.3$ MV m^{-1}; strong nonlinear effects can be expected.

To find an explicit form of the current transient it is easiest to go back to the situation with only a drift current,

$$J(t) = \frac{qp}{L}\frac{\mathrm{d}x(t)}{\mathrm{d}t}. \tag{5.68}$$

The velocity $\mathrm{d}x(t)/\mathrm{d}t$ is given by the product of mobility and local field:

$$\frac{\mathrm{d}x(t)}{\mathrm{d}t} = \mu E(x) = \mu\left[E_{\mathrm{x}} + \frac{qp}{\varepsilon}\left(\frac{2x}{L} - 1\right)\right] \tag{5.69}$$

where E_{x} is the static electrical field, $E_{\mathrm{x}} = V/L$. This equation can be solved analytically and gives

$$x(t) = \frac{L}{2}\left(\frac{\varepsilon E_{\mathrm{x}}}{qp} - 1\right)\left[\exp\left(\frac{2qp\mu}{\varepsilon L}t\right) - 1\right] \tag{5.70}$$

which gives a ToF

$$\tau = \frac{\varepsilon L}{2qp\mu}\ln\left(\frac{\varepsilon E_{\mathrm{x}} + qp}{\varepsilon E_{\mathrm{x}} - qp}\right) \tag{5.71}$$

and a current transient for $t < \tau$ of the form

$$J(t) = \frac{qp\mu}{L}\left(E_{\mathrm{x}} - \frac{qp}{\varepsilon}\right)\exp\left(\frac{2qp\mu}{\varepsilon L}t\right). \tag{5.72}$$

It is easily verified that for small charge densities, the ToF and the current transient return to their original forms,

$$\lim_{p \to 0} J(t) = \frac{q\mu E_x}{L} p \tag{5.73}$$

$$\lim_{p \to 0} \tau = \frac{L}{\mu E_x}. \tag{5.74}$$

Figure 5.15 shows simulations of the transients for such high-charge-density systems. The timescale is normalized for the low-density ToF, $\tau_{ToF} = L/\mu E_x$. The current is scaled by the density p and further divided by this same ratio for low density and at $t = 0$: $[J(t)/p]/J_{00}$, with $J_{00} = \lim_{p \to 0} J(0)/p$. In this way the effects of high-charge density are clearly visible. First of all, the current starts initially rising over time, and secondly, the time it takes for the carriers to reach the other side increases.

It has to be pointed out that these simulations are highly simplistic. First of all, the simultaneously created carriers of opposite sign have been ignored. They were assumed to have been swept away instantaneously, which is unlikely when this would create such large fields.

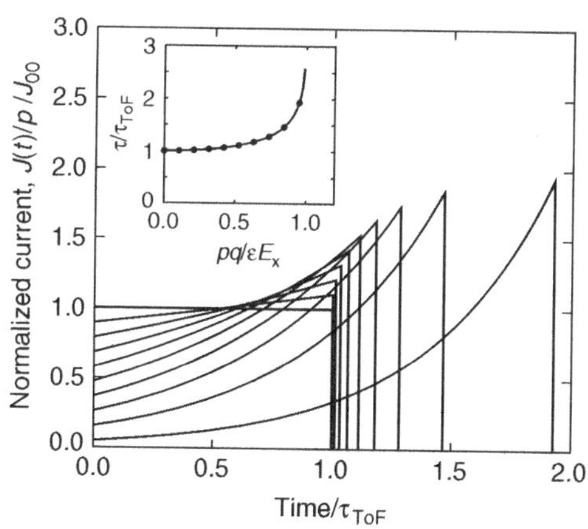

Figure 5.15 Simulation of high-charge-density transients. The time in the plots has been normalized with the low-density ToF, $\tau_{ToF} = L/\mu E_x$. Similarly, the current has been normalized by dividing it by the carrier density (p) and further divided by this ratio for $p \to 0$ at $t = 0$. The inset shows the effective ToF as a function of normalized charge density, p divided by the critical density $\varepsilon E_x/q$. Each dot represents a curve in the main figure. Plasma (space-charge) effects cause a rising current and a retardation of the transient

Furthermore, the superposition principle can no longer be applied; the charge distribution cannot be decomposed into sheets for each of which the transient is calculated. This is because the charge sheets are not independent but feel and disturb each other. Thus, transients where diffusion plays a role, and distributions that are not simple delta functions become highly complex.

Moreover, the simplicity of the calculation is clearly visible in the fact that it has no solution for densities larger than a critical density, $p > \varepsilon E_X/q$, while such densities can easily be induced. For these densities, the electric field at the electrode is larger than the external electric field. Charges will actually move *back* toward the illuminated electrode and are there removed from the device. This will happen in the same stage as the removal of the opposing carriers (for instance in a hole-current transient it occurs with the initial fast removal of the electrons from the device), a complex interplay between the carriers takes place, with a combined dynamics of diffusion, drift and recombination. This will continue until the carrier density becomes so low as to enable the rest of the carriers to drift towards the counter-electrode. The transient and the ToF is thus not governed only by drift mobility, but rather by diffusion and recombination. Researchers are advised to avoid such situations. In cases where the transient shape depends on the excitation intensity (rising currents in part of the transient might be a tell-tale sign), the excitation intensity has to be lowered, in order to guarantee a good extraction of the mobility parameter from the measurements. It is not worth trying to model these space-charge transients, especially when only information about the mobility is sought.

Ideally, researchers should design their experiments to make the analysis simple. Even so, attempts have been made to describe such effects [220]. In view of this, the work of Gross *et al.* is worth mentioning. They observe even a reversal of the current [221–223]. They manage to explain it on basis of the zero-field-plane analysis derived by Lindmayer [224]. The reader is advised to look at these classic papers when confronting problems of this kind.

5.8 SUMMARY OF THE ToF TECHNIQUE

ToF is one of the most challenging measurement techniques to describe and model accurately. Numerous reports have been made. Most apply to some specific conditions; it is impossible to give a general solution to the problem of ToF. The text here may serve as a guideline as to how to think in general terms. It was not intended to give an exhaustive summary of all the possible ways to analyze the ToF experiment, nor to give a summary of the literature. For every experimental situation, it is needed to go back to the drawing board to check which assumption can and must be made to analyze the experimental data or to set up an experiment.

It has been shown here how various transient shapes can result from various effects. The phenomena are summarized in Table 5.2. It is clear from the discussion given here that there can be many reasons for complicated, nonrectangular transients. The main message is that while consistent with a certain model the experimental data often cannot be used to proove it.

Take for example the phenomenon that the current is rising in a transient. This can mean that the initial charge density is high, but equally well that diffusion plays an important role. Another example is that transients can be of the power-law shape. As shown, this can be caused by various effects, ranging from diffusion to an exponentially distributed DOS.

The analysis with an exponential DOS is consistent with the ideas and modeling of Bässler [45]. Where he uses a Gaussian DOS, this can locally be approximated by an exponential DOS when the Fermi level is somewhere in the tail of the distribution. Yet, the observation of a power-law transient by itself does not proove the existence of an exponential DOS nor the model of Bässler.

The analysis here is intentionally simplified. As an example, combinations of the described effects are not treated. Imagine a system

Table 5.2 Summary of concepts and resulting effect on transient as presented in this work

Concept	Transient	Section
Standard (drift)	Rectangular	5.2
+ Low absorption	1 − exp (up to triangular)	5.2
+ Diffusion	Rising current	5.3
+ Diffusion	$t^{-0.5}$	5.3
+ Non-Einstein	see Diffusion	5.4
MTR	see Standard	5.5
Filaments + Exp. DOS	$t^{-1+\alpha}$ $(\alpha = T/T_0)$	5.6
Filaments + Gauss. DOS	Equation (5.62), Figure 5.13	5.6
STR	$\exp(-t/\tau)$	5.6
STR + Exp. DOS	$t^{-1+\alpha}$ $(\alpha = T/T_0)$	5.6
FTR (few-trap-and-release)	$t^{-1+\alpha}$, $t^{-1-\alpha}$	Ref. [219]
High-power (space charge)	Exp. rising current	5.7

with a high density of traps, distributed exponentially in energy, with a low absorption coefficient, under high power illumination. This will result in an extremely complicated transient. Some papers try to describe combinations of two or more of the phenomena. See for example the high-density (space-charge) transients with the inclusion of diffusion by Rosen [225], or the endeavors to model the high-density excess-current transients in conductive materials in the presence of traps [220]. In these cases, the scientist can always opt to revert to Monte-Carlo simulations, as is quite common. (See for instance the works of Scherr and Montroll [202], and Bässler [45] and most works cited in this work.) Experimenters, however, should aim to reduce the complexity of their measurements and isolate or highlight a single process, especially when only knowledge about the mobility parameter is sought.

Other concepts that are not included here, for example, are the effects of the contacts. It is assumed here that the contacts themselves do not disturb the evolution of the system. Electron–hole recombination can take place in the vicinity of the electrodes of the order of the Debeye length. The distribution of charge is then no longer the simple diffusion-broadened Gaussian, as shown in Figure 5.4. The analysis of such a system with inclusion of the effects of the contacts becomes challenging. It is better to avoid such complicated situations; in this case it may serve to increase the device thickness to the point that effects at the extremes become relatively irrelevant, or, alternatively, by making the device so small that the ToF becomes relatively unimportant and the transient is governed exactly by those contact effects in case we want to study these.

In fact, when the contacts become important and they are rectifying, what we wind up with is a Schottky barrier or pn-junction working as a solar cell, namely a device with an internal electric field that separates the electron–hole pairs induced by the light to result in an externally observable current. This is a completely different device architecture. To minimize the effects of the bulk in these devices, the films are made as thin as possible, just thick enough to accommodate the internal depletion regions. In a ToF, however, we want to minimize the effects of the contacts and it is advised to make the bulk as thick as possible. An experimenter is advised to avoid the gray zones, devices that are neither clear solar cells, nor clear bulk samples.

For the same reason, ToF techniques using bias steps instead of light pulses are not described here because, for these systems, two processes play an important role, carrier injection and carrier transport. For this reason the light-pulse ToF technique has preference over bias-pulse

ToF. Likewise, measuring the mobility by the ToF technique in a TFT geometry becomes a challenge (see for instance the recent work of Basu *et al.* in polymeric materials [226]).

Another technique that was not mentioned is ToF using electron bombardment. (See for example the works of Gross [221].) This technique with its higher costs is becoming less popular, especially since (ir)radiation defects and effects are less interesting to the modern post-cold-war scientific community.

Finally, it has been chosen here for all the plots of the simulations to show the current in a normalized form, in order to better show the shape of the transients. In this way, information about the transient amplitude is lost. In cases where such information is important (for instance when changing the temperature), the accompanying formulas give the full analytical expression.

6

Thin-film Transistors

6.1 FIELD-EFFECT TRANSISTORS

Field-effect transistors come in many sorts but can be classified in to four basic types: the junction FET (JFET); the metal–semiconductor FET (MESFET); the metal–oxide–semiconductor or metal–insulator–semiconductor FET (MOS-FET/MIS-FET); and the thin-film transistor (TFT), see Figure 6.1. Organic transistors are predominantly of the latter type. The main reason is that the materials are normally of low purity and, whereas for the other types of transistors high control over purity is needed, for a TFT low requirements exist. Since this book is about organic materials, this chapter deals mainly with TFTs and gives an introduction to MOS-FETs only as a comparison, since most of the literature uses the MOS-FET model to characterize the TFTs. However, this is erroneous, as will be argued.

A little more than a decade after Shockley had shown modulation of conductance in thin-film structures [227], the first TFTs were made of cadmium sulfite and cadmium selenide [228], but the most popular are amorphous silicon, first fabricated in 1979 [229], and, for instance, used as structures to study interface quality [230]. Actually, a TFT can be made of *any* material for the active layer. This, as will be shown, even includes metals! Technologically interesting and worth mentioning is also the 'biological' transistor [231] based on proteins or the 'paper' transistors [232–234] to show that we have come a long way. (See the publications of Reese *et al.* [235] and Newman *et al.* [236] for reviews

Electrical Characterization of Organic Electronic Materials and Devices Peter Stallinga

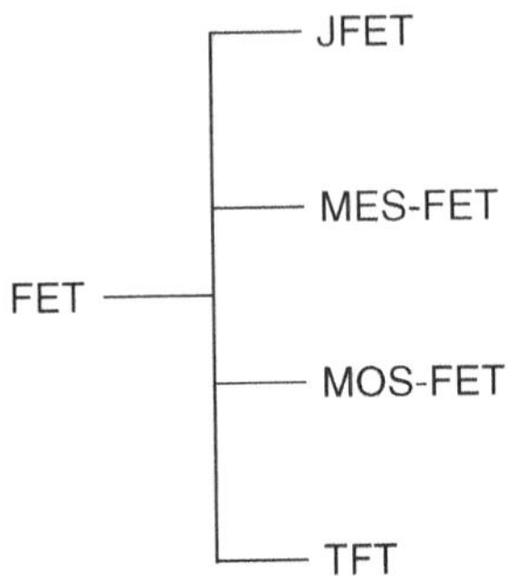

Figure 6.1 Four basic types of field-effect transistors. Organic transistors are mainly of the thin-film (TFT) type

on organic TFTs.) It is envisioned that all-organic integrated circuits are made with organic TFTs driving organic LEDs [237].

Organic TFTs are not of particularly high performance. Where amorphous silicon TFTs have long reached carrier mobilities well above unity [238], organics struggle to reach this level [6, 7, 239], with pentacene one of the few exceptions [240, 241]. The standard benchmark circuit, the ring-oscillator, mostly works in the kilohertz range [242–244], which is only acceptable for low-performance electronics like radio-frequency identification where price is more important than performance. While ink-jet printing of organic electronic circuits is an interesting new technology [12, 242], organics do not even distinguish themselves as the only ones being solution processable. Inorganic TFTs can also be made in this way via nanocrystal colloidal solutions, for instance cadmium selenide [245], zinc oxide [246], or organic/inorganic blends [247]. However, producing n-channel organic transistors [248] is a problem as pointed out by Dodabalapur [249] and they are needed for low-power (complementary MOS; CMOS) logic elements compared with other types of logic [250]. The problem seems to have been tackled; n-type organic materials can be made as shown by many studies [6, 248, 251–255], and integrated CMOS structures of organic transistors can be made (even on a large scale) [243, 256, 257], but n-type channels are still not very common. The reason is that the mobility of electrons is low due to strong electron trapping [248].

Even more technologically interesting are ambipolar materials, since they can be used in logic elements and light-emitting FETs. They can be made (see for example the work of Meijer *et al.* [258]), but ambipolarity remains the exception rather than the rule for organics [259].

The (electrical) behavior of TFTs is remarkably similar to that of the MOS-FET, as will be shown. For the modeling of amorphous-silicon transistors, the works of Powell [260] and Shur and Hack [79] are reference material, while the pioneering work of Horowitz [253, 261–263], Brown *et al.* [264, 265], Gundlach *et al.* [266] and Salih *et al.* [267] in organic TFTs also needs mentioning.

There are, however, important differences between TFTs and MOS-FETs, as will also be discussed. The main difference lies in the dimensionality. A MOS-FET is basically a three-dimensional device with space charge (ionized dopants) and band bendings essential to the workings of a device; a MOS-FET (working in inversion) has a minimum device thickness in the order of 100 nm. The active layer of a TFT can be made as thin as possible; as long as a continuous coverage of the insulating layer exists, the TFT works. Thus, a TFT can be made of a single monolayer [268–270]; any additional layers deposited on top act as passive buffers only for mechanically and chemically stabilizing the layer accommodating the current. For this same reason, contrasting the MOS-FET, the interface between the insulator and the semiconductor is of utmost importance. Surface treatment of the insulator before deposition of the active layer can easily increase the carrier mobility by several orders of magnitude. Traps at the interface, nearly unavoidable because of the lattice discontinuities, have a severe detrimental effect on the electrical behavior. Sze in his book stated that 'To improve device performance, reproducibility, and reliability, the bulk and interface trap densities must be reduced'. All this will be discussed in the following sections. First of all, for comparison the MOS-FET model will be discussed as a starting point.

6.2 MOS-FET

As discussed before, a MOS-FET is an MIS diode with electrodes connected laterally on the semiconductor side. The bias at the gate, relative to the semiconductor, 'programs' the charge in the semiconductor, and the source and drain electrodes are 'measuring' the total free charge in this active layer. No simple charge–voltage relation exists (in contrast to TFTs) and a rather complicated electrical behavior results. For a complete description of a MOS-FET, please refer to Chapters 7.2.1 and 8 of Sze [9].

The local electron and hole concentrations anywhere in the device depend on the energetic depth of the Fermi level and thus by the amount of band bending V_s:

$$n(V_s) = n_0 \exp(qV_s/kT)$$
$$p(V_s) = p_0 \exp(-qV_s/kT) \tag{6.1}$$

where n_0 and p_0 are the electron and hole density in the bulk material, respectively (this assumes a Boltzmann approximation for the distribution). The band bending V_s can be found by Poisson's Equation. For a p-type semiconductor [and setting $V(\infty)$ and $E(\infty)$ to zero]:

$$\begin{aligned}\frac{d^2V_s(x)}{dx^2} &= \rho(x)/\varepsilon_s \\ &= \frac{q}{\varepsilon_s}[p(x) - n(x) - N_A^-(x)]\end{aligned} \tag{6.2}$$

where x is the space coordinate perpendicular to the interface, p and n are the free hole and electron density, respectively, and N_A^- is the ionized acceptor density. Thus, there exists an interdependence between the charge (free electrons and holes and ionized dopants) and the band bending V_s; one causes *and* is the effect of the other. To give an exact solution is not possible, though approximations can be made. Figure 6.2 shows an energy diagram with band bending and charge distributions drawn in a schematic way.

A simple approximation is assuming the charge causing the band bending is caused solely by ionized dopants and the free charge does not contribute to it but is only the result of it. For large doping densities, this approximation seems reasonable. However, it does not explain, for instance, how a MOS-FET can be put into accumulation, since this would need band bending of the sign not possible to make with those dopants (for example, positive space charge is needed while acceptors can only be ionized negatively). However, for a MOS-FET working in depletion and inversion it works quite well. Later it will be shown how the simpler TFT model can be used for a MOS-FET in accumulation.

Thus, for the simple calculation of the band bending the following assumptions are made, which are the same as used for the Schottky and MIS diodes described earlier:

- All acceptors are ionized; $N_A^- = N_A$ everywhere.
- Minority carriers (electrons) do not play a role.
- Inside the depletion (and inversion) zones, all holes have moved out, outside these zones nothing has changed and the free charge is

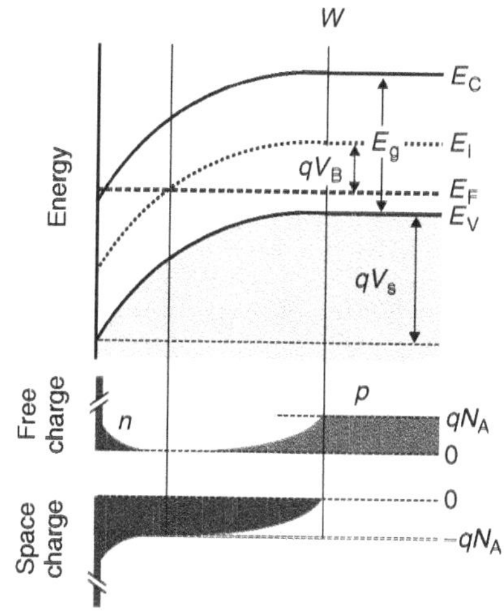

Figure 6.2 Energy diagram, free charge and space charge distributions of an inversion-channel MOS-FET at the threshold voltage $V_g = V_T$ at the onset of the channel, drawn in a rough, schematic way. E_F, Fermi level; E_C, conduction band energy; E_i, intrinsic energy (if $E_F = E_i$ then $n = p = n_i$); E_V, valence band energy; qV_s, band bending (note that the energy bending has opposite curvature compared with voltage, due to the negative sign of the electron charge; electron energy is shown); qV_B, bulk distance between Fermi level and intrinsic level; n, free electron density; p, free hole density; N_A, acceptor density

exactly compensating the ionized acceptors

$$x < W : p(x) = 0$$
$$x > W : p(x) = N_A. \tag{6.3}$$

It results in a quadratic band bending from $x = W$ to $x = 0$, with W the depletion width:

$$V(x) = \frac{qN_A}{2\varepsilon_s}(x - W)^2 \tag{6.4}$$

as shown Section 1.5.2. At the interface this band bending is equal to

$$V_s = V(x = 0) = \frac{qN_A}{2\varepsilon_s}W^2. \tag{6.5}$$

Moreover, the relation between the band bending inside the semiconductor and the total external voltage including the drop over the oxide V_{ox} is

$$V_g = V_s + V_{ox}. \tag{6.6}$$

The voltage drop over the oxide can easily be calculated using the Maxwell continuity equations, more specifically the continuity of displacement $D \equiv \varepsilon E$. On the semiconductor side of the interface the field is given by the derivative of the function $V(x)$ above, namely

$$\begin{aligned} D_s(0) &= -\varepsilon_s \left. \frac{dV(x)}{dx} \right|_{x=0} \\ &= qN_A W. \end{aligned} \tag{6.7}$$

On the oxide side this displacement D_{ox} is therefore equal. Since there exists no charge in the oxide, the field $E_{ox} = D_{ox}/\varepsilon_{ox}$ is constant and the total voltage drop just the field multiplied by the oxide thickness d_{ox}

$$\begin{aligned} V_{ox} &= D_{ox} d_{ox}/\varepsilon_{ox} \\ &= qN_A W \frac{d_{ox}}{\varepsilon_{ox}} \\ &= \frac{\sqrt{2q\varepsilon_s N_A V_s}}{C_{ox}} \end{aligned} \tag{6.8}$$

where the relation between the band bending and the depletion width of Equation (6.5) and the definition of oxide capacitance ($C_{ox} = \varepsilon_{ox}/d_{ox}$) were used. The total external voltage of Equation (6.6) relative to the bulk of the semiconductor side set at zero potential $[V(\infty) \equiv 0]$ then becomes

$$V_g = V_s + \frac{\sqrt{2q\varepsilon_s N_A V_s}}{C_{ox}}. \tag{6.9}$$

In a typical MOS-FET, about half of the voltage is across the interface and the other half is in the semiconductor, see Table 6.1. This means that no simple linear (or quadratic) relation exists between V_g and V_s.

6.2.1 MOS-FET Threshold Voltage

At the threshold voltage, a minority-carrier (electron) channel is formed in the MOS-FET at the interface. According to the literature, by definition, at the threshold voltage the density of minority carriers at the interface is equal to the density of majority carriers in the bulk. In other words, the distance from Fermi level to the conduction band at the

Table 6.1 Parameters of a typical n-type inversion-channel Si-SiO_2 MOS-FET at onset of inversion, $V_g = V_T$

Parameter	Value
N_A	$10^{15}\ cm^{-3}$
d_{ox}	100 nm
E_g	1.15 eV
T	300 K
E_F	$E_V + 225$ meV
C_{ox}	$345\ \mu F\ m^{-2}$
V_s	925 mV
W	630 nm
V_{ox}	512 mV
V_T	1.44 V

interface is equal to the distance from the Fermi level to the valence band in the bulk. Analyzing Figure 6.2 we can see that this is for $V_s = 2V_B$. The threshold voltage is thus [Equation (6.6)]

$$V_T = 2V_B + \frac{\sqrt{4q\varepsilon_s N_A V_B}}{C_{ox}}. \qquad (6.10)$$

The threshold voltage is thus a measure for the doping concentration. In some cases a better definition is a threshold voltage at which the Fermi level is resonant with the minority carrier band, this occurs at

$$V_T = \frac{(E_C - E_F)}{q} + \frac{\sqrt{2\varepsilon_s N_A (E_C - E_F)}}{C_{ox}}. \qquad (6.11)$$

This can be further worked out with the help of the relation

$$(E_C - E_F) = E_g - kT \ln\left(\frac{N_V}{N_A}\right). \qquad (6.12)$$

If the second term in the threshold voltage, which describes the voltage drop in the oxide, is much larger than the first term describing the drop in the semiconductor, the threshold voltage can be approximated by

$$V_T \approx \frac{1}{C_{ox}} \sqrt{2q\varepsilon_s N_A \left[E_g - kT \ln\left(\frac{N_V}{N_A}\right)\right]}. \qquad (6.13)$$

6.2.2 MOS-FET Current

After the threshold voltage is reached and a channel is established, further increments in gate bias do not change the band bendings because the DOS in the conduction band is so huge that it can absorb any additional space charge easily (the free charge itself starts contributing). It also implies that any additional charge is at the interface which thus grows as in a metal plates capacitor, $n(x=0) = (V_g - V_T)/C_{ox}$, and if we neglect the contributions from regions where the Fermi level is below the conduction band the current is proportional to this, $n = n(x=0)$. In the linear region, for small drain-source bias (V_{ds}) the densities are homogeneous from drain to source. The current is then equal to the electron charge density in the channel, $qn = q(V_g - V_T)/C_{ox}$ times the electric field, $E_{ds} = V_{ds}/L$ (where L is the distance between source and drain electrodes), and the electron charge mobility μ_n, scaled with the device dimension, the electrode width W

$$I_{ds} = q\mu_n V_{ds} \frac{W}{L} \frac{(V_g - V_T)}{C_{ox}}. \tag{6.14}$$

Saturation effects, occurring for large V_{ds} will be discussed in Section 6.4.

6.2.3 Exact Solution

The book by Sze [9] gives a semi exact solution that is based on the work of Garrett and Brattain [271]. The equation

$$\rho(x) = q[p(x) - n(x) - N_A^-(x)] \tag{6.15}$$

is for the total charge density inside the material. This assumes all acceptors are ionized over all space ($N_A^- = N_A$. In the bulk, there is charge neutrality, thus $N_A^- = p_0 - n_0$). (Note that this is not valid for the device in accumulation.) With the help of Equation (6.1) (ψ used instead of V_s),

$$p(\psi) - n(\psi) = p_0 \exp(-\beta\psi) - n_0 \exp(\beta\psi) \tag{6.16}$$

substituted in Poisson's Equation we get

$$\begin{aligned} \frac{d^2\psi(x)}{dx^2} &= \frac{\rho(x)}{\varepsilon_s} \\ &= \frac{q}{\varepsilon_s}\{p_0[\exp(-\beta\psi) - 1] - n_0[\exp(\beta\psi) - 1]\}. \end{aligned} \tag{6.17}$$

We can find the electric field $E = \int \psi(x)\mathrm{d}x$ as a function of potential ψ by integrating this, making use of the property of integrating. If

$$\frac{\mathrm{d}^2 y(x)}{\mathrm{d}x^2} = f(y) \tag{6.18}$$

then

$$\left[\frac{\mathrm{d}y(x)}{\mathrm{d}x}\right]^2 = 2\int f(y)\mathrm{d}y. \tag{6.19}$$

This will give a solution for E^2 [with proper boundary condition ($E = 0$ for $\psi = 0$)].

$$E(\psi) = \sqrt{\frac{2qp_0}{\beta\varepsilon_s}} F(\psi, n_0/p_0) \tag{6.20}$$

with

$$F(\psi, n_0/p_0) = \left\{[\exp(-\beta\psi) + \beta\psi - 1] + \frac{n_0}{p_0}[\exp(\beta\psi) - \beta\psi - 1]\right\}^{1/2} \tag{6.21}$$

Now, all space charge inside the device can be found using Gauss's Equation, $Q_s(\psi) = -\varepsilon E(\psi)$,

$$Q_s(\psi) = \sqrt{\frac{2qp_0\varepsilon_s}{\beta}} F(\psi, n_0/p_0). \tag{6.22}$$

Figure 6.3 gives a graphical representation of this space charge as a function of total band bending ψ_s. Note that Q_s is equal to all charge in the device,

$$Q_s = \int_0^\infty q[p(x) - n(x) - N_A^-(x)]\mathrm{d}x. \tag{6.23}$$

What is more interesting to know is how much *free* charge there is in the device, since the current is proportional to the *sum* of the densities of electrons and holes, $I_{ds} \propto n_s + p_s$.

$$\begin{aligned} p_s &= \int_0^d p(x)\mathrm{d}x \qquad (6.24) \\ &= p_0 d + \int_0^\infty [\exp(-\beta\psi) - 1]\mathrm{d}x \end{aligned}$$

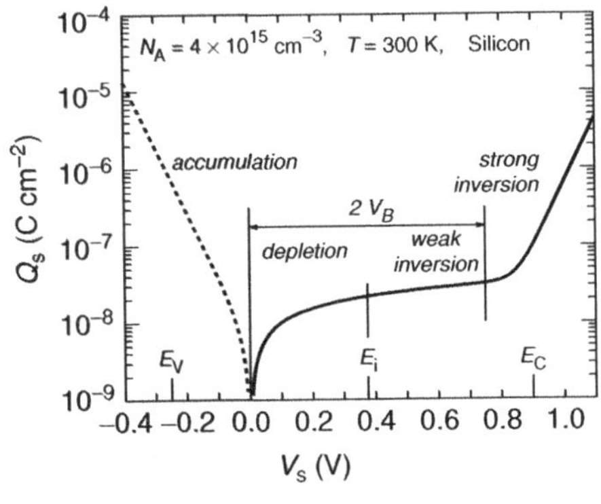

Figure 6.3 MOS-FET space charge Q_s as a function of total band bending in a typical silicon device (assuming the flat-band voltage zero). For completeness the simulation is also shown of the MOS-FET model for accumulation (dashed line), though in this case the TFT model should be used. Based on calculations by Garrett and Brattain [271]

$$= p_0 d + \sqrt{\frac{p_0 \beta \varepsilon_s}{2q}} \int_{\psi_s}^{0} \frac{\exp(-\beta\psi) - 1}{F(\psi, n_0/p_0)} \mathrm{d}\psi$$

$$n_s = \int_0^d n(x)\mathrm{d}x \qquad (6.25)$$

$$= n_0 d + \int_0^\infty [\exp(\beta\psi) - 1]\mathrm{d}x$$

$$= n_0 d + \sqrt{\frac{n_0^2 \beta \varepsilon_s}{2qp_0}} \int_{\psi_s}^{0} \frac{\exp(\beta\psi) - 1}{F(\psi, n_0/p_0)} \mathrm{d}\psi$$

where d is the thickness of the active layer. These equations are difficult to solve. One important observation is that the density of free charge is never zero, implying that a MOS-FET cannot be switched off completely.

An additional parameter is defined for a MOS-FET that is called the flat-band voltage, V_{FB} [272]. As the name implies, V_{FB} is that voltage that causes zero band bending and space charge in the semiconductor. Deviations from zero are caused by an electrochemical potential (or workfunction difference) between the gate material and the semiconductor layer.

Once again, the solution given here does not apply to the situation of a MOS-FET in accumulation because the assumption that $N_A^- = N_A$ everywhere is not true. In accumulation the acceptors get neutralized in a region close to the interface. Note that also for materials with deep levels, such as organic materials, the assumption that all impurities are ionized is not valid. A much more complicated $Q_s - \psi$ relation can emerge. However, as we will see, in accumulation the picture is much simpler.

6.2.4 MOS-FET Subthreshold Current and Subthreshold Swing

The threshold voltage is given by Equation (6.10). Above this voltage, the density of charge is huge and depends linearly on the gate and drain biases (until saturation effects kick in); a relatively homogeneous distribution of charge exists along the channel, more or less linearly dropping. On account of this, the current is comprised primarily of drift current.

Below the threshold voltage, in weak inversion, the current is not zero, since the charge density along the channel is not zero. Owing to the relative large gradients of charge density and low absolute densities, the subthreshold drain current in this regime is dominated by diffusion. These currents are proportional to the gradient of the density and, because the current is constant along the channel, the density is linearly dropping. In other words, the channel current is

$$I_{\mathrm{ds}} \propto D_{\mathrm{n}} \frac{\mathrm{d}n}{\mathrm{d}x} = \frac{n(0) - n(L)}{L}. \tag{6.26}$$

The density at the source, $n(0)$, exponentially depends on the band bending V_s, as shown in Equation (6.1) and at the drain the Fermi level is qV_{ds} deeper. The current in Equation (6.26) can thus be written as

$$\begin{aligned} I_{\mathrm{ds}} &\propto \exp\left(\frac{qV_s}{kT}\right) - \exp\left[\frac{q(V_s - V_{\mathrm{ds}})}{kT}\right] \\ &= \exp\left(\frac{qV_s}{kT}\right)\left[1 - \exp\left(-\frac{qV_{\mathrm{ds}}}{kT}\right)\right]. \end{aligned} \tag{6.27}$$

See the book by Sze [9] for a more complete equation that includes also the effective thickness of the channel (assumed to be constant here). For drain-source biases exceeding about $3kT/q$ ($\approx 80\,\mathrm{mV}$ at room temperature), the term in square brackets is close to unity and independent of V_{ds}, leaving behind an exponential dependency of I_{ds} on band bending V_s. The relation between V_s and the external gate bias V_g is nonlinear [Equation (6.9)] and a simple exponential behavior of the current on the gate bias cannot be expected. The subthreshold swing, defined as the gate-voltage needed to change the current by one decade,

$$S \equiv \ln(10)\frac{\mathrm{d}V_\mathrm{g}}{\mathrm{d}\ln(I_\mathrm{ds})} \tag{6.28}$$

and for an inversion-channel MOS-FET, Brews gives an expression approximately [273]

$$S = \frac{kT}{q}\ln(10)[1 + C_\mathrm{D}(V_\mathrm{s})/C_\mathrm{ox}] \tag{6.29}$$

where C_D is the depletion capacitance that still depends on band bending and thus bias. Normally, however, the subthreshold swing is small (in the order 10–100 mV per decade) and consequently S does not vary much over a decade. Since the depletion capacitance depends on the acceptor concentration, the subthreshold swing gives some information about the doping. It has also been suggested that the interface traps can impact the subthreshold swing, but there might be other effects of similar magnitude that alter S [273].

It has to be well understood that the expression for S is valid only for MOS-FETs working in inversion. Since TFTs commonly work in accumulation, the expression cannot be used. Later a more appropriate expression for the subthreshold current will be given. As will be shown, the current is no longer exponential and the definition of a subthreshold swing does not make sense for TFTs; semi-log plots of the subthreshold current of TFTs are invariably nonlinear.

6.3 INTRODUCING TFTs

The Algarve approach to the TFT, developed in the last decade (1997–2007), treats it purely as two-dimensional. In other words, a TFT is like a metal-plates capacitor filled with insulator material. The model is a great simplification of reality, as reality is more complicated. However, it describes the reality very well and any modeling of reality should always find an optimum balance between simplicity and descriptive power. I set out on this journey not to make as good a description of devices as possible (good textbooks exist, for example the book by

Sze [9]), but to come up with one that is easy to understand by a large community. I tried to find the minimum requirements for a model to explain our (quite complicated) data measured in our laboratory. The search for simplicity was inspired by my great heroes, William of Ockham (in the 14th century probably the first modern scientist when describing his Ockham's Razor) and Saint-Exupéry, both stating that models to describe nature should be as simple as possible. William of Ockham wrote (Wikipedia) '*Entia non sunt multiplicanda praeter necessitatem*' 'Entities should not be multiplied beyond necessity'. This is freely translated as 'the explanation of any phenomenon should make as few assumptions as possible, eliminating those that make no difference in the observable predictions of the explanatory hypothesis or theory' (Wikipedia), thereby advocating using a 'razor' to cut away all the nonessential parts of the theory. Saint-Exupéry said more-or-less the same when he stated that 'An engineer should be happy not when there is nothing left to add, but when there is nothing left to take out'. Thus, the best models are those that are simple while still maintaining an adequate description of reality. As an example, we assume all charge to be directly at the interface between the insulator and the semiconductor. In reality, the charge induced by the gate can be several layers distant from the interface (see the works of Horowitz, for example ref. [274]). These effects can be absorbed by the insulator capacitance C_{ox}, which is then slightly smaller than the one calculated on the basis of the insulator thickness and can even depend on the bias. However, these are minor perturbations and should be looked at only when the rest of the characterization of the device measured is under control. One should not get bogged down in insignificant details even if they are correct. On the other hand, the simplicity of the model can directly show us some strong conclusions such as the impossibility to make a Schottky barrier at the contacts, as will be shown here, and a correct modeling of possible contact resistance, where these two are often used in the literature to write off any undesired and not well understood behavior of the devices. Moreover, all sorts of gate-bias- and temperature dependent field-effect mobilities and distortions in I–V curves are very easily found. Summarizing, while the modeling is simple, it is able to draw some very powerful conclusions. I hope that the model can rapidly identify why your device is behaving in a particular peculiar way and that it may lead the way to removing the unwanted effects and paving the way for better devices. 'If you don't know what is the problem you cannot find a solution'.

6.4 BASIC MODEL

Figure 6.4 shows a schematic diagram of a TFT and the labeling of the parameters used. Unless specified otherwise, the parameters are as in Table 6.2.

The single postulate in the model is to treat the device as a simple parallel plate capacitor. All charge on both sides of the insulator is immediately at the interface, and there are no band bendings or other ways to distribute the charge in another way. As such, the charge at any place in the device is directly proportional to the voltage drop at the insulator, like in a classical capacitor ($C = Q/V$):

$$\rho(x) = C_{ox}[V(x) - V_g] \tag{6.30}$$

where $\rho(x)$ is the charge density at position x along the channel (see Figure 6.4), C_{ox} is the oxide capacitance density ($= \varepsilon_{ox}/d_{ox}$), $V(x)$ is the local potential in the channel and V_g is the uniform potential at the gate.

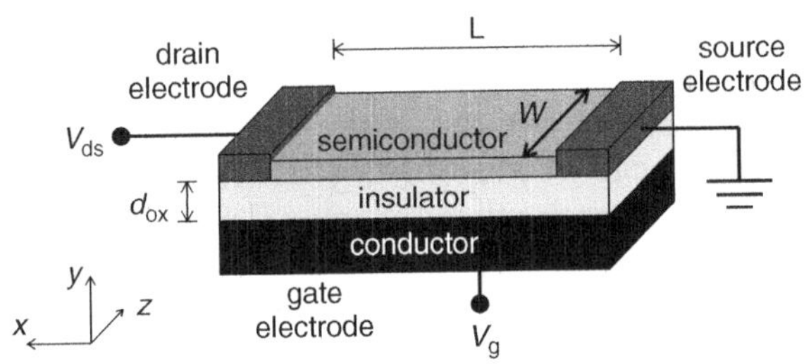

Figure 6.4 Schematic diagram of a TFT device and its parameters

Table 6.2 Parameters of TFTs used presenting the basic model

Parameter	Value	Unit
d_{ox}	216	nm
ε_{ox}	3.9	ε_0
C_{ox}	160	$\mu F\ m^{-2}$
ε_s	11.9	ε_0
L	10	μm
W	1	cm
μ_p	3×10^{-4}	$cm^2\ V^{-1}s^{-1}$
μ_n	0	–

In this chapter all capacitances are capacitance densities.

Note that the density of charge is 'per square meter', thus effectively treating the active layer as two-dimensional. For pure materials (without traps, doping or other electronically active impurities), all this charge is free charge,

$$\rho(x) = qp(x)$$
$$p(x) = C_{ox}\left[V(x) - V_g\right]/q. \qquad (6.31)$$

(Later we will show cases in which this is different when traps are introduced into the system.) Now, at any point in the channel, the current I_x is then given by the product of the local free charge density qp, the local electrical field $[E_x = -dV(x)/dx]$, the charge-carrier mobility μ multiplied by the device width W:

$$I_x(x) = -qp(x)\mu W dV(x)/dx. \qquad (6.32)$$

This differential equation can easily be solved when appropriate boundary conditions are used, namely $V(0) = 0$, $V(L) = V_{ds}$ and $I_x(x) = I_{ds}$ for all x (indicating there are no current sources in the channel). The solution is then (positive biases; n-channel)

$$I_{ds}(V_g, V_{ds}) = \mu C_{ox}\frac{W}{L}\left(V_g V_{ds} - \frac{1}{2}V_{ds}^2\right) \qquad (6.33)$$

which is equal to the behavior of a MOS-FET, explaining the persistence of the application of the MOS-FET model to TFT devices; it simply seems to be adequate. Further examples will be shown where the output curves deviate from simple MOS-FET behavior. It will be shown how the effective – as-measured – mobility can be different from μ and how it can depend on the bias (V_g and V_{ds}) and the temperature.

Equation (6.33) is valid for drain-source biases up to V_g, the so-called 'linear region'. After that point saturation starts. The same reasoning can be applied to TFTs as is common for MOS-FETs. When at the drain the potential is equal to the gate bias, no voltage drop exists there and no free charge (6.30) which can be said to be pinched-off. The current remains constant at the saturation value,

$$I_{ds}^{SAT}(V_g) = \mu C_{ox}\frac{W}{L}\frac{1}{2}V_g^2. \qquad (6.34)$$

Figure 6.5 shows simulations of the above equations. The thin lines are the linear regime and follow Equation (6.33), whereas the thick

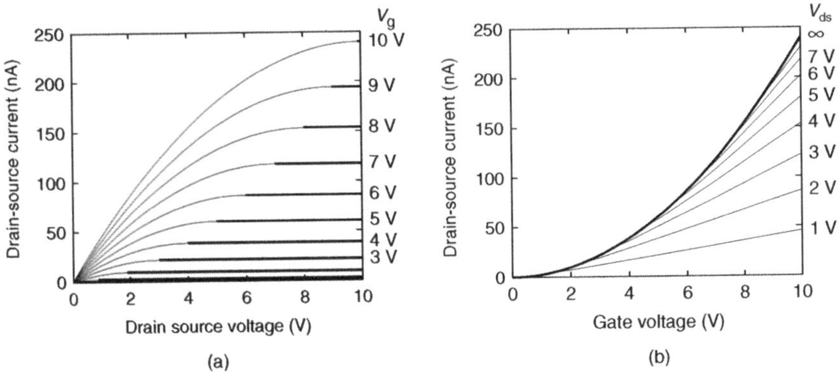

Figure 6.5 Simulations of n-channel TFT (a) I–V and (b) transfer curves in the linear regime (thin lines) and the saturation regime (thick lines). Parameters as in Table 6.2, with μ_n and μ_p exchanged

lines are the saturation currents according to Equation (6.34). Special attention goes to the point of onset of saturation, $V_{ds} = V_g$, the locus point, which in some cases can be used to minimize (measuring) the leakage through the insulator.

For tiny V_{ds} the quadratic term in Equation (6.33) is negligible and the current can be approximated by

$$I_{ds}(V_g, V_{ds}) = \mu C_{ox} \frac{W}{L} V_g V_{ds} \tag{6.35}$$

This allows for the definition of the as-measured field-effect mobility via the derivative of the transfer curve,

$$\mu_{FET} \equiv \frac{\partial I_{ds}}{\partial V_g} \frac{1}{C_{ox} V_{ds}} \frac{L}{W}. \tag{6.36}$$

This is the standard definition used in the literature. It is also possible to define the mobility in the saturation regime, via Equation (6.34), but for various reasons this is less reliable.

The mobility μ has only slight dependence on temperature ($\propto T^{-3/2}$, $\propto T^{3/2}$ or $\propto T^{1/2}$ for acoustic phonon scattering, ionized impurity scattering and optical phonon scattering, respectively). Thus, the current in a TFT is basically independent of temperature. This is what one would expect; if the current is proportional to the density of charge and this density follows a simple capacitor behavior, the current is expected to be independent of temperature. Later it will be shown how mobility and

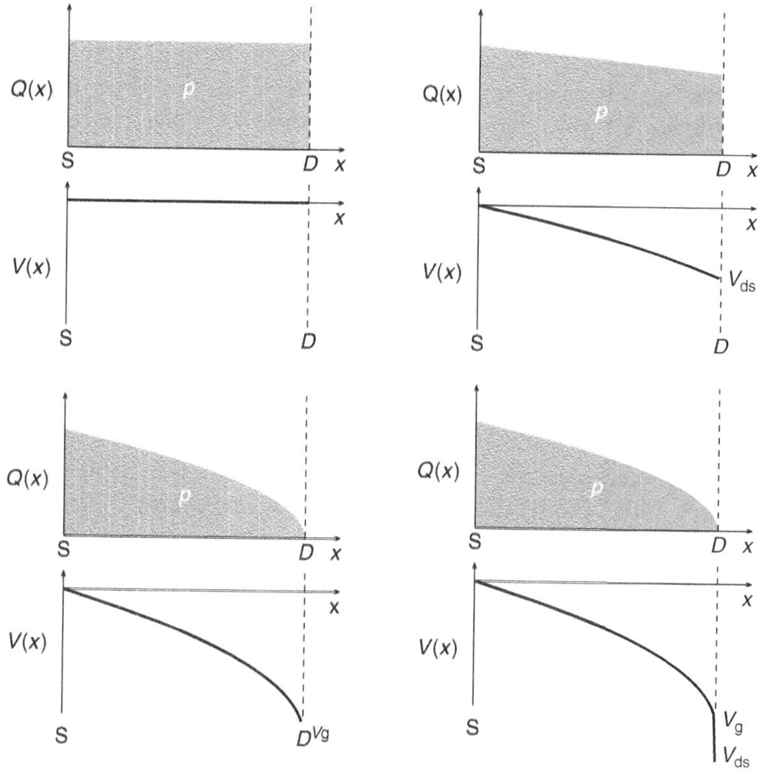

Figure 6.6 Distribution of charge $Q(x)$ and potential $V(x)$ along the channel for various biases V_{ds} for a p-channel TFT (negative gate bias)

current can be temperature (and drain-source and gate-bias) dependent in the presence of traps.

Figure 6.6 shows the distribution of charge and potential for several drain-source biases for a p-channel TFT (negative biases). Before saturation, these follow

$$V(x) = V_g - \sqrt{V_g^2 + \frac{V_{ds}^2 - 2V_{ds}V_g}{L}x} \tag{6.37}$$

and

$$p(x) = \frac{C_{ox}}{q}\sqrt{V_g^2 + \frac{V_{ds}^2 - 2V_{ds}V_g}{L}x}. \tag{6.38}$$

Note that after saturation the distribution is basically constant except for an infinitesimally thin region close to the drain. Because the potential and charge distribution are equal, the current is equal [Equation (6.32)] and is independent of V_{ds}, hence the term 'saturation'. Note also that the charge in all cases is two-dimensional. The plot does not show the *thickness* of the channel, but the *density* of the charge in the two-dimensional channel.

6.4.1 Threshold Voltage and Subthreshold Current

It is common practice to use the threshold voltage and the subthreshold current as device-evaluation parameters [9]. They are often used to extract information about impurity concentrations, traps and interface states. The equations used for the parameter extraction, however, have been derived for MOS-FETs working in inversion. Therefore, these parameters cannot be applied to TFTs. In the context of the two-dimensional model described above, it is important to understand the physical meaning of these device parameters. In this section, we will analyze the trap-free device based on intrinsic materials and show that the threshold voltage and subthreshold current do have different behavior compared with the conventional MOS-FET models.

In a TFT of a pure material, without traps, or a transistor in accumulation, all charge induced in the channel is free charge and the threshold voltage is therefore zero. [In the presence of traps, or dopants, this can be different, as will be shown later, with the general threshold voltage being given by Equation (6.84)]. In MOS-FETs, the subthreshold current is exponentially depending on the gate bias as well as the drain-source bias. The reason for this is that below threshold the free carrier density is exponentially dependent on the local bias. (The energetic distance between band edge and Fermi level is linearly dependent on the voltage drop across the insulator and the free carrier concentration is dependent exponentially on this distance.) In the linear region, the potential at the drain is slightly smaller than at the source. Therefore, p is exponentially smaller at the drain compared with the source. Such a high gradient in density causes the diffusion current to dominate. (Drift currents are still insignificant because the densities are still too small.) The gradient and the current thus depends exponentially on V_{ds} and V_g. The current is proportional to the difference in density at the source and the drain, $I_{ds} \propto \exp(V_g) - \exp(V_g - V_{ds}) \approx \exp(V_{ds})\exp(V_g)$, which leads to the equation normally found in textbooks. Above threshold, the densities depend linearly on the potential and drift currents exceed the diffusion currents.

The charge density does not depend exponentially on the potential as in MOS-FETs, but always linearly. To make a crude analysis of the sub-threshold current in TFTs we have to bring electrons into the model and make three assumptions: (1) The charge, electrons and holes, is homogeneously distributed over the channel thickness. This is either the thickness of the active layer or the diffusion length, whichever is smaller. For the calculation we use a thickness of $d = 1$ nm. (2) The thermal equilibrium equation of electrons and holes is maintained:

$$n \times p = n_\mathrm{i}^2 = N_\mathrm{V} N_\mathrm{C} \exp(-E_\mathrm{g}/kT) \tag{6.39}$$

where n_i is the intrinsic electron density (note again that all units of density are 'per area', including n_i, N_V and N_C) and E_g is the electronic band gap. For small currents this holds. (3) The electrons are immobile and do not contribute to current [248] (for the ambipolar devices discussed later this third condition will be abandoned). Then we have to imagine that the induced charge is caused by the the difference between p and n, Equation (6.30) becomes

$$n - p = V_\mathrm{g} C_\mathrm{ox}/q. \tag{6.40}$$

The solution of Equations (6.39) and (6.40) is

$$p = -\frac{V_\mathrm{g} C_\mathrm{ox}}{2q} + \sqrt{n_\mathrm{i}^2 + \left(\frac{V_\mathrm{g} C_\mathrm{ox}}{2q}\right)^2}. \tag{6.41}$$

For large negative voltages $p \gg n_\mathrm{i}$ this reduces to the original form; the currents are proportional to the gate voltage. For large positive voltages, the subthreshold current can be shown to be

$$I_\mathrm{ds} \propto p = 2\left(\frac{q n_\mathrm{i}}{V_\mathrm{g} C_\mathrm{ox}}\right)^2 \tag{6.42}$$

i.e., not exponential; a representation of the subthreshold data in a semi-log plot does not make sense for TFTs since they are not linear and no subthreshold swing parameter S can be extracted. To give an idea, the interval at which the currents goes from two times to half the zero-bias current is

$$V_\mathrm{g} = \pm \frac{3 q n_\mathrm{i}}{C_\mathrm{ox}} \tag{6.43}$$

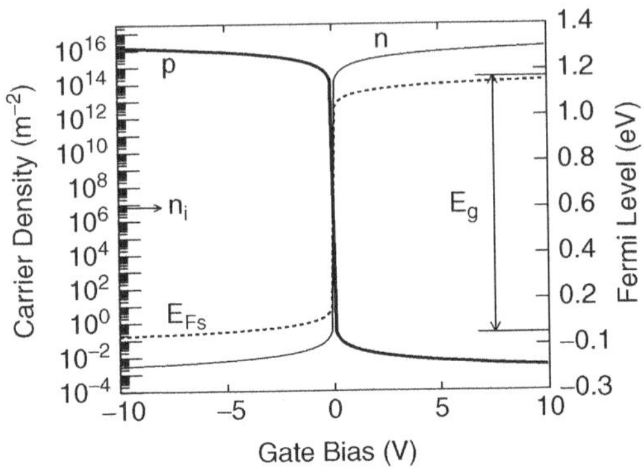

Figure 6.7 Hole and electron density and position of Fermi level as a function of bias with $V_{ds} = 0$. Reproduced with permission from Synthetic Metals, Modeling electrical characteristics of thin-film field-effect transistors: I. Trap-free materials by P. Stallinga and H. L. Gomes, 156, 21–24 Copyright (2006) Elsevier

and is in the order of nanovolts. Figure 6.7 shows the position of the Fermi level and the charge densities as a function of bias. The example is for a silicon TFT. For materials with wider band gap, such as most organic materials, the subthreshold current is much less. Therefore, for all practical purposes, the subthreshold current can be ignored. In any case, the subthreshold swing is a parameter for a weak inversion MOS-FET and not for an accumulation TFT.

6.5 JUSTIFICATION FOR THE TWO-DIMENSIONAL APPROACH

The basic question is: Is the assumption that the active layer can be treated as two-dimensional justified? For instance, how thin can the active layer be and still behave as a TFT? Empirically it has been shown that a single mono-layer is enough. As long as a continuous film exists between the drain and source electrodes the device works properly, with a constant mobility [268]. Any additional layers only serve as stabilization of the device, either in chemical or mechanical terms, for instance to prevent diffusion of impurities to the interface.

Theoretically, a single monolayer has a large enough density of states (N_V) to accommodate the bias induced for all reasonable voltages. To give an example, for a silicon-based device, with -1 V at the gate relative to drain and source (both at ground) and an oxide thickness of $d_{ox} = 200$ nm, the induced charge is $0.17\,\mathrm{mC\,m^{-2}}$. With a DOS of N_V

of $1.04 \times 10^{19}\,\mathrm{cm}^{-3}$ and assuming continuity, this can fit into 1 Å; less than the height of a monolayer.

A TFT – or a MOS-FET in accumulation for that matter – has all the induced charge at the interface, for the simple reason that it is mobile (enough) to migrate to the interface. It is composed of free holes that have no way to resist the pulling force of the field caused by the (negative) gate bias. Even if the charge is trapped on deep localized states, thermal excitation will liberate the charge from the states occasionally and eventually it will be able to slowly reach the interface. At steady state, all induced charge is at the interface. In other words, the relation in Equation (6.30) is valid everywhere in the device.

One important observation that needs to be made is that a MOS-FET in accumulation is also best described by a two-dimensional TFT model. The reason is that the dopants do not play any role, since they are not ionized in accumulation. For instance, the charge that is attracted to the interface by a negative gate bias in a p-type material is positive. Since acceptors can only be negatively charged or neutral, they will all be neutral at the interface and will not contribute to space charge. Any gate-induced charge therefore has to be of free-hole type and the above reasoning of a two-dimensional TFT applies. In contrast, the MOS-FET model [9] assumes all dopants to be ionized and space charge can consist of ionized dopants not compensated by their emitted charge.

In reality, the charge layer is not truly two-dimensional. Even though the energy for a charge is lowest close to the interface, there are two reasons why charge can be found at a distance from the interface:

- thermal excitation;
- diffusion.

Thermal excitation can cause charges to also occupy some of the states further away from the interface even when they have a little more energy. There is therefore a thermodynamic distribution over the levels. To find the final distribution in space is very difficult. It is the result of solving equations as in Section 6.2.3 (without N_A). Horowitz describes a nice procedure for finding the distribution by dividing the device in thin layers and let them iteratively establish a thermal equilibrium between each other [274].

Another effect that makes charges move away from the interface is diffusion. Diffusion current is proportional to the gradient of density. The total current in the y-direction, perpendicular to the interface, is zero,

$$J_y = q\left(\mu_p p E_y + D_p \frac{dp}{dy}\right) = 0 \tag{6.44}$$

where D_p is the hole diffusion coefficient, expressed in the mobility through the (Nernst)–Einstein relation, $D = \mu kT/q$. Therefore, the free charge p can resist the pulling field of the gate because of the quasi-force of diffusion. There is a balance between enthalpy and entropy, one trying to place all charges at the interface and the other trying to equalize the density all over space. In this way, the charges can spread out away from the interface in a distance characterized by the Debye length

$$\lambda_D = \sqrt{\frac{\varepsilon_s kT}{q^2 N_V}} \tag{6.45}$$

which is about 1.3 nm for silicon at room temperature, about twice the lattice constant.

Even if the charge layer is thick it would not necessarily imply that the two-dimensional approach is invalid. As long as there exists a linear relation between voltage drop across the insulator and induced charge, the device will effectively behave as two-dimensional. In contrast, a MOS-FET in inversion has a rather complicated voltage–charge relation because of the character of the induced charge. While in a TFT all charge is assumed to be mobile (enough to reach the interface), in a MOS-FET some of the charge is in the form of ionized dopants (for instance acceptors that have received an electron) which, by their nature, have zero mobility. Not all charge is then directly at the interface and a complicated Q–V relation emerges. While charge at the interface 'feels' a capacitance C_{ox}, charge further away feels a reduced capacitance and Equation (6.30) is no longer valid.

The effects of a thicker charge layer in undoped materials (and devices in accumulation) are minimal. As an example, a silicon TFT device will have the charge exponentially spread out by diffusion with a characteristic Debye length, $\lambda_D = 1.3$ nm and will have its oxide capacitance C_{ox} replaced by an effective capacitance C'_{ox} because the average distance of the charge to the gate is increased, by an amount λ_D,

$$C_\lambda = \frac{\varepsilon_s}{\lambda_D} \tag{6.46}$$

$$C'_{ox} = \left(\frac{1}{C_{ox}} + \frac{1}{C_{\lambda}} \right)^{-1}. \qquad (6.47)$$

For a Si-SiO_2 device with a 200 nm insulator, the oxide capacitance is about $C_{ox} = 173\,\mu F\,m^{-2}$ and the Debye capacitance caused by the 1.3 nm diffusion layer, $C_{\lambda} = 81\,mF\,m^{-2}$, will only lower it by about 0.2%. Unless (until) all the other parts of the modeling and measurements are under control, 0.2% is not something to worry about. Moreover, as long as this capacitance is independent of gate bias, the only effect is a slight overestimation of the capacitance and thus underestimation of the field-effect mobility, μ_{FET}.

One issue that remains is that if traps or any other electronic levels are present in the device apart from the band states, and they are not discrete in energy but rather distributed, realizing that the Fermi level is shallower compared with the deeper layers, different states play a role in the electrical behavior. Once again, these effects are not expected to be major and should be addressed only when the rest is under control. For inorganic materials this stage is far from reached and a simpler hands-on theory is needed at this moment. The secondary effects can be studied as an academic hobby to dot the i's and cross the t's. What this text aims to do is to explain a basic understanding of the behavior of the device, rapidly identifying the limiting factors of the product to help remove the obstacles. Getting bogged down in details is not advisable.

In the theory presented here, the charge is considered to be purely two-dimensional and all feels the same oxide capacitance and follows a simple charge–voltage relation as in Equation (6.30). Moreover, the material is considered to be homogeneous in all respects. There are no grains, or special regions at the contacts, etc.

6.6 AMBIPOLAR MATERIALS AND DEVICES

Ambipolar materials are materials which have a high electron *and* hole mobility and can thus conduct both. Various organic materials show ambipolar conduction [146, 275–280]. Because the electrons and holes can be made to meet somewhere in the middle of the channel, far away from the electrodes, they can be made to produce light when they recombine in so-called light-emitting field-effect transistors (LEFETs) [281–284]. It becomes technologically even more interesting when these LEFETs are mechanically flexible [285]. Recently, gate-controlled electroluminescence has also been observed in TFTs based on silicon nanoparticles [286]. Light-emitting or ambipolar field-effect transistors can also be made from two separate layers [251, 287, 288], one with high electron mobility and one with high hole mobility, but the description here will assume that the material itself is ambipolar. For such materials, saturation cannot occur. The reason for this is quite simple. In unipolar devices, when the density of the holes goes to zero, a large field can exist while maintaining a small current. At this point the opposite carrier (electrons) gets a high density (because $np = n_i^2$ still holds, as long as the currents are low) but because they have low mobility, they cannot contribute to current. However, in ambipolar materials the mobility is also relatively high and no strong fields can be maintained. In the equation replacing Equation (6.32),

$$I_x(x) = -q[p(x)\mu_p + n(x)\mu_n]W\mathrm{d}V(x)/\mathrm{d}x \tag{6.48}$$

where μ_p and μ_n are the field-effect mobility of holes and electrons, respectively. This equation reduces to the original Equation (6.32) when $\mu_n = 0$. It is obvious that always a hole current or an electron current exists.

For the distributions we make use of the basic equation [Equation (6.30)], and we assume that when there is one type of carrier present, the other type is absent (valid for small n_i^2)

$$V(x) > V_g: \quad p(x) = C_{ox}\left[V(x) - V_g\right]/q \tag{6.49}$$
$$n(x) = 0$$
$$V(x) < V_g: \quad p(x) = 0 \tag{6.50}$$
$$n(x) = C_{ox}\left[V_g - V(x)\right]/q$$

substituting in Equation (6.48). When the gate bias is outside the range $0 - V_{ds}$ it has only one type of charge throughout the device and it can be treated as a p-channel or n-channel device. A more complicated case exists when the gate bias is in the range between the drain and source potential. In this case, there exists a region of length L_n with free electrons and a region of length L_p with free holes, see Figure 6.8. At the junction point, the potential is equal to the gate bias. Diffusion carries the electrons to the hole channel and vice versa. There therefore exists a zone approximately of length equal to the sum of the individual diffusion lengths of holes and electrons where both these carriers coexist and where the recombination takes place and light originates. The device thus behaves as a LEFET [281–283, 287]. By playing with the gate and

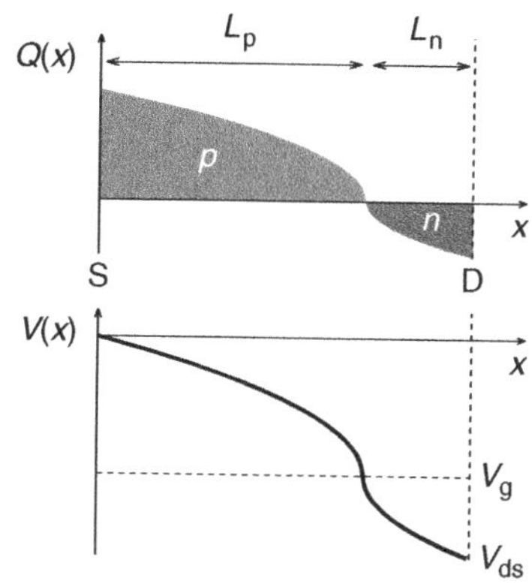

Figure 6.8 Distribution of charge $Q(x)$ and potential $V(x)$ along the channel for an ambipolar device for a bias where otherwise saturation would have occurred for a unipolar device replacing panel 4 of Figure 6.6 for unipolar devices

drain voltages the position of the cross-over point can be programmed, as well as the total current (and thus light intensity).

When we assume that at this junction the electron–hole recombination is not the limiting factor, we can separately treat the electron and hole regions as an FET in saturation, with effective channel length L_n and L_p, respectively. Then, demanding equal current in both regions and using the fact that the total length must equal the channel length $L_p + L_n = L$, we find for $V_{ds} > 0$

$$I_{ds} = \frac{1}{2}\frac{W}{L}C_{ox}\left[\mu_n V_g^2 + \mu_p\left(V_{ds} - V_g\right)^2\right]. \tag{6.51}$$

In the same way, for $V_{ds} < 0$

$$I_{ds} = -\frac{1}{2}\frac{W}{L}C_{ox}\left[\mu_n\left(V_{ds} - V_g\right)^2 + \mu_p V_g^2\right]. \tag{6.52}$$

Figure 6.9 shows output and transfer curves of ambipolar devices. The thick parts of the curves indicate the dual-injection regime.

Such curves can serve as rapid evaluation tools. The minimum of a particular transfer curve can be found by taking the derivative of Equation (6.51) or (6.52) and putting to zero:

$$V_g^{min}|_{V_{ds}<0} = \frac{\mu_p}{\mu_p + \mu_n}V_{ds} \tag{6.53}$$

$$V_g^{min}|_{V_{ds}>0} = \frac{\mu_n}{\mu_p + \mu_n}V_{ds}. \tag{6.54}$$

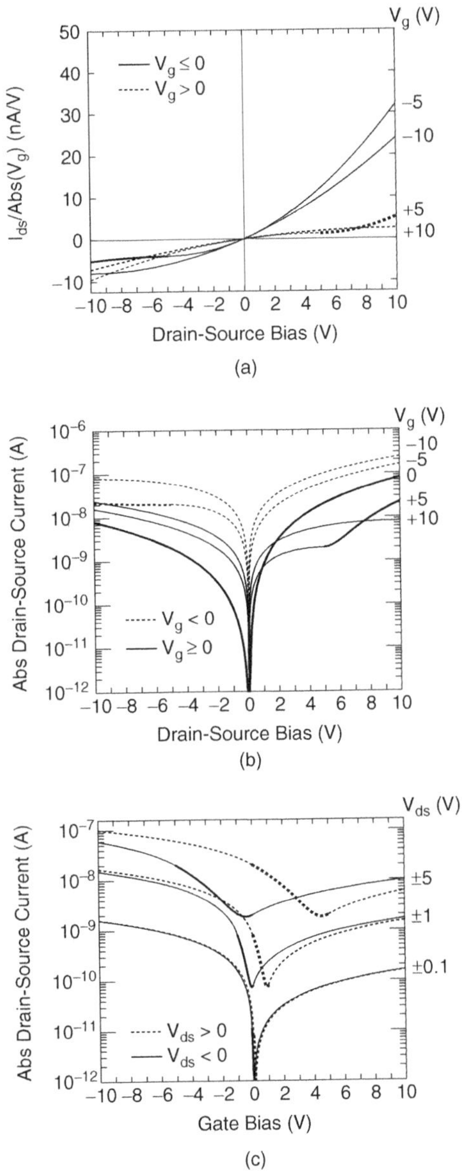

Figure 6.9 Simulation of I–V curves in linear scale (a), logarithmic scale (b) and transfer curves in log scale (c) for ambipolar devices, $\mu_p = 10^{-4}\,\mathrm{cm^2\,V^{-1}s^{-1}}$. $\mu_n = 0.1\,\mu_p$ (b and c) or $0.3\,\mu_p$ (a). The thick parts of the curves are for currents in the dual-injection regime, as shown in Figure 6.8, whereas the thin traces are for unipolar behavior with a single type of carrier in the channel. Reproduced with permission from Synthetic Metals, Modeling electrical characteristics of thin-film field-effect transistors: I. Trap-free materials by P. Stallinga and H. L. Gomes, 156, 21–24 Copyright (2006) Elsevier

The ratio of the positions of the minima in positive and negative bias thus directly yields the ratio of electron and hole mobility.

The position of the recombination zone, where the electron region and hole regions touch, is for positive biases, $0 < V_g < V_{ds}$, given by

$$\frac{L_n}{L} = \frac{\mu_n V_g^2}{\mu_p (V_{ds} - V_g)^2 + \mu_n V_g^2} \tag{6.55}$$

and for negative biases, $V_{ds} < V_g < 0$ (see Figure 6.8)

$$\frac{L_p}{L} = \frac{\mu_p V_g^2}{\mu_n (V_{ds} - V_g)^2 + \mu_p V_g^2}. \tag{6.56}$$

In other words, the position of the region with light emission can be controlled by the combination of drain and gate biases when the relative mobilities for electrons and holes are known.

The device described above is based on a material that itself is ambipolar, with high values for both the electron and hole mobility. In other geometries, this limitation can be overcome. In one design, two layers are deposited, one on top of the other. One layer has high electron mobility while the other is a good hole conductor. The model described above is adequate for this geometry with a small modification. For the layer closest to the interface, the insulator capacitance is the insulator capacitance C_{ox}. For the top layer, however, the effective capacitance is reduced and is the series capacitance of the insulator and the first layer. This capacitance has to substitute C_{ox} in Equation (6.49) or (6.50), depending on which layer is the top layer. A slightly more complicated solution compared with Equations (6.51) and (6.52) is then found. Care has to be taken to include in the design a barrier for carriers of the top layer to be prevented from moving into the bottom layer. This determines the preferred stacking order of the materials. Another geometry is a deposition of parallel or interdigitated tracks of hole and electron transport material. The fabrication of this device is rather complicated, but the above theory will work without modifications.

6.7 CONTACT EFFECTS AND OTHER SIMPLE NONIDEALITIES

Often nonlinearities are encountered in the electrical characteristics and the I–V and transfer curves deviate from the ideal cases shown in

Figure 6.5. Most TFT literature opts to classify them as contact effects [289], or use the generic term 'contact resistance' (or 'series resistance') [290–296] to cover all such nonlinearities. Since often the found value of the contact resistance depends on bias or current, it is a clear misnomer and 'contact effects' at first seems better, leaving room for both contact effects being purely resistive and Schottky-barrier-like.

Sometimes, overwhelming evidence is found that the nonlinearities are caused by some kind of barrier at the contact. In Kelvin-probe profilometry, Nichols *et al.*, for instance, found that the choice of contact metal changes a voltage drop at the contact that correlates to the nonlinearities in the output curves [297]. The question in this work is, however, if the material can still be considered homogeneous in the channel and the metal has not chemically reacted with the semiconductor material in the channel to produce a region with lowered mobility. The most reactive metals appeared to produce the most nonlinearities. In contrast, huge voltage drops at the drain are observed without any sign of nonlinearities in the work by Tanimura *et al.*, undermining the Schottky-barrier ideas [298]. Lin *et al.* observe strong nonlinearities when they fabricate Schottky-barrier TFTs on purpose [299]. However, Wang *et al.*, in a similar work with planned Schottky barriers programmed by the gate find close to no nonlinearities [300]. This just shows how difficult it is to get conclusive results. Theoretical work sometimes also shows strong effects of the contacts, see for instance the work of Bolognesi *et al.* [301]. Other numerical work does not seem to manage to reproduce non-linearities with contact barriers and need to introduce a field-dependent mobility [302]. The nonlinearities also seem to increase when the channel length is decreased, where the contact regions are gaining importance [303]. These are all very suggestive towards a conclusion of contact effects.

Strong doubt on the interpretation of contact effects is cast by a myriad of observations:

- The fact that the contact effects depend on the mobility of the material [295, 304] (huge 'contact effects' are seen for trap-ridden pentacene [305], yet not in single-crystal pentacene transistors [306]).
- The contact effects depend on the bias conditions [295].
- The contact barrier height of the proposed Schottky contact is negative [307], or in general is not equal to the barrier height predicted by thermionic emission theory [308].

- A contact barrier exists between T6 or pentacene and gold [309, 310], while gold is known to form low-ohmic contacts well with organic materials.
- The contact effects depend on the choice of gate insulator, see for example the pentacene transistors by the same group of Klauk *et al.*, where TFTs with SiO_2 as insulator give large nonlinearities [311], while those based on PVP show nearly ideal characteristics [312]. (Note that silicon oxide surfaces are full of dangling bonds and other electrically active centers, whereas organic materials do not have this property.)
- Another group, Lee *et al.* [313, 314], fabricated top-contact pentacene transistors on sapphire dielectrics with varying (insulator and pentacene) film thickness and found perfect output curves for all of them.

Generally, it seems strange that the contact effects in the form of nonlinearities seem to disappear for modern devices that have higher purity and mobility, for instance in single-crystal devices [306, 315]; the Schottky-barrier height, for example, is supposed to be (nearly) independent of doping concentration and carrier mobility and depends mostly on the workfunctions of the metal and semiconductor.

The conclusion is that the nonlinearities are caused by the semiconductor material properties and not so much by the contacts. As will be shown, following this reasoning will result in an adequate description of the nonlinearities, see Section 6.10. This leaves room for a proper description of the effects of the metallic contacts, see Section 6.8.

First, the real effects of contact resistance and contact Schottky barriers are discussed. This section will discuss:

- insulator leakage currents;
- contact resistance at source and drain;
- Schottky barriers at source and drain;
- parallel conductance.

6.7.1 Insulator Leakage

In the above description it was assumed that the insulator is perfect and has infinite resistivity. No charge can pass from the channel to the gate, or in other words, the gate current is zero. The most used insulator material, silicon oxide, however, has a finite resistivity of about 10^{14}–10^{16} Ω cm. For high-mobility materials, such as silicon in the MOS-FET geometry, even the low-end silicon oxide is good enough and gate currents are negligible compared with channel currents. For most organic materials and for TFTs of low-mobility materials in general, the channel currents are smaller and requirements on the insulator stronger if distortions of the ideal characteristics are unwanted.

Leakage to the gate comes in many forms but can be divided in to three basic types: (1) Leakage below the electrodes; (2) leakage from gate to source over the entire plate; (3) leakage to the gate in the channel area. Not all leakage current is measured, as Figure 6.10 clearly shows.

The first type, leakage to the gate below the contacts, is always present. It can easily be modeled by placing a shunt resistor R_{dg} bridging the

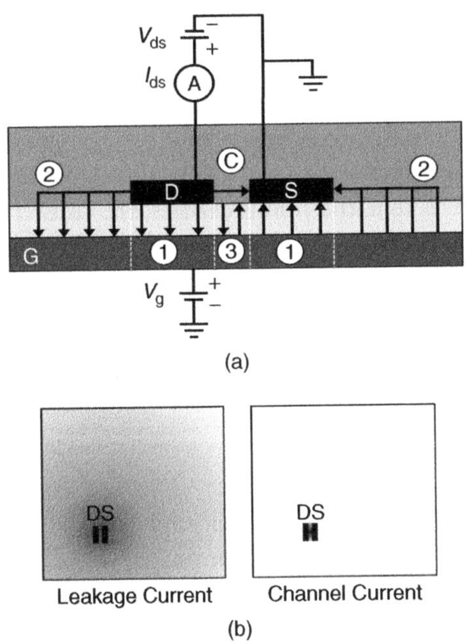

Figure 6.10 (a) Types of insulator leakage, assuming V_g between 0 and V_{ds}: (1) Leakage below the electrodes. (2) Leakage from gate to source over the entire plate. This needs free charge in the active layer and thus only exists when the channel is open. (3) Leakage to the gate in the channel area. (C) Channel current. The arrows pointing down are measured by the amperimeter at the drain, arrows pointing up represent currents not measured. (b) Graphical impression of density of (type 2) leakage current (left) and channel current (right) ($V_{ds} = 0$)

drain and the gate, as shown in Figure 6.11. (This assumes the current meter is placed at the drain.) In principle, a similar resistor has to be placed at the source, but this will not influence the drain current. Figure 6.12 shows simulations of a device with such a resistor at the drain. Note the crossing point of the I–V curves which is a tell-tale sign of leakage. (Experimentally this is for example seen by Torsi [62, 316, 317], Babel and Jenehke [318], Peng [319], or Dodabalapur and coworkers [303] in TFTs based on thiophene, or devices with ultrathin insulators [320].) The crossing voltage and current are given by $V_{ds} = 1/aR_{dg}$ and $I_{ds} = 1/(2aR_{dg}^2)$, respectively ($a = C_{ox}\mu W/L$). It is obvious that this type of leakage is reduced by diminishing the size of the contact pads, or placing them on elevated layers (with a trade-off of increased production cost). This further has the advantage that the device capacitance is reduced.

The second type of leakage is over the entire plate including to neighboring transistors. For this type of leakage, free charge has to be induced in the active layer to make it have finite resistivity. The leakage

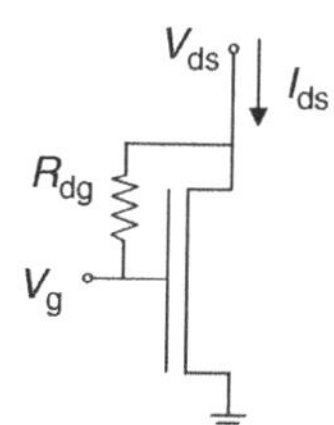

Figure 6.11 Simple modeling of type 1 insulator leakage

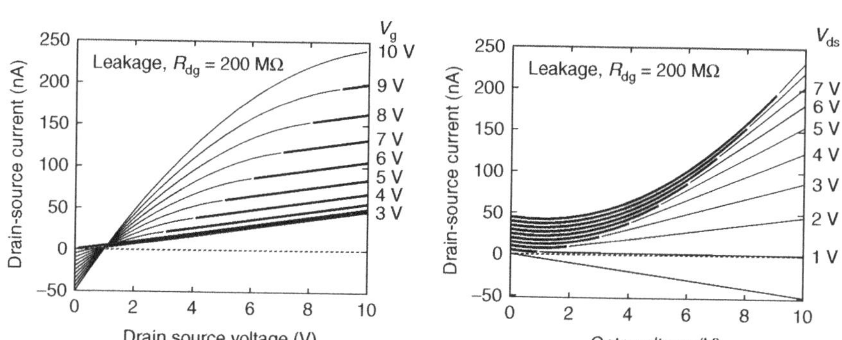

Figure 6.12 Simulation of I–V and transfer curves of an ideal TFT with a resistor bridging the gate and the drain, as shown in Figure 6.11, type 1 of Figure 6.10. The thin lines are the linear regime and the thick lines are the saturation regime

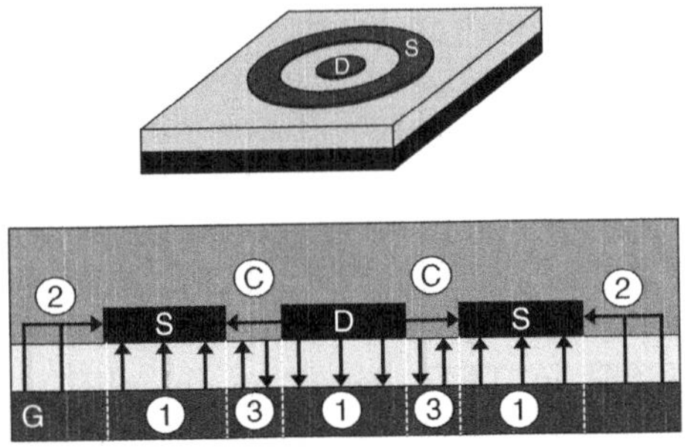

Figure 6.13 Circular FET design in which the source entirely surrounds the drain as a solution to eliminate the measured type 1 leakage current (compare Figure 6.10); the measured insulator-leakage current (arrows pointing down) has been greatly reduced

equivalent resistance thus depends on bias and the current cannot be modeled easily. Sometimes the leakage is caused by material deposited on the edges of the plate thereby shorting the top and bottom layers. To minimize (measuring) this kind of leakage, a guard ring could be placed around the device which is then grounded, or alternatively, the source electrode, whose current is not measured, should be encircling entirely the drain electrode, as shown in Figure 6.13, this eliminates most of the leakage current. A more obvious solution is to not deposit active material outside the channel region, or remove it after depositing [321], again with a trade-off of fabrication complexity.

Another way of eliminating the measured leakage current is to measure with V_{ds} always equal to V_g, in what we call 'locus curves' since they follow the locus points of Figure 6.5. This reduces the power of electrical measurements as an analytical tool, though. Moreover, it reduces the device effectively to a simple nonlinear two-terminal component, like a diode.

The last type of leakage is that one below the channel. This is unavoidable (save increasing the insulator thickness or quality). Figure 6.14

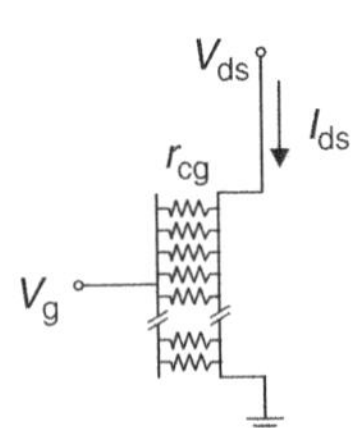

Figure 6.14 Modeling of type 3 leakage from channel to gate

shows how 'channel-leakage' current (type 3 leakage) could be modeled. The result might be that the crossing point of output curves is no longer independent of gate bias, as in type 1 leakage, see for example the p-channel T6 transistors presented by Dodabalapur *et al.* [322].

6.7.2 Contact Resistance

As stated in the introduction of Section 6.7, the term 'contact resistance' is often used in TFT literature to indicate any deviation from the ideal MOS-FET behavior. There exist a multitude of works describing the effects of contacts, for example changing the material used for the electrodes. An effect normally ascribed to the contacts is superlinear growth of the current with bias in the linear region of the I–V curves. Horowitz has spent a substantial amount of effort on designing ways to characterize and extract the contact resistance on the basis of electrical measurements [323]. Another rather straightforward way of finding the (apparent) contact resistance is by measuring the channel resistance for a set of transistors with different channel lengths; extrapolating the channel to zero length then yields the contact resistance [293, 324]. This scheme seems rather dubious, in view of the results that in this way sometimes negative contact resistance is found. What is more, it is intrinsically assumed that the channel length of each device is well known, whereas the literature in the CMOS community of MOS-FETs describes ways of extracting the effective channel length L_{EFF} via, for example, a resistance derivative method [325].

As will be shown later, a basic device, with a homogeneous active layer, and with contact electrodes touching the channel (Figure 6.15a) cannot have any contact resistance, or other substantial contact effects. However, in top contact devices (Figure 6.15b) the source and drain electrodes do not make direct contact with the accumulation channel. Charges injected into the active layer have to travel a substantial distance through a (possibly) low-conductivity material [326]. Similar problems occur for bottom contact devices where the deposition was not optimal. In such cases, for instance when using spin-coating techniques, gaps can exist between electrodes and active layer at the interface (Figure 6.15c). Charges then either have to tunnel through these gaps or find an alternative path through low conductive materials.

However, in many cases, the found values for the contact resistance often depend on bias conditions [302], thereby immediately invalidating the term 'resistance', which is a word reserved for purely linear elements following Ohm's Law. In view of this, before continuing, it is interesting

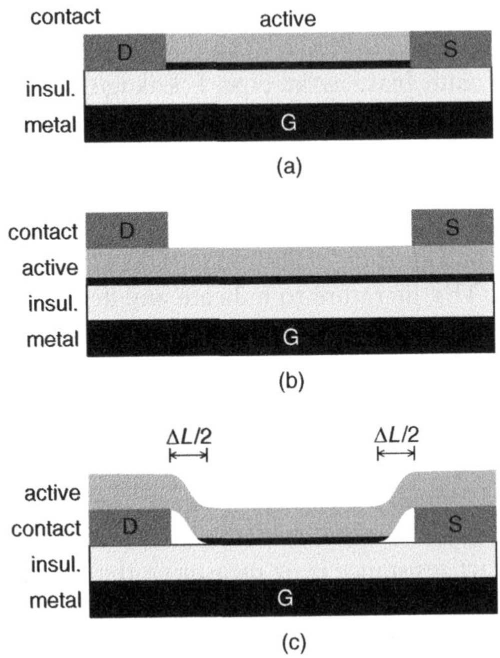

Figure 6.15 A basic (bottom contact) TFT device (a), a top contact device (b), and a device with deposition problems (c). The first has no noticeable contact effects, while the other two can have a substantial contact resistance or contact barriers because the electrodes do not touch the channel (dark layer); a resistive distance has to be traveled by the charges before they reach the channel. Only the top structure is a true TFT; the other two are TFTs in series with other components fabricated on a single plate

to determine what a pure resistor at the contacts would do to the electrical characteristics. Figure 6.16 shows a circuit of a TFT with resistances at the contact. The equations to solve are the following

$$I_{ds} = \mu C_{ox}\frac{W}{L}\left[(V_d - V_s)(V_g - V_s) - \frac{(V_d - V_s)^2}{2}\right] \tag{6.57}$$

$$V_d = V_{ds} - I_{ds}R \tag{6.58}$$

$$V_s = I_{ds}R. \tag{6.59}$$

Figure 6.17 shows the I–V curves of such a system for various values of the contact resistance. The effect of the resistance is that for increased current I_{ds} it will lower the voltage drop across the insulator $V_g - V_s$ and across the channel $V_d - V_s$, thereby closing the channel and reducing

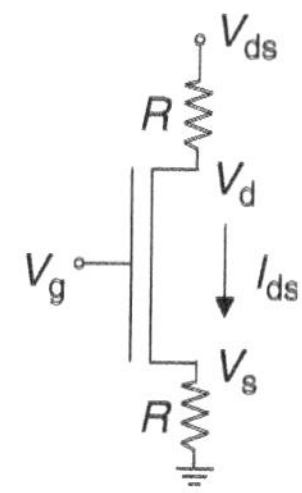

Figure 6.16 A TFT with contact resistance

the current. This kind of negative feedback tends to stabilize the current and make it independent of bias. This is visible in the figures. For high contact resistance, the currents become independent of gate bias, the I–V curves for different gate biases 'pinch' together. Yet, the curves remain (sub)linear [327]. The same effect can be seen in the transfer curves of Figure 6.18. For large contact resistance, the current becomes independent of gate bias. This is indeed what one would expect. For large contact resistance values, the channel resistance becomes negligible and the current becomes equal to $V_{ds}/2R$, independent of V_g.

In conclusion, contact resistance causes 'current crowding' or 'curve pinching' in I–V curves and saturation in transfer curves [327], sometimes observed in the literature [328]. No superlinear curves are observed in I–V or transfer curves. For this we have to look for other sources, as will be discussed later in Section 6.10.2.

6.7.3 Contact Barriers

Because a classical MOS-FET is based on pn-junctions of p- and n-doped same-material, and in view of the success of the MOS-FET model to explain the basic behavior of a TFT, it is logical to continue the reasoning and apply the barrier ideas to TFTs. In most TFTs, the electrodes are made of metals, and it seems logical to expect Schottky barriers at the source and drain contacts. It is even more so, since Schottky-barrier MOS-FETs are a well studied subject [329] (see also the thesis of Vega [330] for a good summary on Schottky-barrier FETs). Yet, a TFT is not simply a thin MOS-FET. It does not make sense to place barriers at the contacts for reasons that will be discussed in the next section where an adequate description of the metallic contacts will be given. Still, it is interesting to consider what would happen if we were to place barriers at the contacts.

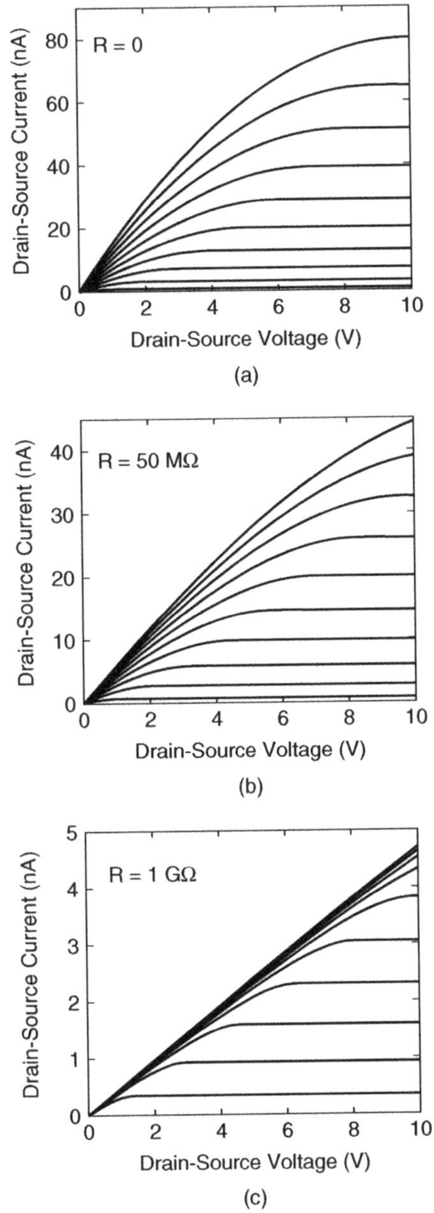

Figure 6.17 I–V curves of a TFT with various values for the contact resistance, as shown in Figure 6.16: (a) 0; (b) 50 MΩ; (c) 1 GΩ. Reproduced with permission from Synthetic Metals, Modeling electrical characteristics of thin-film field-effect transistors: I. Trap-free materials by P. Stallinga and H. L. Gomes, 156, 21–24 Copyright (2006) Elsevier

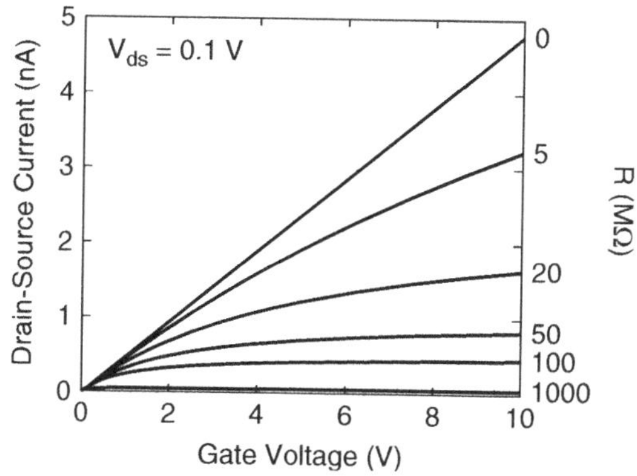

Figure 6.18 Transfer curves for various values of contact resistance (shown on the right). Reproduced with permission from Synthetic Metals, Modeling electrical characteristics of thin-film field-effect transistors: I. Trap-free materials by P. Stallinga and H. L. Gomes, 156, 21–24 Copyright (2006) Elsevier

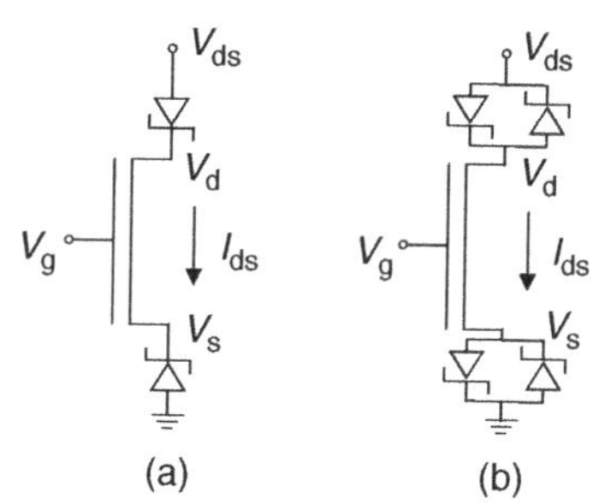

Figure 6.19 Schottky diodes at the contacts. (a) Single Schottky pair. (b) Double Schottky pair

The first observation that has to be made is that the Schottky diodes come in pairs. If a forward-biased Schottky diode is present at one electrode (source or drain), then a reverse-biased Schottky diode is present at the counter-electrode. Figure 6.19(a) shows the situation of Schottky diodes at the contacts. In the literature often double-Schottky-diode contacts are used, as shown in Figure 6.19(b) [331]. Although this can model the irregularities observed by these authors, it lacks physical meaning. If we are to search for the elimination of the observed effects, a good understanding of the physical origin of the effects is needed first. Before continuing, it is interesting to see the effects of diodes placed at the electrodes.

The current through a diode is equal to [9]

$$I(V) = I_0\left[\exp\left(\frac{qV}{nkT}\right) - 1\right] \tag{6.60}$$

where n is the ideality factor (equal to 1 for ideal diode and normally between 1 and 2) [9]. If the diodes are the limiting factor, the two diodes back-to-back can pass a maximum current equal to the reverse-bias diode current I_0. The total current–voltage relation is

$$I_{\text{ds}} = I_0 \tanh\left(\frac{qV_{\text{ds}}}{nkT}\right) \tag{6.61}$$

which is a linear function close to the origin. Figure 6.20 shows simulations of such systems with single Schottky diodes at both barriers. For comparison a standard ideal TFT curve is also shown. For an ideality factor equal to 1, the currents saturate at a bias approximately kT/q, or 26 mV (at room temperature). Larger ideality factors can stretch this voltage somewhat but the I–V curves remain linear for small voltages.

Sometimes two antiparallel diodes at both source and drain electrodes are used to explain the TFT data [331]. It can easily be shown that a set of antiparallel diodes have an I–V relation given by

$$I = 2I_0 \sinh\left(\frac{qV}{nkT}\right). \tag{6.62}$$

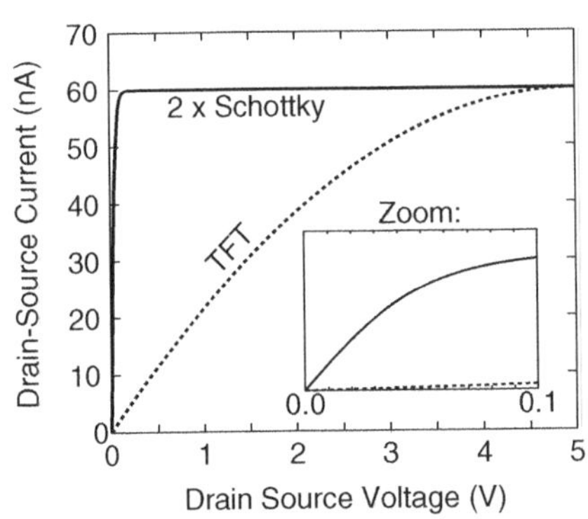

Figure 6.20 Simulation of I–V curves for a circuit as shown in Figure 6.19(a), based on mirrored Schottky diodes at the contacts. The drain-source voltage is the external bias V_{ds} and not the internal voltage drop $V_{\text{d}} - V_{\text{s}}$. Reproduced with permission from Synthetic Metals, Modeling electrical characteristics of thin-film field-effect transistors: I. Trap-free materials by P. Stallinga and H. L. Gomes, 156, 21–24 Copyright (2006) Elsevier

When these are placed at both source and drain contacts of a TFT, they have some interesting effects on the I–V curves, as evidenced by Figure 6.21. When the parameter I_0, the reverse-bias saturation current of an individual diode, is reduced, the I–V curves become nonlinear in the origin. This effect is amplified when the product of ideality factor and temperature nT is increased. Although these curves resemble some of the I–V characteristics reported in the literature, there is no physical basis for the circuit and it can thus not be an explanation for the observed nonidealities. When an elimination of the effects is sought we have to

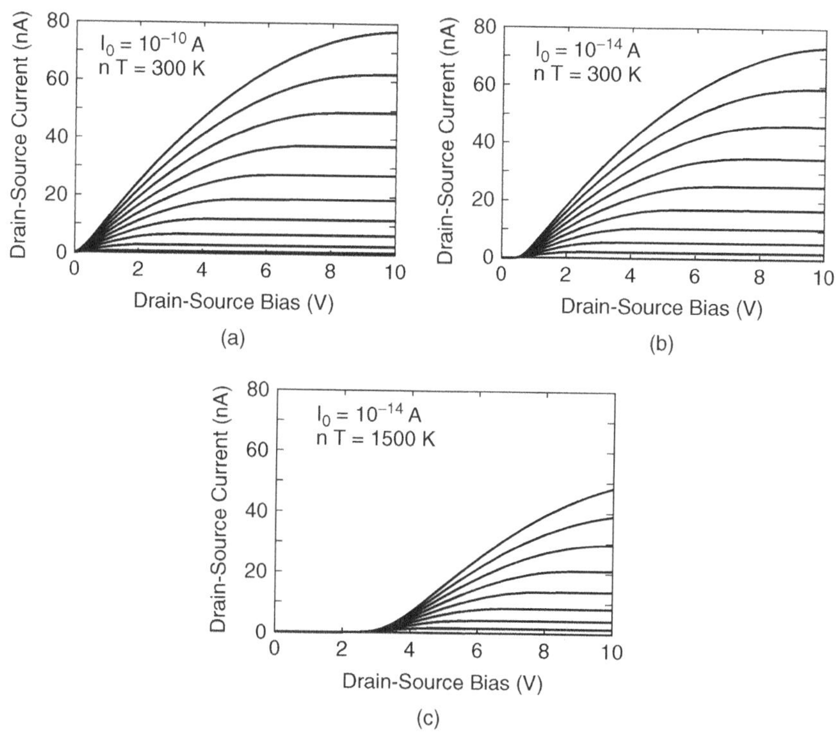

Figure 6.21 Simulation of I–V curves for a circuit as shown in Figure 6.19(b), namely based on antiparallel Schottky diodes at both electrodes. (The drain-source voltage is the external bias V_{ds} and not the internal voltage drop $V_d - V_s$.) (a) I–V curves for large I_0. (b, c) I–V curves for small I_0 causing the diodes to become the limiting factor. For increased ideality factor or temperature the curves become highly nonlinear. While being able to simulate some interesting characteristics, the physical meaning of the antiparallel diodes remains to be explained. Reproduced with permission from Synthetic Metals, Modeling electrical characteristics of thin-film field-effect transistors: I. Trap-free materials by P. Stallinga and H. L. Gomes, 156, 21–24 Copyright (2006) Elsevier

look elsewhere for the explanation. If you do not identify the problem correctly you cannot hope to eliminate it. Later we will show some more examples of physically correct explanations for nonidealities.

6.7.4 Grain Boundaries

The barriers might also be embedded in the material itself, rather than at the contacts. In polycrystalline materials, at grain boundaries the discontinuities may locally cause a change in the electronic structure, i.e., traps, resulting in band bendings and barriers [6, 332–334]. Levinson has calculated for such systems a linear dependence of current on V_{ds}, but a nonlinear dependence on V_g. For hole channels (negative V_g) [335]:

$$I_{ds} = -\mu C_{ox} \frac{W}{L} V_{ds} V_g \exp\left(a/V_g\right) \tag{6.63}$$

where a is a temperature- and trap-density-dependent parameter. A Levinson plot of $\ln(I_{ds}/V_g)$ vs. $1/V_g$ is a straight line and this can be used to test the model. Street *et al.* have applied this theory to organics but found it to be inadequate for their pentacene TFTs [336]. Horowitz developed a model that uses tunneling rather than thermionic emission at the grain boundaries and found a gate-dependent mobility [332]. In numerical simulations, Bolognesi and coworkers likewise found gate-bias-dependent mobilities depending on the grain size [334]. Though not all grain boundaries necessarily cause nonlinear effects, as demonstrated by the measurements of Wang *et al.* on pentacene-island TFTs [306]. A junction between the same materials (same Fermi levels) will not by itself cause any barriers. A substantial amount of different material (metal/trap-rich material with different Fermi level) has to be sandwiched in between, enough material to store the charge needed for the band bending and the barrier effects. Depending on the total thickness of the contact region, the barrier will act as a back-to-back Schottky barrier (very thick) or as a single tunnel barrier (very thin). Never will repulsive Coulomb forces of the accumulated charge play a role; it was the attractive force of the region that put the surplus charge there in the first place. This will continue until the total force is zero.

Since most organic materials are amorphous and this book is primarily about this type of material, the grain boundaries are only mentioned here as a brief aside. Moreover, none of the interface layers (where the channel resides) are expected to be of any form of crystallinity, even when the morphology of the film, when viewed from the top by atomic force microscopy (AFM) or the like, indicates that the film is (poly)crystalline. At the interface any material will be amorphous because of the lattice mismatch between the materials of insulator and semiconductor, or the underlying insulator itself is amorphous, and this will inhibit crystal growth. Only after some layers can the growth become crystalline. As will be shown later, traps alone can adequately explain nonlinearities in output and transfer curves.

6.7.5 Parallel Conductance

As discussed in Section 6.4, a TFT can be as thin as one monolayer. In reality, it is difficult to make devices as thin as that with techniques other than vacuum deposition. Spin coating, or – even worse – drop-casting will produce layers easily some hundreds of nanometers thick. This causes no problem if the material itself is undoped and has a low free charge density. However, if the material is conductive, the device will suffer seriously (see for instance the work of Sehil *et al.* on poly-Si TFTs [337]). Since basically only the charge in the first monolayer is modulatable, the other layers cause parallel conductance and will reduce the on/off ratio of the device dramatically. This is best seen in transfer curves, which will basically look as shown in Figure 6.22. The on/off ratio is determined by the mobility of the charges which is proportional to the slope of the transfer curves [Equation (6.36)]. The conductance through the top layers is determined by the product of mobility, charge density p_0 and layer thickness d. Obviously, minimizing the layer thickness (reducing d) or purifying the material (reducing p_0) will increase the device performance. Normally-on devices, such as the one given here and others, will be discussed in Section 6.9.

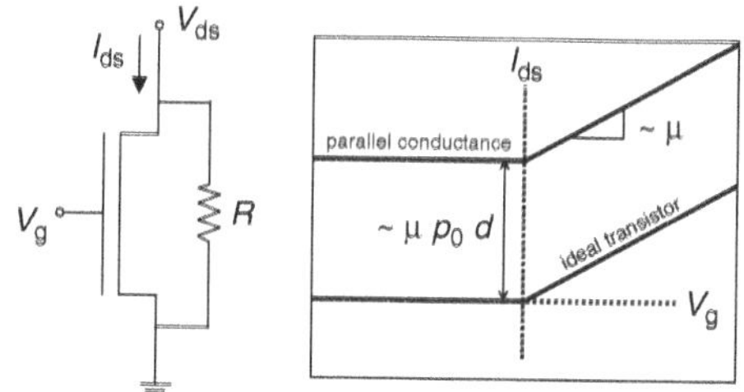

Figure 6.22 Schematic illustration of the effect of thick film on transfer curves. The parallel conductance is determined by the product of charge mobility μ, bulk charge density p_0 and layer thickness d

6.8 METALLIC CONTACTS IN TFTs

In contrast to MOS-FETs, where the drain and source electrodes are made of the same material (different doping type) semiconductor, most contacts in TFTs are made of metal. In fact, based on the experience with OLEDs, the metal of choice is normally gold, since gold is proven to be a good hole-injector for most hole-conducting organic materials. Other materials form a so-called Schottky barrier when used in a diode geometry. It is therefore common sense to imagine Schottky barriers are also generally formed when metals are used for the contacts in TFTs. Such ideas are normally used in the literature to account for non-idealities in the electrical characteristics of TFTs. However, as shown in the previous section, Schottky barriers at the contacts are not capable of explaining the nonidealities such as superlinear I–V curves. This, in fact, was to be expected, since in the prototype of an FET with barriers, namely the MOS-FET (containing pn-junctions), no nonlinearities are observed. Once a current flows, the barriers have disappeared: In a MOS-FET of type p-n-p (drain-channel-source), by the time the hole density equals the hole density in the drain and source, by definition the bands have aligned and no barrier exists. Barriers exist when the transistor is switched off by the gate. Since then no charge exists in the channel and no current flows, naming them 'barriers' is somewhat misplaced. Substituting the pn-junctions by Schottky barriers only slightly changes this viewpoint, only a tiny residual barrier of limited size can result when the channel contains charge, as will be demonstrated here.

Moreover, there is another, fundamental reason why Schottky barriers are not an adequate description of a metal–semiconductor interface in TFTs and this entails the symmetry argument. The symmetry of a Schottky barrier is not the same as the symmetry of the metal contacts in the TFT geometry, see Figure 6.23. Because of this, they have to be treated differently.

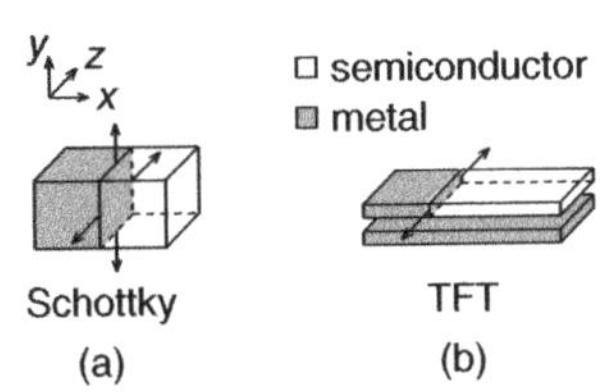

Figure 6.23 Comparison of a Schottky barrier (a) and metal contacts in TFTs (b). Symmetry invariance directions are indicated by arrows. The different symmetry and geometry warrants a different description

We start with the basic Maxwell equation,

$$\nabla \cdot \mathbf{D}(x, y, z) = \rho(x, y, z) \tag{6.64}$$

where $\mathbf{D} = \varepsilon\mathbf{E}$ is the displacement vector and ρ is the charge density scalar, both functions of space coordinates x, y and z. For a Schottky diode, the device has invariance to any translation, rotation and mirroring that leaves the plane of the interface invariant, see Figure 6.23. The interface is imagined to stretch to infinity in these two directions (y, z) for the sake of the calculation. The true interface area then only enters as a scaling parameter in the final equation. Therefore, **D** is a vector pointing in the x-direction whose magnitude only depends on x, and ρ is a function of x only. In other words,

$$\frac{\mathrm{d}D(x)}{\mathrm{d}x} = \rho(x). \tag{6.65}$$

This is a form of Poisson's Equation which results in a good description of a Schottky barrier with its band bendings and electric fields, as discussed earlier.

For a FET, the only symmetry is along z, along the length of the electrodes because of the symmetry-breaking presence of the gate and the thinness of the device. Maxwell's Equation therefore only reduces to

$$\frac{\partial D_{\mathrm{x}}(x, y)}{\partial x} + \frac{\partial D_{\mathrm{y}}(x, y)}{\partial y} = \rho(x, y). \tag{6.66}$$

Thus, the use of Poisson's Equation and the Schottky-barrier approach is not permitted for FETs (MOS-FET and TFT alike). Contrasting a Schottky diode, the distribution of charge is *highly* dependent on y. In fact, as we have argued, in a TFT or MOS-FET in accumulation the charge ρ resides in only a very thin layer, as in a capacitor. This is caused by the near presence of the gate (field). (Temporarily) ignoring this gate in the analysis of the contacts will lead to wrong conclusions. There is only one situation where Schottky barriers can exist in a TFT-like device, namely in a situation of top electrodes on a thick active layer (see Figure 6.15b). In this case, the gate is far away enough from the contacts to justify a Schottky-diode analysis. Such a device is not a pure TFT but rather a TFT with Schottky barriers in series and has been fabricated on purpose to improve performance [338]. The behavior of such a device has been simulated in the previous section and does not show nonlinearities.

In another study of top-electrode TFTs the distant contacts resulted in Fowler–Nordheim tunnel diodes reducing the conduction and apparent mobility [339], anticipated in MOS-FETs by Tucker *et al.* [340], where they expect these diodes to be programmable by the gate. In a real TFT, with the contacts in direct contact with the channel, no barrier can exist.

Even if we were to continue using the Schottky barrier approach for the analysis of a pure TFT we have to do it with care. The extent of the influence of a normal Schottky barrier is determined by the band bending in the depletion zone, which in turn is determined by the acceptor or donor concentration, and is given by $W = \sqrt{2\varepsilon_s V_{bi}/qN_D}$. For normal doping levels this is in the order of some hundreds of nanometers. For a TFT or a transistor in accumulation there are no donors and the only charges that can be positive are the free holes themselves. Substituting N_D for N_V will give a 'depletion width' in the order of nanometers. In this range it is unlikely that thermionic emission is the limiting process for transport across the barrier. Tunnel currents dominate at this distance and are not likely to be the limiting factor in the current through the transistor. Experimentally it has been shown that a TFT channel length can be scaled down to the 10-nm range, without short-channel effects [341]. Short-channel effects are expected when the depletion widths at the drain and source start overlapping [9]. The absence of short-channel effects in the sub-10-nm TFTs thus means that barrier regions in TFTs are indeed below this size.

For these reasons, a better approach is to continue treating the transistor as a metal plates capacitor and to analyze what the effect will be of the contacts. In other words, maintaining the simple charge–voltage relation $\rho(x) = C_{ox}[V(x) - V_g]$ everywhere along the device. Although it is also an approximation, since it ignores any band bendings possibly existing at the contacts, this approximation is a better one than the Schottky barrier approximation for reasons given above. Since the densities of electronic levels playing part in the description are very high, close to metallic, it is best to compare it with metal–metal connections. This approach has been presented by us in a recent publication [342] and is summarized here.

What will happen when two materials with different Fermi levels are brought into intimate contact is that charge will flow from one material to the other. As an example, Figure 6.24 shows a schematic energy diagram of a metal–semiconductor–metal system, with, before contact, the Fermi level for electrons in the metal E_{Fm} lower than the Fermi level in the semiconductor E_{Fs}. Because of this difference ΔE_F, the system can

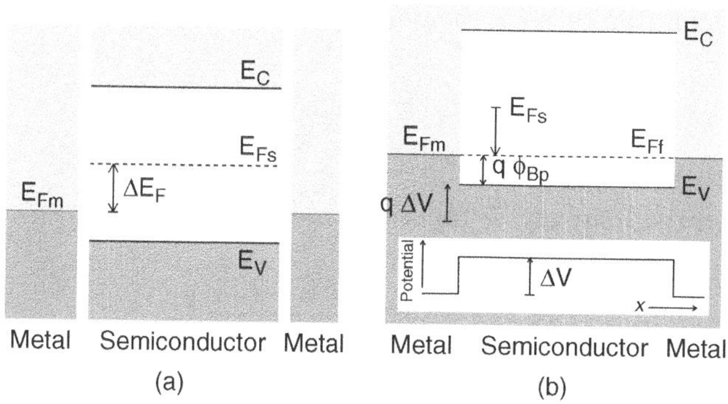

Figure 6.24 Metal–semiconductor–metal system in a TFT (a) before and (b) after contact. Reproduced with permission from Organic Electronics, Metal contacts in thin-film transistors by P. Stallinga and H. L. Gomes, 8, 4, Copyright (2007) Elsevier

gain energy by transferring electrons from semiconductor to metal (or holes from metal to semiconductor).

The conventional buildup of space charge by uncompensated ionized acceptors or donors via Poisson's Equation as in Schottky barriers does not take place due to the nonexistence of such impurities ionizable with the correct sign. Instead, the charge flowing into the channel is solely comprised of free holes. As a first effect, the holes cause a change of the Fermi level from E_{Fs} to E_{Ff} according to the Boltzmann relation

$$p = N_V \exp\left(\frac{E_V - E_{Ff}}{kT}\right) \tag{6.67}$$

where N_V is the effective (two-dimensional) density of valence band states, E_V is the energy of an electron in the valence band, and E_{Ff} is the final position of the Fermi level. As a second effect, the new free holes are adding a potential to the channel (and to all energy levels in the channel) via Equation (6.30)

$$\Delta V = qp/C_{ox}. \tag{6.68}$$

The free-hole density and thus the potential are constant over the entire length of the channel, because any gradient in either will cause a redistribution of the charges until the gradients disappear. At steady state, the only solution is a homogeneous distribution of holes and a constant voltage in the channel. The two effects, changing the Fermi level and increasing the in-channel potential, continue until the Fermi levels have lined up, $E_{Ff} = E_{Fm}$. Combining the above two equations, with the definition of ΔE_F, the potential in the channel will be given by

$$\Delta V = \frac{kT}{q} W\left(\frac{q}{kT} A\right) \tag{6.69}$$

where W is the Lambert-W function, and A is given by

$$A = \frac{qN_V}{C_{ox}} \exp\left(\frac{E_V - E_{Fm}}{kT}\right). \tag{6.70}$$

At the same time, a residual barrier $q\phi_{Bp}$, equal to the difference in Fermi level and valence band in the semiconductor, exists at the contacts. Combination of Equations (6.30) and (6.67) yields

$$q\phi_{Bp} = -kT \ln\left[\frac{C_{ox}(\Delta V - V_g)}{qN_V}\right]. \tag{6.71}$$

This barrier height thus depends on the bias V_g. When the bias is changed, the density of free holes changes and the Fermi level depth changes and the barrier height with it.

The easiest way to measure the height of the barrier is via the temperature dependence of the drain-source current. However, this necessitates the existence of a channel and thus free holes. Free holes imply that the Fermi level in the active layer is close to the band edge. The Fermi level depth being equal to the barrier height (see Figure 6.24b) then tells us that the measured barrier height is always rather small, in the order of tens of millielectronvolt. We can thus give an estimate for the measured barrier height by substituting typical values for the parameters: $N_V = 1.06 \times 10^{16}\ \mathrm{m^{-2}}$, $C_{ox} = 170\ \mu\mathrm{F\ m^{-2}}$, $V_g = -1$ V and $T = 300$ K yields about 60 meV for the barrier height. For the same reason, the barrier height, when measured by current, is independent of the type of metal used for the contact, since it depends only on the final Fermi level depth; at most, a slightly different threshold voltage results.

An observation can also be made for the in-channel potential ΔV. It can take values ranging from zero up to the volts range. For intrinsic materials, even a tiny amount of charge in the channel can shift the Fermi level substantially and it is thus possible that the effect is nearly completely absorbed by a shift of the Fermi level, $E_{Fs} - E_{Ff} \approx \Delta E_F$, and $\Delta V \approx 0$. However, in a system where the Fermi level is pinned, for instance in doped semiconductors, the effect is mainly a voltage drop at the contact, $\Delta V \approx \Delta E_F/q$ and this can easily be in the volts range.

Summarizing, the metal contacts of a TFT working in accumulation cause two effects: (i) The building up of an in-channel potential that can be in the volts range; and (ii) the formation of a residual barrier, with a height that depends on the gate bias and, when measured, is of the order of some tens of millielectronvolts.

Interestingly, both these effects have been reported in the literature. Yagi *et al.* reported a gate-bias-dependent barrier with, for a certain bias, an activation energy of 80 meV [290], which is very close and easily within the range of the prediction above. Note that the devices of Yagi have top electrodes, a slightly different geometry compared with the bottom-electrode devices presumed here. Apparently, the difference between the two geometries is not important enough and the model still applicable. Similarly, Bürgi and coworkers, in bottom-contact devices, find an activation energy of 60–140 meV for the contact barrier [308], which they themselves find surprisingly small compared with the conventional Schottky-barrier analysis value.

Concerning the in-channel potential predicted above, relevant is the voltage profiling experiment of Bürgi *et al.* [343]. They measured a potential in the channel, in the absence of bias, with an order of magnitude as described in the current work. Others report similar Kelvin probe experiments, but without presenting the most relevant zero-bias results.

Some observations have to be made about the limitations of the model. First of all, in Equation (6.67), the Boltzmann approximation to the full Fermi–Dirac distribution has been used. This simplification is done to make the idea more understandable and make it analytically solvable. For Fermi levels close to – or resonant with – the valence band, this approximation is no longer valid. Note however that for such systems, the barrier height goes to zero and the effects of the contacts disappear completely.

Even a negative barrier height is predicted for large biases. The model thus loses applicability for such conditions. What happens for large biases is that all available states N_V are occupied by holes and saturation is thus expected for the transfer curves. As long as such saturation is not observed, the model is apparently applicable. In practice what will happen is that when the states in the first monolayer are depleted, the second layer (if present) starts getting filled. This layer is slightly more distant from the gate electrode and feels a slightly different capacitance and a different charge–voltage relation exists, compared with Equation (6.30). To a good approximation, when the insulator is much thicker than a monolayer, it can be imagined that the DOS is

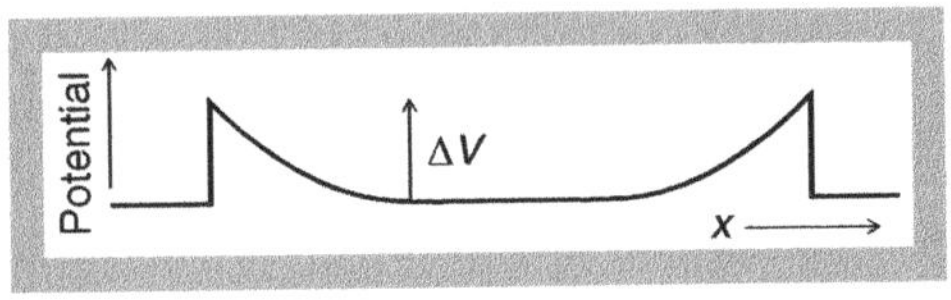

Figure 6.25 Effect of a less than perfect insulator. When the channel is still highly resistive, the channel potential is determined more by the gate then by the contacts and the band bends as shown (for zero bias)

simply multiplied by a factor 2 and then 3, etc. The simplified model presented only describes nondegenerate cases for low biases. Moreover, in practice it is expected that burning through the oxide occurs long before depletion of the valence band sets in.

Other nonidealities are expected (no real system is ideal). For instance, when the insulator has finite resistance. In this case, far away from the contacts – especially when there is still little charge in the channel – the potential is more defined by the gate then by the source and drain contacts. This is just a matter of resistance paths. What is then expected is that, instead of the flat bands in the semiconductor as shown in Figure 6.24, the bands will bend in such a way as to make the potential in the channel equal to the gate, see for example Figure 6.25.

6.9 NORMALLY-ON TFTs

In the basic model some assumptions were made to come up with the basic workings of a TFT. In reality, some of these assumptions can and should be reanalyzed. In this section, features will be introduced into the model that will result in so-called normally-on TFTs, in other words, TFTs that already conduct at zero bias. It is a summary of our work published in 2008 [344].

The reason for normally-on TFTs is easy, namely there are already free charges in the device at zero bias. In the basic model, *all* charge was assumed to be bias induced. However, as shown in Chapter 1, there can be more sources of free charge.

- Thermal excitation. In the case of narrow band gap materials, thermal energy is enough to excite carriers across the gap and have free charge without bias.
- Doping. Like in any semiconductor material, doping causes free charges in TFTs not associated with fields.

- Metals. As will be shown, a TFT can be made from a metal. These materials have free carriers without bias or doping even at low temperatures. However, this charge can be modulated by the bias and thus a transistor can result.

To include these effects and give a description of these devices, the following components have to be used:

1. The active layer is neutral at zero bias. This implies that there is no charge transfer between the active layer and the gate or insulator layer. In other words, they all have the same electrochemical potential.
2. All charge is free charge, free holes p and free electrons n, implying that the material is intrinsic without traps or other ways of storing immobile charge; all induced charge contributes to current. For doped semiconductors this limitation will be relaxed and charge can also be stored on the dopants by ionizing them.
3. The total charge, the sum of negative charge and positive charge follows the basic charge equation, Equation (6.30), namely

$$p(x) - n(x) = \left[V(x) - V_g\right] C_{ox}/q. \tag{6.72}$$

   For doped semiconductors an extra term is added to the left-hand side of this equation to include ionized impurities, while for metals a term is added to the right-hand side representing the electron sea, as will be discussed later.
4. The relative densities of electrons and holes follow a Fermi–Dirac distribution defining a Fermi level. For all normal currents (small drain-source biases), this Fermi level is equal for electrons and holes (there is thermal equilibrium at all places and no electron–hole recombination currents exist). The product of electron density and hole density thus depends on the temperature and the band gap of the material [9]

$$pn = n_i^2 = N_V N_C \exp(-E_g/kT) \tag{6.73}$$

   where n_i is the intrinsic electron density, N_V is the effective density of states of the valence band and N_C is the conduction band states. E_g is the electronic band gap. Note that all units of density are 'per area'.

5. The current can have contributions from both electrons and holes, Equation (6.32) is replaced by

$$I_x(x) = -qW\left[p(x)\mu_p + n(x)\mu_n\right]\frac{dV(x)}{dx}. \qquad (6.74)$$

The difference between these narrow gap materials and the wide band gap devices described earlier is that in these materials electrons and holes can be present in substantial amounts *at the same place and time*, because Equation (6.73) allows for that; n_i is large where for wide band gap materials it is negligible. Even if the wide band gap materials are ambipolar (with both electron and hole mobility large), they can only have electrons and holes and their associated currents at separate regions of the device. In narrow gap materials, however, electrons and holes can coexist at the same time and place in the channel.

In the following subsections, transfer curves ($I_{ds}-V_g$) and output curves ($I_{ds}-V_{ds}$) are discussed. The transfer curves are based on the calculated zero-V_{ds} conductance, which is defined as

$$G \equiv \left.\frac{\partial I_{ds}}{\partial V_{ds}}\right|_{V_{ds}=0} = q\mu_p pW/L. \qquad (6.75)$$

For the figures, this conductance is multiplied by a small potential, for example $V_{ds} = -0.1$ V to result in transfer curves ($I_{ds}-V_g$). For the I–V curves ($I_{ds}-V_{ds}$), the differential equation, defined by Equations (6.72), (6.73) and (6.74), is solved. This is done numerically. See Table 6.3 for the parameters used.

Table 6.3 Simulation parameters used in the section on normally-on TFTs (unless otherwise specified)

Parameter	Value	Unit
N_C	2.8×10^{16}	m^{-2}
N_V	1.04×10^{16}	m^{-2}
C_{ox}	160	μF m^−2
V_{ds}	−0.1	V
W	1	cm
L	10	μm
μ_p	3	$cm^2\ V^{-1}s^{-1}$
μ_n	1	$cm^2\ V^{-1}s^{-1}$
E_g	0.1	eV
E_A	0.2	eV
T	300	K
n_i	2.5×10^{15}	m^{-2}

6.9.1 Narrow Gap Semiconductors

Narrow gap materials, in the context of the current work, are defined as materials in which both electrons and holes exist in thermal equilibrium. Silicon, for instance, is here considered a wide band gap material since the room temperature density of free charge in intrinsic silicon is much less than the charge induced by typical operating voltages. Unipolar devices are made of semiconductors where the mobility of one type of carrier is much higher than that of the other type. For most organic semiconductors, it is normally found that the hole mobility is much higher than the electron mobility. As we have shown, low mobility in organic materials is due to a high density of traps; the intrinsic mobility of pure crystalline materials is expected to be as high as their technologically more advanced inorganic counterparts [345]. In this section we will give an example of a narrow gap material with high hole mobility ($\mu_p = 3\,\mathrm{cm^2V^{-1}s^{-1}}$) and zero electron mobility ($\mu_n = 0$).

The first direct obvious result of the description in the previous section is that there is free charge (both electrons and holes) and thus nonzero conductivity of the channel even at zero bias. As an example, for a device with parameters as in Table 6.1, with a bad gap of 0.1 eV, the room temperature zero-bias density of electrons and holes according to Equation (6.73) is $p = n = 2.5 \times 10^{15}\ \mathrm{m^{-2}}$ uniform over the entire channel. The conductance in the linear region (tiny V_{ds}), defined in Equation (6.75) is 120 μS, resulting in a current of approximately −12 μA for a bias of $V_{ds} = -0.1$ V. In other words, the device is of the so-called 'normally-on' type because a current can exist even in the absence of a (gate) bias.

The holes can be driven out of the channel by a positive gate bias. However, holes cannot be completely removed from the channel, since this would need an infinite density of electrons [Equation (6.73)] and thus an infinite gate bias [Equation (6.72)]. For a normally-on intrinsic device, the hole density along the channel can thus be calculated from Equations (6.72) and (6.73):

$$p(x) = \frac{[V(x) - V_g]C_{ox}}{2q} + \sqrt{n_i^2 + \left\{\frac{[V(x) - V_g]C_{ox}}{2q}\right\}^2}. \tag{6.76}$$

It is easily verified that for small n_i this reproduces the basic model charge equation [Equation (6.31)]. In the linear regime [$V(x) \approx 0$ and a distribution of p uniform in the channel], the bias interval at which p

and thus the current goes from two times to half the zero-bias value is

$$V_g = \pm\frac{3qn_i}{C_{ox}} = \pm\frac{3q}{C_{ox}}\sqrt{N_C N_V}\exp\left(-\frac{E_g}{2kT}\right). \tag{6.77}$$

For the device given here, this is in the order of ± 1.1 V. Figure 6.26(a) shows a simulation of the transfer curves for such a unipolar normally-on

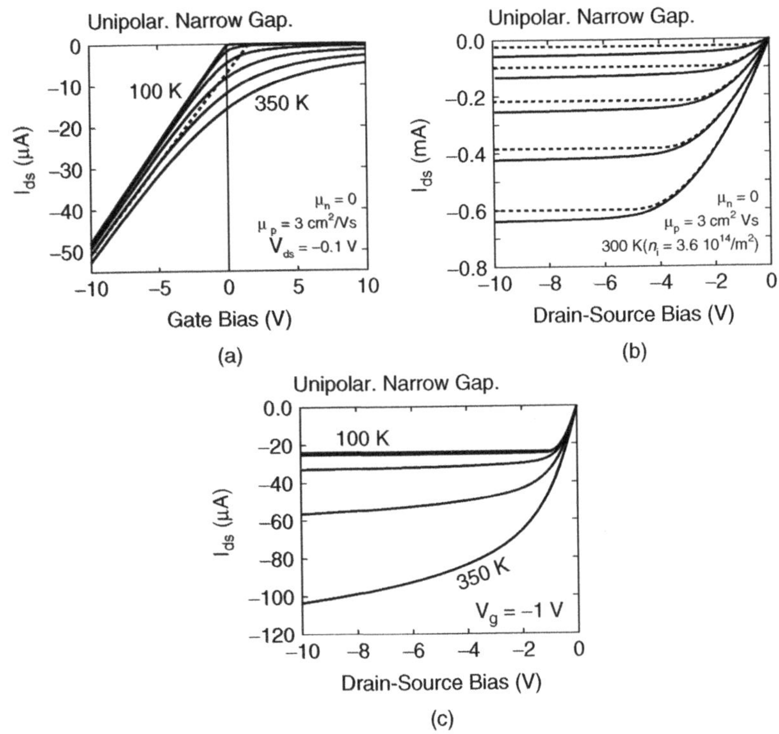

Figure 6.26 Simulation of a narrow band gap unipolar device with parameters as in Table 6.3 ($\mu_n = 0$): (a) Transfer curves ($I_{ds}-V_g$) for temperatures ranging from 100 to 350 K in steps of 50 K. The dashed line shows an attempt-to-fit a classical MOS-FET model to the behavior at $T = 300$ K, resulting in $\mu_{FET} = 2.7\ \text{cm}^2\ \text{V}^{-1}\text{s}^{-1}$ and $V_T = +1.64$ V. (b) Output curves ($I_{ds}-V_{ds}$) at 300 K for gate biases from -1 to -5 V in steps of 1 V. The dashed lines show the behavior for a wide band gap material ($n_i = 0$). (c) Output curves ($I_{ds}-V_{ds}$) at $V_g = -1$ V as a function of temperature (100 K to 350 K in steps of 50 K). For the lowest temperatures, the device behaves like a wide band gap device with a pinch-off voltage equal to V_g. For higher temperatures, the device stops showing saturation. Reproduced with permission from Synthetic Metals, Modeling electrical characteristics of thin-film field-effect transistors: III. Normally-on devices by P. Stallinga and H. L. Gomes, 158, 11 Copyright (2008) Elsevier

device. Interesting to observe is that were the device analyzed with a classical MOS-FET model in the bias range shown, a threshold voltage of approximately 1.6 V would have been estimated at room temperature (dashed line). Yet, this estimation depends on the selected bias range and tends to zero for large gate voltages, as Equation (6.76) easily demonstrates (extrapolate to $p = 0$ with n_i small compared with the other terms). The threshold voltage is not a device parameter for pure materials and only makes sense when traps are present [345].

Figure 6.26(b) shows simulations of the output curves at room temperature. They are based on a substitution of Equation (6.76) into Equation (6.32) and numerically solving this differential equation. Note the absence of saturation in the current. For the same reason that the channel is open at $V_g = 0$, it is not possible to completely close the channel ('pinch off'); this would need an infinite density of electrons [$p = 0$ implies $n = \infty$, according to Equation (6.73)]. For comparison, the dashed lines show the equivalent wide band gap device ($n_i = 0$). Figure 6.26(c) shows the temperature dependence of a typical output curve at various temperatures. For the lowest temperatures, the device behaves like a standard TFT with pinch-off at $V_{ds} = V_g$, while for higher temperatures, no saturation takes place.

In ambipolar materials, mobilities of electrons and holes are comparable. Such materials have already been described at the beginning of this chapter (Section 6.6), but they were limited to wide band gap materials. In this section we describe the narrow gap ambipolar materials. The solution can be found by combining Equations (6.72), (6.73) and (6.74), now with a nonzero mobility for both electrons and holes. The complexity of the differential equation prohibits its analytical solution, but a numerical solution is readily found (as for all of the systems presented here). Figure 6.27(a) shows transfer curves at various temperatures.

The channel in narrow band gap ambipolar devices cannot be completely closed, since driving out holes with the field means accumulating electrons and increases the conduction. However, the transfer curves have a minimum in the linear region. It can easily be shown that the minimum in conductance [Equation (6.75)] occurs for a bias

$$V_{g,min} = \frac{qn_i(T)}{C_{ox}}\left(\sqrt{\frac{\mu_p}{\mu_n}} - \sqrt{\frac{\mu_n}{\mu_p}}\right) \tag{6.78}$$

and thus depends on the relative mobilities and the temperature [Equation (6.73)]. This point of minimum conductance is indicated in Figure 6.27(a). As such, these minima can serve as rapid evaluation tools of the relative mobilities.

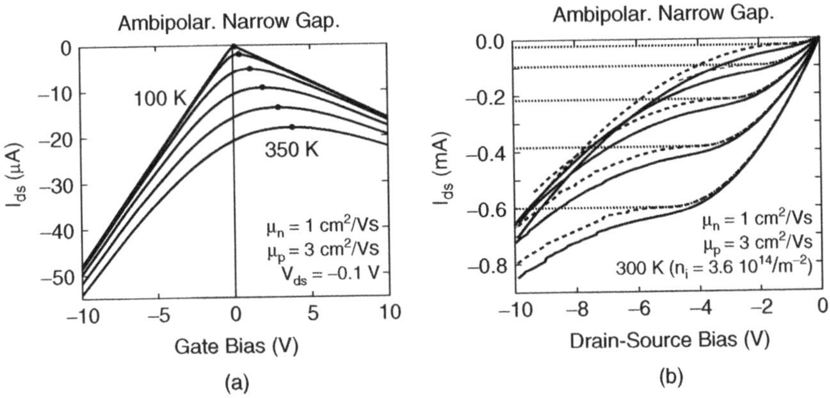

Figure 6.27 Simulation of a narrow band gap ambipolar device with parameters as in Table 6.3: (a) Transfer curves ($I_{ds}-V_g$) for temperatures ranging from 100 to 350 K in steps of 50 K. The full circles indicate the minimum conductance according to Equation (6.78). (b) Output curves ($I_{ds}-V_{ds}$) at 300 K for gate biases from −1 to −5 V in steps of 1 V. The dashed lines show the behavior for a wider band gap ambipolar material (n_i 1000 times smaller) and the dotted lines represent output curves for wide band gap unipolar devices. Reproduced with permission from Synthetic Metals, Modeling electrical characteristics of thin-film field-effect transistors: III. Normally-on devices by P. Stallinga and H. L. Gomes, 158, 11 Copyright (2008) Elsevier

Again, the lack of the possibility of closing the channel goes hand in hand with the nonsaturability of I–V curves, as can be seen in Figure 6.27(b).

6.9.2 Thick TFTs

As discussed before, a TFT only needs a single monolayer to function well [327]. A special case occurs when the transistor works in the thin-film regime – when not working in inversion as in a MOS-FET, so either intrinsic material, as discussed above, or working in accumulation in doped semiconductors, to be discussed later – but the film itself is thicker than one monolayer. When the material is not conductive, nothing will change and normal electronic TFT behavior will be observed, as for wide band gap intrinsic materials. Conduction will always be in the first monolayer and the other layers are passive buffers. However, when the material is conductive, either because of the narrow band gap, or because of doping of the material, or the layers are metallic, things change. The field effect still occurs in the first monolayer, but the adjacent monolayers

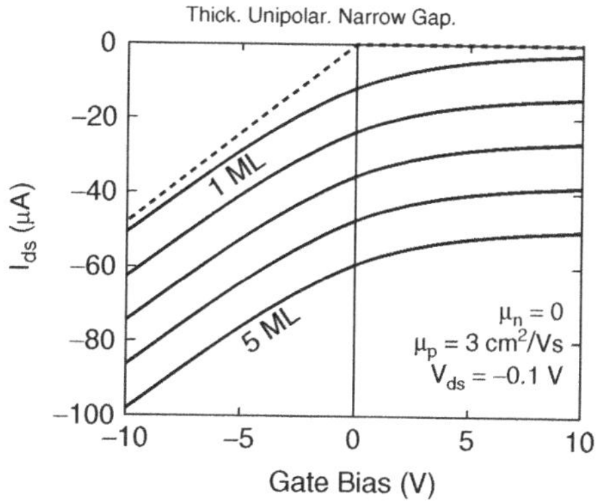

Figure 6.28 Simulation of a thick narrow band gap unipolar device with parameters as in Table 6.3 ($\mu_n = 0$) for different thicknesses ranging from one monolayer (1 ML) to five monolayers (5 ML). The dashed line shows the behavior of similar devices made of wide band gap materials, independent of film thickness. Reproduced with permission from Synthetic Metals, Modeling electrical characteristics of thin-film field-effect transistors: III. Normally-on devices by P. Stallinga and H. L. Gomes, 158, 11 Copyright (2008) Elsevier

now form parallel conduction paths, not influenced by any bias at the gate. The field effect is thus drowned in the parallel conductance. Once this is realized, the simulation of the behavior is obvious, yet it is still useful to illustrate this in a figure, for a quick overview. Figure 6.28 shows the output and transfer characteristics for a device as used in the previous section at 300 K with various numbers of monolayers. In this simulation, each monolayer contributes a conductance of 120 μS (or a parallel resistance of 8.4 kΩ), unaffected by the gate. As can be seen, multilayer devices can easily be confused with narrow band gap devices; both cause a noncompletely closing of the device at reverse bias. For comparison, the dashed line shows the behavior of similar films made of wide band gap materials, which is independent of film thickness since additional layers carry no free charge and do not contribute to conductance.

In the discussion above, it was assumed that the second layer (and consecutive layers) cannot be affected by the gate. This is true up to a certain extent. Once the first layer has been depleted, the second layer starts being modified. The bias at which this starts being important however is quite large, as can be seen by checking when the induced charge becomes comparable with the density of band states. For the above device this is expected to happen at approximately $V_g = -qN_V/C_{ox} = -10$ V.

6.9.3 Doped Semiconductors and Inversion-channel TFT

Another class of conductive materials is doped semiconductors. For the same reasons as for the narrow gap materials, devices based on these semiconductors will show conductance, even in the absence of a gate field and will thus result in a normally-on device. In the case of a material doped with donors with density N_A at energetic position E_A this implies substituting Equation (6.72) by

$$p(x) - N_A^-(x) - n(x) = \left[V(x) - V_g\right] C_{ox}/q. \quad (6.79)$$

The system is not analytically solvable, but it can easily be solved numerically. For the linear region the approximation can be made that the charge densities of Equation (6.79) are constant in space (independent of x). The solution for each bias is then found by applying a zero-finding algorithm to the difference of the left-hand side and the right-hand side of Equation (6.79) as a function of the position of the Fermi level, using Fermi–Dirac distributions for the occupancy of the levels. The conductance is then found by Equation (6.75) (assuming a unipolar behavior, $\mu_n = 0$). Figure 6.29 shows a simulation of transfer curves for such a wide band gap doped semiconductor device. The effects are similar to the effects of narrowing the band gap (Figure 6.26), namely a normally-on behavior, but, as can be seen, for doped semiconductors

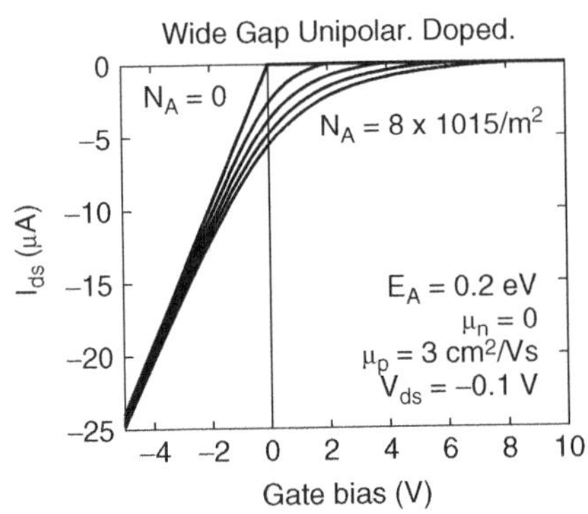

Figure 6.29 Simulation of a doped wide band gap unipolar device with parameters as in Table 6.3 ($\mu_n = 0$) for different acceptor concentrations ranging from 0 to 8×10^{15} m^{-2} in four equal steps with energy $E_A = E_V + 0.2$ eV. The current is found as the conductance of Equation (6.75) multiplied by V_{ds}. Reproduced with permission from Synthetic Metals, Modeling electrical characteristics of thin-film field-effect transistors: III. Normally-on devices by P. Stallinga and H. L. Gomes, 158, 11 Copyright (2008) Elsevier

the channel can be closed, namely for a threshold voltage equal to $V_T = qN_A/C_{ox}$.

Finally, it is interesting to see what happens when the devices are driven in inversion (with majority carrier mobility zero). In this case the dopants are no longer compensated by free charges. This can cause band bending. If the 'film' is very thick, the device behaves like a normal MOS-FET, with band bendings perpendicular to the interface and at the contacts (pn junctions). In a regular TFT, no space is available for the band bendings perpendicular to the interface, as described before. Also, the band bendings at the contacts are no longer regular pn junctions or Schottky barriers because the symmetry is not adequate for that kind of analysis (see the discussion about the contacts in Section 6.8). Still, some form of banding will occur. Regions exist without free charge and the current is no longer given by drift only. Some form of thermionic emission theory or tunneling current theory can be applied, although not exactly of the classic forms for two-dimensional devices or for standard MOS-FET carrier injection. This is an unexplored area.

When the gate bias is strong enough so that the induced charge is larger than the dopant density, once again a channel will open, this time of minority carriers in an inversion channel. This thus occurs for a threshold bias (n-channel in p-type material) equal to

$$V_T = qN_A/C_{ox} \tag{6.80}$$

where N_A is the ionized acceptor concentration (normally equal to the acceptor concentration). Everywhere in the calculations V_g has to be replaced by $V_g - V_T$. First of all, note that this threshold voltage is much simpler than the threshold voltage of a MOS-FET inversion, the latter having a square-root dependence on N_A [see Equation (6.10)], since it comes from depletion and band-bending calculations, an approach that in a TFT does not make sense. Next, note that, to make sure that the entire channel has free charge, including the region at the drain, the drain bias has to be added to this threshold voltage. For gate biases between V_T and $V_T + V_{ds}$, the behavior will be peculiar, with a bias-dependent depletion region at the drain. Since all organic TFTs without exception work in accumulation (either because they are undoped and either polarity causes accumulation, or because only p-channels are made in p-type materials), this issue is not yet studied.

Finally, note that the threshold voltage for starting an inversion channel is equal to the threshold voltage for closing an accumulation channel and vice versa. This is something that we have already seen in the basic model. Doping only shifts the behavior along the gate-bias axis [by the amount given in Equation (6.80)]. The threshold voltage is also equal to the one found for trapped charge, with N_T substituting N_A. In fact, in this respect, ionized dopants are nothing more than trapped charges. They are defined by the fact that this charge has zero mobility.

If one type of 'free' carrier has zero mobility, the threshold voltage is the voltage needed to turn on the transistor, either in accumulation or inversion. If both types of carriers have nonzero mobility, the material is ambipolar and the threshold voltage is that bias that makes the device commutate between an n-channel and a p-channel. (If both types of carrier have zero mobility, no current will flow in any case.)

6.9.4 Metal–Insulator–Metal TFT

One interesting aspect of the TFT model is that the materials are not limited to semiconductors only. As long as the active layer is thin, so that parallel conductance (current that is unaffected by the gate bias) is minimal, a device can be made with a metal for the active layer. As an example we will describe here a metal–insulator–metal TFT with a monolayer of gold as the channel. In practice, such ideal devices are difficult to make in reality because the gold deposited on top of the oxide tends to form islands instead of a uniform coverage of the insulator. However, it may serve as a prototype for the model.

Gold has a density of 19 300 kg m^{-3} and an atomic mass of 197 g mol^{-1} (1 mol is $6.02214199 \times 10^{23}$ particles). With one free electron per atom, this gives a free-electron density of 5.9×10^{28} m^{-3}. The electrical resistivity of 22.14 nΩ m then translates into an electron mobility of 47.8 cm^2 V^{-1}s^{-1} (values taken from ref. [346]), a value we will use for this exercise. A film with a thickness equal to one unit cell of the gold FCC crystal, 2 Å, has a two-dimensional electron density of $n_0 = 1.2 \times 10^{19}$ m^{-2} independent of temperature. (Neutrality is maintained by the positively charged matrix of the metal.) The local electron density in the presence of a field is then equal to

$$n(x) = n_0 + C_{ox}\left[V_g - V(x)\right]/q. \tag{6.81}$$

Such a film, used in a device with parameters as in Table 6.1, will have a zero-bias conductance [Equation (6.75)] of 9.2 S. Figure 6.30 shows a simulation of a transfer curve. Because of the high zero-bias density of electrons, the relative effect of the gate field is small. However, the absolute effect, the slope of the transfer curve, yields the same mobility given

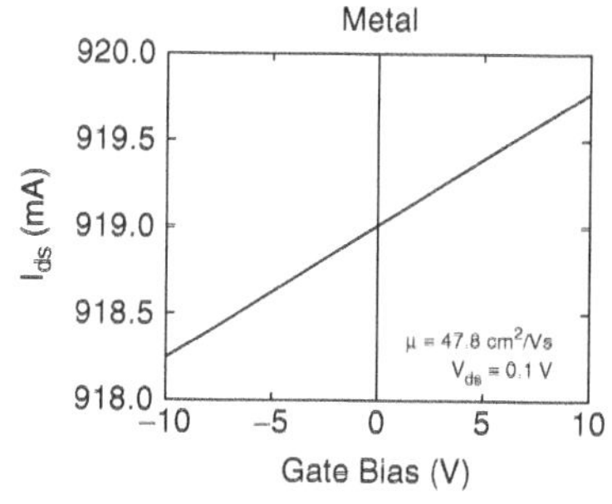

Figure 6.30 Simulation of a gold–insulator–gold TFT transfer curve with parameters as in Table 6.3 ($n_0 = 1.2 \times 10^{19}$ m^{-2} and $\mu = 47.8$ cm^2 V^{-1}s^{-1}). Reproduced with permission from Synthetic Metals, Modeling electrical characteristics of thin-film field-effect transistors: III. Normally-on devices by P. Stallinga and H. L. Gomes, 158, 11 Copyright (2008) Elsevier

above. For the same reason, the bias needed to close the channel is large, $V_T = -qn_0/C_{ox}$, and is of the order of 12 kV; impossible to attain in practice. The same voltage is needed at the drain to pinch-off the channel there. In other words, output (I–V) curves do not saturate and remain linear for a very large range of voltages, with relative tiny field effect.

Organic devices may have DOS that are much lower than pure gold devices, even when the materials are metallic. Next we make a simulation of metallic behavior (half-filled band), but with much smaller electronic state density, though with the same charge carrier mobility of 47.8 cm^2 V^{-1}s^{-1}. We call this a metalloid behavior to distinguish it from pure metallic behavior. As can be seen in Figures 6.31 and 6.32, when the

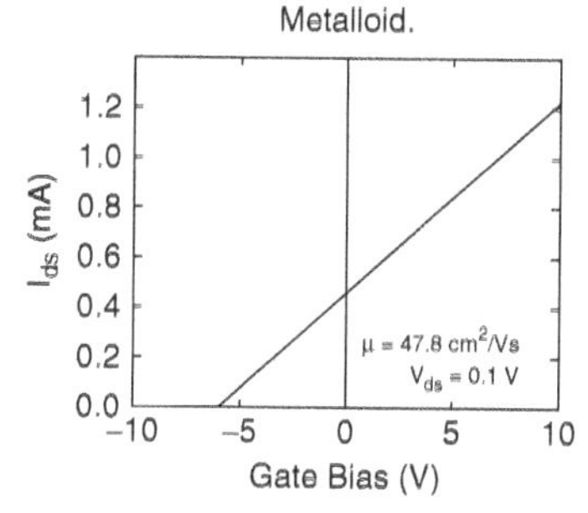

Figure 6.31 Simulation of a metal–insulator–metal ('metalloid') TFT transfer curve with parameters as in Table 6.3 ($n_0 = 6 \times 10^{15}$ m^{-2} and $\mu = 47.8$ cm^2 V^{-1}s^{-1}). Reproduced with permission from Synthetic Metals, Modeling electrical characteristics of thin-film field-effect transistors: III. Normally-on devices by P. Stallinga and H. L. Gomes, 158, 11 Copyright (2008) Elsevier

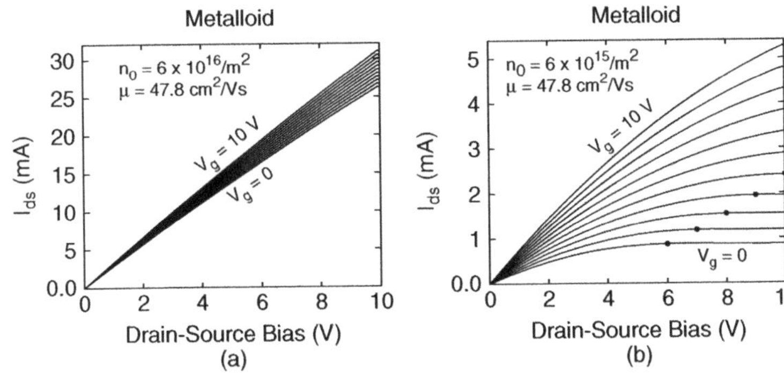

Figure 6.32 Simulation of metal–insulator–metal ('metalloid') TFT output curves with parameters as in Table 6.3 ($\mu = 47.8$ cm^2 V^{-1}s^{-1}) for (a) $n_0 = 6 \times 10^{16}$ m^{-2} and (b) $n_0 = 6 \times 10^{15}$ m^{-2}. The full circles indicate pinch-off at $V_{ds} = V_g - V_T$. Reproduced with permission from Synthetic Metals, Modeling electrical characteristics of thin-film field-effect transistors: III. Normally-on devices by P. Stallinga and H. L. Gomes, 158, 11 Copyright (2008) Elsevier

density of states diminishes, the device behaves more and more like typical semiconductor TFTs. For a zero-bias electron density of $n_0 = 6 \times 10^{15}$ m^{-2}, the bias to close the channel is $V_T = -qn_0/C_{ox} = -6$ V (Figure 6.31). The output curve (Figure 6.32b) thus saturates at a pinch-off voltage of $V_P = V_g - V_T$.

The metal transistor was recently demonstrated by us [347].

6.10 EFFECTS OF TRAPS

In the previous sections, all charge induced by the gate was assumed to be free charge (with a brief reference to doping). In particular, organic materials, or amorphous materials in general, have a large density of traps. This means that not all charge induced by the gate will contribute to current. As will be shown in this chapter, these traps can cause a remarkably intricate behavior of the device even for simple systems. The effective mobility, as defined by the derivative of the transfer curve, can become temperature and/or bias dependent, a phenomenon often observed [262, 348, 349]. Moreover, the output curves can become highly nonlinear in the linear region (small V_{ds}), an observation that is often mistakenly attributed to contact effects (resistance or rectifying Schottky barriers). Note that the real contact effects are described in Section 6.7. In the context of this work, a trapped charge is any charge that has zero mobility and can thus not contribute to current. This is very similar to dopant levels. The difference between a trap and a dopant level is the charge states. Both are derived from (or 'communicate to') a band, but whereas a dopant becomes neutral when capturing the charge from the band, a trap becomes charged when a charge is captured. Apart from this, the behavior is similar; charges become unavailable for conduction when captured.

In this section our work is summarized based on our publications in *Organic Electronics* [350] and *Synthetic Metals* [345].

6.10.1 Traps and Threshold Voltage

The immediate effect of traps is similar to doping. In a regular semiconductor, ionized dopants are compensated by free charge, thereby maintaining charge neutrality. In the same way, a charge on a trap will be compensated by a charge of opposite sign in the bands. The difference between a dopant and a trap is that they communicate to different bands. For instance, a positive charge on a trap empties to the valence band, while a positively ionized donor receives an electron from the conduction band:

$$\begin{aligned} T^0 + h &\leftrightarrow T^+ \\ D^0 &\leftrightarrow D^+ + e \end{aligned} \tag{6.82}$$

In other words, charged immobile species are compensated by mobile charges in the conductive bands. Filled traps will thus change the threshold voltage because an extra bias has to be applied to drive out these compensating charges before the channel can be opened of the appropriate carrier type. As an example, imagine a TFT of a material that has low electron mobility and high hole mobility, but there exists a trap that can capture and immobilize these holes. Imagine that at zero bias for some reason (probably the history of the device) there exists a certain trapped positive charge density, N_T^+. In the conduction band these are compensated by 'free' electrons (of low mobility). To remove these charges, a negative bias has to be applied to the gate. Only biases beyond this threshold voltage will start attracting holes to the valence band. From the basic charge–voltage relation [Equation (6.31)] it can easily be found that

$$V_T = -qN_T^+/C_{ox} \tag{6.83}$$

The situation becomes more complex when the occupancy of the dopants (or traps in general) depends on the bias. In this case, the effect is not simply an introduction of a threshold voltage. This is easily understood when we realize that the gate bias can sweep the Fermi level through the energy diagram and thus change the occupancy of any states passing under it. The curves become curved and the effective mobility bias-dependent. An example for doped materials was given earlier. More complicated examples will be given later in Section 6.10.3.

The threshold voltage is equal to the one found for doping, with $-N_T^+$ substituting N_A. In fact, in this respect, ionized dopants are nothing more than trapped charges. Both species are charges with zero mobility. In all cases, the general expression for threshold voltage is

$$V_T = -\rho_{immobile}/C_{ox} \tag{6.84}$$

where $\rho_{immobile}$ is the immobile charge density. This can be charge trapped on deep (slow) levels, for instance qN_T^+, or charge of ionized dopants, qN_D^+ or $-qN_A^-$. Once again, the threshold voltage for opening a channel of one type is equal to the threshold voltage for closing a channel of the opposite type. When the material is ambipolar and both types of carriers have nonzero mobility, current will flow on either side of V_T. When only one carrier type has nonvanishing mobility, the threshold voltage is similar to the conventional threshold voltage. When both types of carriers have zero mobility, obviously no current will flow under any condition.

6.10.2 Traps and Output Curves

Nonlinearities are often observed in I–V curves. They can be described as supralinear close to the origin, and are often attributed to the effects of the contacts. As shown in earlier sections, this in not an adequate physical picture; resistances at the contacts will cause an effect called 'current crowding', while rectifying contacts cause saturation at a few times the thermal voltage (26 mV at room temperature). A more likely cause for nonlinearities in I–V curves is the presence of abundant traps causing a Poole–Frenkel-like behavior [55, 345].

A Poole and Frenkel approach shows how the current, and thus the effective mobility ($\mu \propto I_{ds}/V_{ds}$), of a trap-ridden material can depend on the temperature and electrical field [9, 54]:

$$\mu_{PF} = \mu_0 \exp\left\{-\left[(E_T - E_V) - q\sqrt{q\,|E_x|\,/\pi\varepsilon}\right]/kT\right\} \tag{6.85}$$

where μ_0 is the free carrier mobility, $E_T - E_V$ is the discrete trap depth, and $E_x = -dV(x)/dx$ is the in-plane electric field along the direction of the current. The field used in the calculation is only the in-plane field E_x and not the field of the gate. This is because the perpendicular field E_y is shielded by the build-up of charges. Since there is no current in this direction, the total field is necessarily zero: $0 = J_y = qp\mu E_y$.

The differential equation for TFTs now becomes

$$
\begin{aligned}
I_x(x) &= Wqp(x)\mu(x)\frac{dV(x)}{dx} \qquad (6.86)\\
p(x) &= C_{ox}\left[V_g - V_T - V(x)\right]/q\\
\mu(x) &= \mu_{00}\exp\left(\frac{a}{kT}\sqrt{|dV(x)/dx|}\right)
\end{aligned}
$$

where the same boundary conditions apply as before $[V(0) = 0,\ V(L) = V_{ds}$ and $I_x(x) = I_{ds}]$ and $\mu_{00} = \mu_0 \exp[-(E_T - EV)/kT]$ depends on the temperature and the trap depth and $a = \sqrt{q^3/\pi\varepsilon}$ is a constant that depends only on the permittivity of the material (ε). The above equations do not have a simple analytical solution, however, they are not difficult to solve numerically. Figure 6.33 shows simulations of the I–V curves for different temperatures.

The curves are clearly supralinear for small V_{ds}, especially for lower temperatures, because a larger portion of the charges is trapped, whereas for higher temperatures the charges are excited to the mobile band and the mobility approaches μ_0, independent of field. It is also predicted that the nonlinearities become more pronounced when the channel length is reduced. This is because the factor in the exponent of Equation (6.86) scales with the inverse of the channel length and takes an increasingly important role. As to the effect on the shapes of the curves, it can be said that the effect of reducing the channel length is equal to lowering the temperature. This will be technologically important when the devices are reduced to nanoscale dimensions. Severe nonlinearities can be expected when abundant traps are present.

The saturation behavior is rather abrupt. At this point, the numerical simulations encounter problems, since a singularity is approached, with the charge density tending to zero and the field to infinity, maintaining the product constant. It has to be pointed out once again that diffusion is not incorporated in this simple model. Diffusion would avoid such singularities because large gradients in charge are removed.

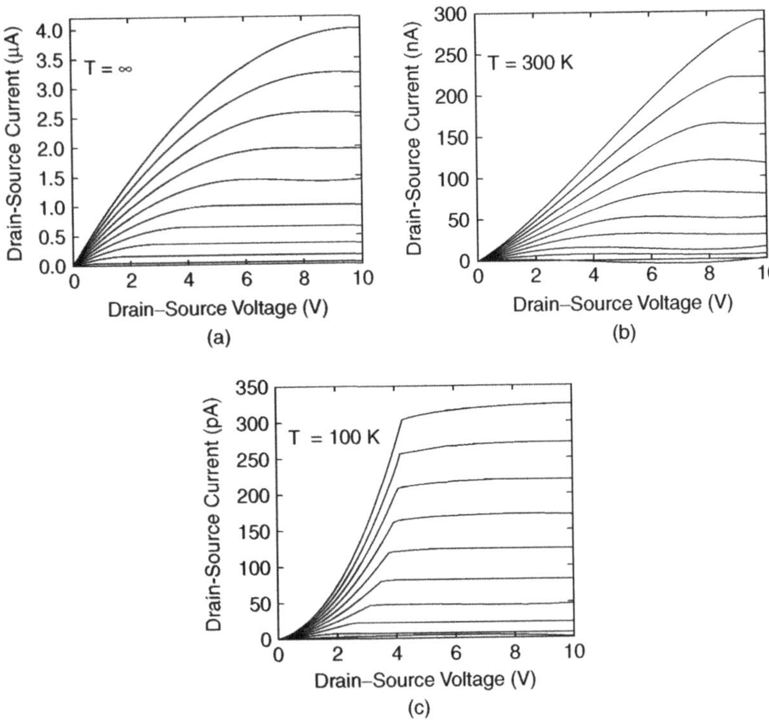

Figure 6.33 Simulations of I–V curves for field-dependent mobility as described by Poole and Frenkel for temperatures as indicated. Gate bias ranging from 1 to 10 V. Parameters as in Table 6.3 and $E_T = 100$ meV and $\varepsilon = 5\varepsilon_0$. For small field dependence (infinite temperature), the familiar FET curves emerge. However, for larger field dependence, the curves become pronouncedly concave (supralinear). Reproduced with permission from Synthetic Metals, Modeling electrical characteristics of thin-film field-effect transistors: II: Effects of traps and impurities by P. Stallinga and H. L. Gomes, 156, 21–24 Copyright (2006) Elsevier

Thus, it is concluded that nonlinearities are readily explained in the framework of traps by assuming a field-assisted thermal excitation from these deep levels, as proposed by Poole and Frenkel. The model described above has, apart from the scaling parameter E_T, only one adjustable parameter, namely the permittivity of the material, ε. For the simulations in Figure 6.33, an ε equal to $5\varepsilon_0$ was assumed; a reasonable value. For lower ε the effects are more pronounced. As shown before, in standard theory the currents are independent of the ratio W:L. When Poole–Frenkel conduction is important, this is no longer true, since the current depends nonlinearly on L. Severe short-channel effects thus result.

6.10.3 Traps and Transfer Curves

In the basic model described in Section 6.4 all induced charge was assumed to be free charge, see Equation (6.31). In that case, standard field-effect transfer curves emerge. Because of the similarity between TFT and MOS-FET characteristics, the latter are often used to define an effective 'as-measured' mobility. The field-effect mobility μ_{FET} is defined via the derivative of the transfer curve (I_{ds} vs. V_g) in the linear regime.

$$\mu_{FET} \equiv \frac{L}{W} \frac{1}{C_{ox} V_{ds}} \frac{\partial I_{ds}}{\partial V_g}. \tag{6.87}$$

Equally standard is applying this definition of μ_{FET} to TFTs, which then sometimes become bias or temperature dependent [6]. Furthermore, it can be stated that at low drain-source bias, in the so-called linear regime, the charge density and electric field can be considered homogeneous along the channel. In this case, the current is proportional to the free charge density, holes (p) in the case of p-channel FETs. The mobility in the linear regime is thus proportional to the derivative of the function of the hole density as a function of gate bias.

$$\mu_{FET} = -\frac{q\mu_0}{C_{ox}} \frac{\partial p(V_g)}{\partial V_g}. \tag{6.88}$$

For intrinsic TFTs this relation is linear [see Equation (6.31)] and the as-measured mobility is equal to the intrinsic charge mobility μ_0. The as-measured mobility is therefore bias independent in pure crystalline materials. [For this analysis the intrinsic (band) mobility μ_0 is considered to be temperature independent; effects of optical-phonon scattering, etc., are not included. In any case, these are slowly varying functions of temperature, such as $T^{1/2}$ [9]]. The function becomes nonlinear and the transfer curves with it, when – and only then – the material is full of traps.

The basic idea is that the traps are differently distributed in energy (see Figure 13.1). Changes in Fermi level in the channel as caused by changes of gate-bias therefore have different effects on the filling of traps and conduction states. For instance, the conduction charge density can grow faster than the trapped-charge density, while the sum still follows the charge–bias relation of a parallel plates capacitor. In this case, the current grows faster than linear (superlinear transfer curves in the linear region) and the effective as-measured field-effect mobility, defined via the derivative of the transfer curve is gate-bias dependent, in this case increasing with bias. The exact relation between mobility and bias depends on the distribution functions of trap states and conduction states. This is comparable with the work of Shur and Hack [79], but opposes the ideas of, for example Vissenberg and Matters, that see the origin of a bias dependence of mobility in (variable-range) hopping conduction [351].

In this subsection on transfer function for systems with traps the following assumptions are made. [Some of them have already been mentioned and used in earlier (sub)sections]:

1. All charge is directly at the interface; the active layer is effectively two-dimensional.
2. Charge can be free charge (holes), with high mobility, μ_0, or trapped charge with zero mobility.
3. Sum density, $\rho = q(p + N_T^+)$, depends on the local local bias, $\rho(x) = C_{ox}\left[V(x) - V_g\right]$.
4. Relative density (can) depend on temperature and bias; they follow the Fermi–Dirac distribution.
5. Thermal equilibrium is reached instantaneously.
6. Local current is proportional to the free charge and field: $I_x(x) = qp(x)\mu_0 E_x(x)$, with $E_x(x) = -dV(x)/dx$.
7. Field-effect mobility is defined via the derivative of the transfer curve, see Equation (6.87).

The parameters used in this section are summarized in Table 6.4. For a trap-free device the system is easily solvable. The numerical solution presented in Figure 6.34 is equal to the analytical solution given earlier.

Introducing traps into the system the first type is a discrete trap. In this case, the mobility is lowered significantly by the reduced ratio of free-to-total charge, and becomes temperature dependent, but remains

Table 6.4 Simulation parameters used (unless otherwise specified)

Parameter	Value	Unit
N_V	1.04×10^{16}	m^{-2}
N_T	1.04×10^{17}	m^{-2}
C_{ox}	160	$\mu F\ m^{-2}$
N_{T0}	10^{18}	$m^{-2}\ eV^{-1}$
N_{V0}	10^{17}	$m^{-2}\ eV^{-1}$
T_2	800	K
T_1	300	K
μ_0	3	$cm^2\ V^{-1}s^{-1}$

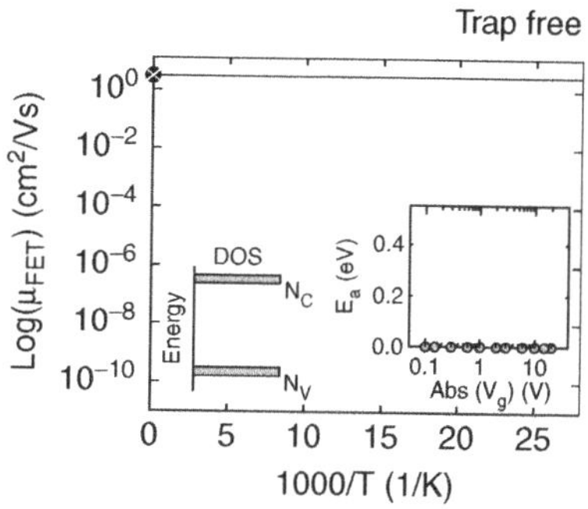

Figure 6.34 Temperature dependence of mobility represented in an Arrhenius plot for the trap-free case. The insets show the schematic DOS and the activation energy of mobility (and current) as a function of bias. The symbol ⊗ on the mobility axis indicates the free-hole mobility μ_0. Reproduced with permission from Organic Electronics, Thin-film field-effect transistors: The effects of traps in the bias and temperature dependance of field-effect mobility, including the Meyer-Neldel rule by P. Stallinga and H. L. Gomes, 7, 6, 592–599 Copyright (2006) Elsevier

independent of bias. This result is similar to the model of Poole and Frenkel [9] and the reasoning is as follows: Free holes (p) in the conduction band, originally induced by the gate bias, can be captured by the traps, making these positively charged. At thermal equilibrium, the ratio of densities of holes and charged traps N_T^+ is determined by the energetic distance $E_T - E_V$ between them, the relative abundance of the levels, N_V and N_T respectively, and the temperature T (Note, since the active layer is treated as purely two-dimensional, all densities have units 'per square meter'):

$$\frac{p}{N_T^+} = \frac{N_V}{N_T} \exp\left(\frac{E_V - E_T}{kT}\right) \tag{6.89}$$

where the Boltzmann statistics function was used presupposing that the Fermi level is far away from both the conductive as well as trap state levels. (For the simulations, however, the full Fermi–Dirac distribution was used.) The total charge induced in the channel is proportional to the gate bias [the basic charge equation, Equation (6.30)]:

$$p + N_T^+ - n = -C_{ox} V_g / q. \tag{6.90}$$

The current is only proportional to the free hole density because the trapped states, by definition, do not contribute to current and the density of electrons is insignificant. The solution of the above equations is that the current is linearly proportional to the gate bias and that the effective, as-measured mobility of Equations (6.87) and (6.88), defined via the derivative of the transfer curve, is therefore depending on temperature, but not on bias

$$\mu_{\mathrm{FET}} \approx \mu_0 \frac{N_{\mathrm{V}}}{N_{\mathrm{T}}} \exp\left(-\frac{E_{\mathrm{T}} - E_{\mathrm{V}}}{kT}\right). \tag{6.91}$$

In other words, the Arrhenius plots of mobility are straight lines; independent of bias, the slope of the plot reveals the activation energy of mobility, which is then equal to the depth of the trap level, $E_{\mathrm{a}} = E_{\mathrm{T}} - E_{\mathrm{V}}$, see Figure 6.35. This result is similar to the Poole–Frenkel model [9], or the MTR model of Horowitz [261]. Note that extrapolation of the curves to $T = \infty$ gives an effective prefactor in the mobility equal to $\mu_\infty = \mu_0 N_{\mathrm{V}}/N_{\mathrm{T}}$, which can be well below the free-hole band value μ_0 when the traps are abundant. For Figure 6.35, $N_{\mathrm{T}} = 10N_{\mathrm{V}}$ was used.

The assumption was made here that the trap states are truly abundant, effectively unlimited: $N_{\mathrm{T}} \gg N_{\mathrm{V}}$. When this is not the case, the trap states can be exhausted and, once all filled, the induced charge is necessarily free charge (holes) and the mobility returns to the band value μ_0. In the above calculations, it involves replacing the Boltzmann distribution approximation by the full Fermi–Dirac distribution function. The traps

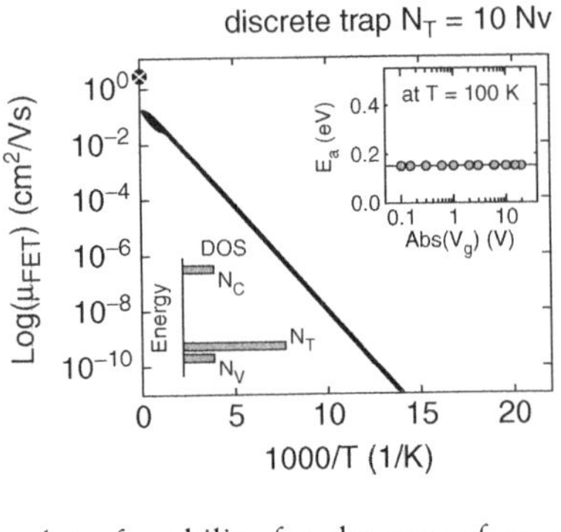

Figure 6.35 Arrhenius plot of mobility for the case of an abundant discrete trap. In this particular case, $N_{\mathrm{T}} = 10N_{\mathrm{V}}$. The mobility is strongly temperature dependent and the plot reveals the trap depth of 150 meV, independent of bias, see Equation (6.91). The insets show the schematic DOS and the activation energy of mobility (and current) as a function of gate bias. The symbol ⊗ on the mobility axis indicates the free-hole mobility μ_0. Reproduced with permission from Organic Electronics, Thin-film field-effect transistors: The effects of traps in the bias and temperature dependance of field-effect mobility, including the Meyer-Neldel rule by P. Stallinga and H. L. Gomes, 7, 6, 592–599 Copyright (2006) Elsevier

become depleted when the induced charge density is comparable with the trap density. This defines the trap-free-limit voltage for the gate bias

$$V_{\text{tfl}} = -qN_T/C_{\text{ox}}. \tag{6.92}$$

In Figure 6.36, a transition case is shown with the trap density equal to the effective density of valence band states, $N_T = N_V$. For this specific case, $V_{\text{tfl}} = -10.4$ V. For gate biases below this voltage, the activation energy is equal to the trap depth (150 meV), while above it, the mobility rapidly becomes independent of temperature and settles at the free-hole

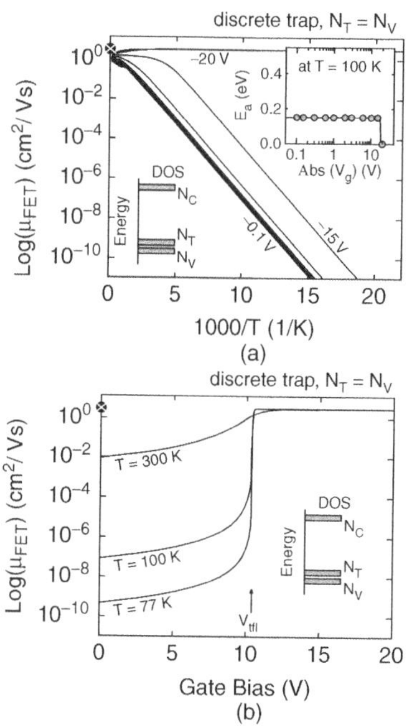

Figure 6.36 (a) Arrhenius plot of mobility for a discrete trap with density equal to the effective density of valence band states, $N_T = N_V$. The activation energy is equal to the trap depth for biases below the trap-free-limit voltages $V_g < V_{\text{tfl}}$. For larger biases, the mobility is no longer thermally activated and the plots resemble those of trap-free devices. The insets show the schematic DOS and the activation energy of mobility (and current). The symbol ⊗ on the mobility axis indicates the free-hole mobility μ_0. (b) The bias dependence of mobility at three different temperatures. The trap-free limit voltage, V_{tfl}, of Equation (6.92) is indicated. Reproduced with permission from Organic Electronics, Thin-film field-effect transistors: The effects of traps in the bias and temperature dependance of field-effect mobility, including the Meyer-Neldel rule by P. Stallinga and H. L. Gomes, 7, 6, 592–599 Copyright (2006) Elsevier

value μ_0, as can be seen in Figure 6.36. In this case, it is not easy to give an algebraic solution. Figure 6.36(b) shows the bias dependence of the mobility for different temperatures, which is based on numerical simulations: For a certain bias V_g, the Fermi level that zeros the total charge minus induced charge is found by a numerical algorithm as described by ref. [78]. Once the Fermi level is found, the free charge is determined by substituting this energy in the free hole distribution $p(E_F)$. The voltage is stepped by a tiny amount and the free charge is calculated again. The mobility is then the derivative according to Equation (6.88). (Note: this technique was also used for Figure 6.35.) As can be seen, for $-V_g > -V_{tfl}$ the mobility is equal to the free-hole value indicated by ⊗ on the mobility axis. In fact, Equation (6.92) gives a fast way of determining the trap density, provided the rapid transition in the transfer curves is observed that allows for a determination of V_{tfl}.

Inspired by the model of Shur and Hack [79] is the next attempt of a model in which the trap states are distributed in energy. This, as shown now, can give rise to a bias- and temperature-dependent mobility. Using a normal valence band and trap states N_T exponentially distributed in energy,

$$N_T(E) = N_{T0} \exp\left(\frac{E_V - E}{kT_2}\right) \tag{6.93}$$

where E is the energy of an electron, N_{T0} is the DOS of traps at the valence band E_V, k is Boltzmann's constant, and T_2 is a parameter describing the distribution (the slope of a logarithmic plot of the DOS, see Figure 6.37). When following the same reasoning as followed for the discrete trap, but with a convolution over trap states in Equations (6.89) and (6.90), it can be shown that the drain-source current is (see Appendix A, with $\rho = -C_{ox}V_g$)

$$p(V_g) = N_V \left[\frac{-C_{ox}V_g}{qN_{TT}(T)}\right]^{T_2/T} \tag{6.94}$$

giving

$$I_{ds} = q\mu_0 \frac{W}{L} V_{ds} N_V \left[\frac{-C_{ox}V_g}{qN_{TT}(T)}\right]^{T_2/T} \tag{6.95}$$

where

$$N_{TT}(T) = \alpha(T) N_{T0} \frac{(kT_2)^2}{kT_2 - kT} \tag{6.96}$$

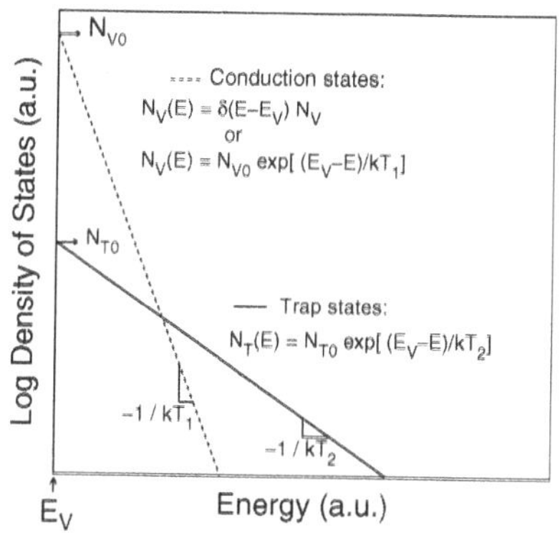

Figure 6.37 Density of states used in the calculations for Figures 6.38 and 6.39. The trap states (solid line) are exponentially distributed in energy. For the conduction states, two distributions are tried: A discrete band at $E = E_V$ and an exponentially decaying function (dashed line). The former results in the observation of the MNR with bias and temperature dependent current and mobility, see Figure 6.38. For the latter, only the bias dependence remains and the temperature dependence nearly vanishes, see Figure 6.39. Reproduced with permission from Organic Electronics, Thin-film field-effect transistors: The effects of traps in the bias and temperature dependance of field-effect mobility, including the Meyer-Neldel rule by P. Stallinga and H. L. Gomes, 7, 6, 592–599 Copyright (2006) Elsevier

where $\alpha(T)$ is a slowly varying function of temperature (and therefore irrelevant for the discussion), oscillating between 1 and 0.8 in the temperature range $0-T_2$, with a minimum half way. The as-measured mobility is proportional to the gate-bias derivative of this equation [according to Equation (6.87) or Equation (6.88)]

$$\mu_{FET} = \frac{T_2}{T}\mu_0\frac{N_V}{N_{TT}(T)}\left[\frac{-C_{ox}V_g}{qN_{TT}(T)}\right]^{T_2/T-1}. \tag{6.97}$$

It is immediately clear that: (i) The mobility depends on gate bias; (ii) the dependence disappears at a temperature $T = T_2$, thus following the MNR, with $T_{MN} = T_2$. Figure 6.38 shows simulations of the above equation with parameters as in Table 6.1. From Figure 6.38 it can be seen that, because of the effects of the factor T in the denominator of Equation (6.97), as well as the temperature dependence of α, the isokinetic temperature falls slightly below T_2 and the curves do not exactly extrapolate to a single point. However, in most cases the instrumental resolution will be too low to accurately determine this small deviation. Note also the sharp drop in current when the temperature approaches T_2. This is due to the factor $1/(kT_2 - kT)$ in Equation (6.96), which

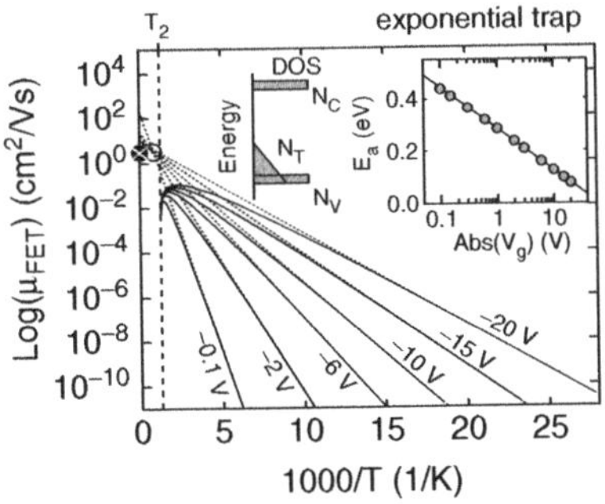

Figure 6.38 Graphical representation of temperature-dependent as-measured mobility of Equation (6.97) of a system with a DOS as in Figure 6.37, namely an exponential distribution of trap states and discrete conduction states. Parameters as in Table 6.4, with gate biases from −0.1 to −20 V as indicated. The open circle (○) represents the Meyer–Neldel point (T_{MN}, μ_{MN}). The inset shows the effective activation energy as a function of bias. Reproduced with permission from Organic Electronics, Thin-film field-effect transistors: The effects of traps in the bias and temperature dependance of field-effect mobility, including the Meyer-Neldel rule by P. Stallinga and H. L. Gomes, 7, 6, 592–599 Copyright (2006) Elsevier

diverges for $T \rightarrow T_2$. To our knowledge, no reports in the literature exist for measurements at or in the vicinity of the isokinetic temperature; in all cases, T_{MN} is found by extrapolation.

Analyzing Equations (6.95) and (6.97) it is easily shown that the activation energy of the field mobility (and current alike), as measured via the slope of an Arrhenius plot, depends on the bias in the following way

$$E_a \equiv -\frac{d\ln(\mu_{FET})}{d(1/kT)} = kT_2\ln(N_{TT}) - kT_2\ln(-C_{ox}V_g/q). \tag{6.98}$$

This is shown in the inset of Figure 6.38. Thus, the activation energy of mobility or current does not reveal the depth of an energetic level. Rather, it depends on the parameters of the distribution (T_2 and N_{T0}) and the bias.

It is interesting to point out the difference between this model and the model of Shur and Hack [79]. Where they have a factor of $2T_2/T - 1$ in the exponent in the current, the above has T_2/T [Equation (6.95)]. This results in an infinite isokinetic temperature for the current, whereas they have $T_{MN} = 2T_2$ [352]. For mobility, both models arrive at $T_{MN} = T_2$;

the dependence of mobility on bias disappears at this temperature, as shown by Equation (6.97).

A more fundamental difference is that Shur and Hack use exponential distribution for both the trap states as well as the valence band states (so-called 'tail states'), where here only a distributed trap state is used, while maintaining a Dirac-delta function for the DOS of the valence band. When exponentially distributed tail states are included,

$$N_V(E) = N_{V0} \exp\left(\frac{E_V - E}{kT_1}\right) \tag{6.99}$$

with similar reasoning a strongly bias-dependent, but temperature-independent mobility is found (see the derivation in Appendix A, with $\rho = -C_{ox} V_g$),

$$p(V_g) = N_{VT}(T) \left[\frac{-C_{ox} V_g}{q N_{TT}(T)}\right]^{T_2/T_1} \tag{6.100}$$

The mobility via Equation (6.88) then becomes

$$\mu_{FET} = \frac{T_2}{T_1} \mu_0 \frac{N_{VT}(T)}{N_{TT}(T)} \left[\frac{-C_{ox} V_g}{q N_{TT}(T)}\right]^{T_2/T_1 - 1} \tag{6.101}$$

with

$$N_{VT}(T) = \beta(T) N_{V0} \frac{(kT_1)^2}{kT_1 - kT} \tag{6.102}$$

where T_1 is the parameter describing the distribution of band-tail states, see Figure 6.37, and β is a function equal to α but scaled with T_1 instead of T_2. Figure 6.39 shows a simulation of the mobility as a function of temperature and bias. Interesting in this respect is the observation by us of exactly such a behavior [47], something that is inexplicable in the theory of Shur and Hack. Figure 6.40 plots this behavior. Note that the curves were not taken consecutively and stressing may have occurred in between the measurements, causing the curves to shift.

Figure 6.41 compares the various models described in this work and the model of Shur and Hack [79]. For the latter a value of 0.484 eV was used for their parameter E_{F0} and the value for their parameter g_{F0} (defining the DOS at E_{F0}) can then be found by extrapolating $N_T(E)$ to $E = E_{F0}$ in Equation (6.93) and dividing it by 1 nm. This gives $g_{F0} = 8.93 \times 10^{23}\ \text{m}^{-3}\text{eV}^{-1}$. As can be seen, the model of Shur

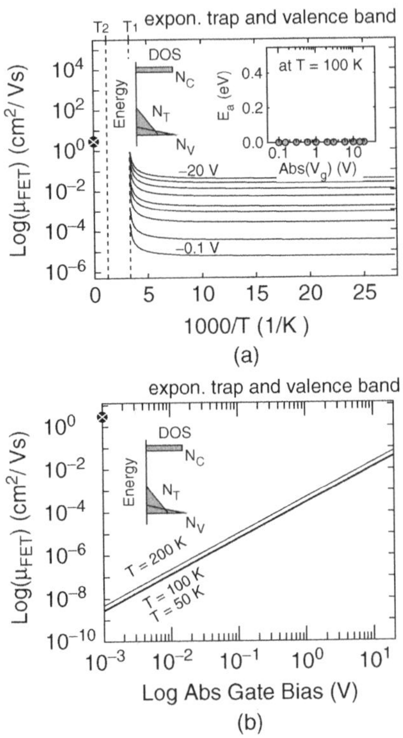

Figure 6.39 (a) Graphical representation of temperature-dependent as-measured mobility of Equation (6.101) of a system with a DOS as in Figure 6.37, namely an exponential distribution for both the trap states as well as the conduction states. Parameters as in Table 6.4, with gate biases from −0.1 to −20 V as indicated. The mobility is strongly bias dependent, but independent of temperature; the inset shows the effective activation energy as a function of bias. (b) The bias dependence of mobility at three different temperatures, which clearly follow a power law. Reproduced with permission from Organic Electronics, Thin-film field-effect transistors: The effects of traps in the bias and temperature dependance of field-effect mobility, including the Meyer-Neldel rule by P. Stallinga and H. L. Gomes, 7, 6, 592–599 Copyright (2006) Elsevier

and Hack with conduction and donor states exponentially distributed in energy behaves much like our model with only the trap states distributed exponentially in energy. Note, the model of Shur and Hack was developed for amorphous silicon devices based on the MOS-FET model with a three-dimensional active layer, while our model is based on the two-dimensional accumulation-channel TFT model.

As a final remark, it has to be pointed out that in this analysis thermal equilibrium is assumed at all times. Especially for deep traps

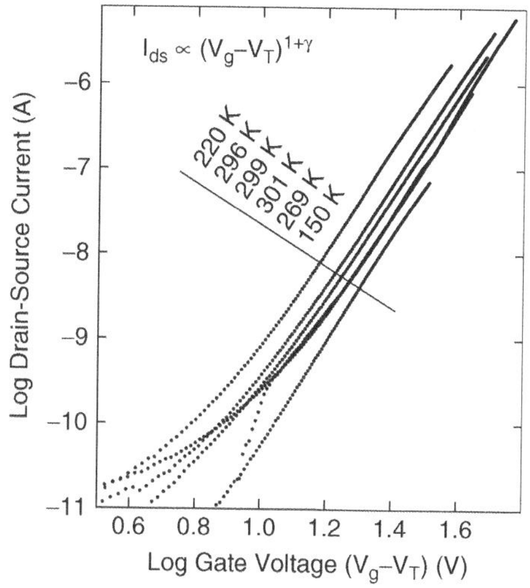

Figure 6.40 Experimental observation of the behavior predicted by the assumption of experimental distributions for both the trap states and the conduction states. The sample measured was a TFT based on sexithiophene (T6) [47]. (Note, the difference between slopes of mobility and current plots in log–log scale is 1). Reproduced with permission from Journal of Applied Physics, Electronic transport in field-effect transistors of sexithiophene by P. Stallinga, 96, 9 Copyright (2006) American Institute of Physics

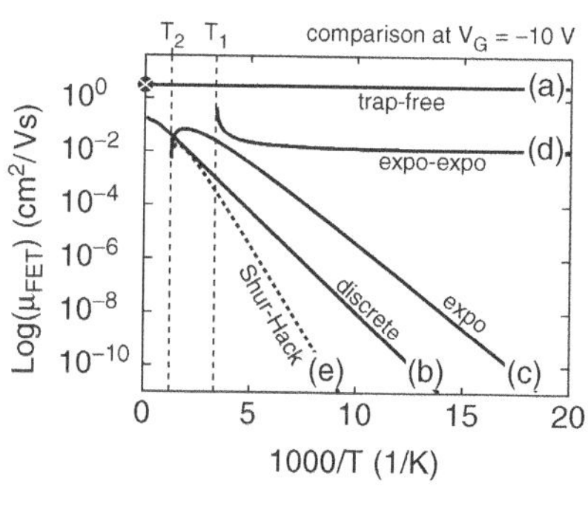

Figure 6.41 Comparison of the various models. (a) Trap-free model, (b) abundant discrete trap, (c) exponentially distributed trap, (d) exponentially distributed trap and conduction band and (e) model of Shur and Hack (with $E_{F0} = 0.484$ eV, $g_{F0} = 8.93 \times 10^{23}$ $m^{-3}eV^{-1}$, $N_V = 1.04 \times 10^{25}$ m^{-3}, $\varepsilon = 11.9\varepsilon_0$ and other parameters as in Table 6.4). Reproduced with permission from Organic Electronics, Thin-film field-effect transistors: The effects of traps in the bias and temperature dependance of field-effect mobility, including the Meyer-Neldel rule by P. Stallinga and H. L. Gomes, 7, 6, 592–599 Copyright (2006) Elsevier

this equilibrium can take a very long time to establish, in which case the electrical characteristics will depend on things such as the scanning speed and even the history of the device in the case of extremely deep electronic levels. An effect often observed and named 'stress' then might result. This is discussed in the next section.

6.10.4 Traps and 'Stressing' (Threshold-voltage Shift)

When the traps are very deep and the capture of free charge is very slow, the transfer curves are undistorted; the time it takes to measure a transfer curve is much shorter than the relaxation time of charges. However, sustained application of bias will eventually cause the charges to be removed from conductive states and to be captured on the trap states. In a subsequent measurement of the transfer curve, the newly trapped charge will be visible in a shifted threshold voltage, as given by Equation (6.83). For slow traps this equation becomes

$$V_T(t) = -qN_T^+(t)/C_{ox} \tag{6.103}$$

where t is the time the gate bias was applied. This phenomenon of continually shifting of the threshold voltage as caused by the prolonged application of the gate bias is called 'stressing' and is well known in the amorphous silicon community [85, 353–357] and organic community alike [358–368]. This behavior is summarized in Figure 6.42. Since behind this threshold-voltage shift lies the same process of trapping as

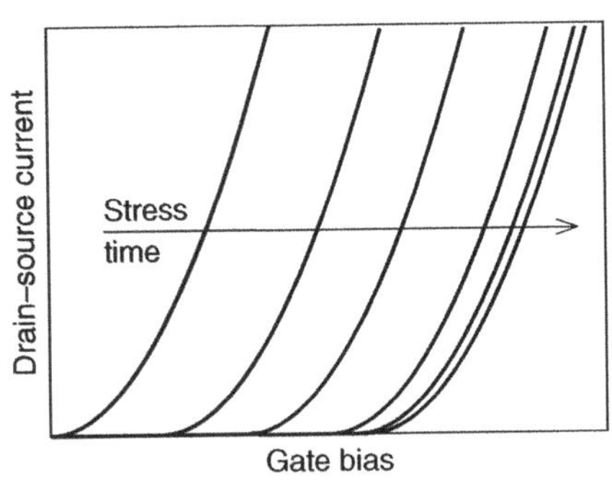

Figure 6.42 Example of stressing. A sustained application of a gate bias (except when rapidly scanning a transfer curve) causes the transfer curves to shift along the gate-bias axis. In modeling terms this means a monotonous increase of threshold voltage V_T. The cause for this shift in threshold voltage is an increase in amount of trapped charge, as given in Equation (6.103)

responsible for other transient behavior, the same kinetics are expected, albeit with time constants in a different range. Indeed power-law functions and stretched exponentials have been proposed for the transient stressing behavior. In the framework of the traps presented in this book, a power law is expected. Experimentally deciding between a stretched exponential and power law is difficult due to the small difference between them and the limitations of experimental data. Rahal *et al.* for instance plot their data in both forms, without clearly deciding between them.

In the presence of abundant traps, all free charge induced by the gate can effectively be removed from the channel and converted into trapped charge. This implies that the final current is zero and thus that the threshold voltage is equal to the stressing gate bias. In this case the channel is marginally closed,

$$I(t=\infty)=0 \tag{6.104}$$

and

$$V_{\mathrm{T}}(t=\infty)=V_{\mathrm{g}}. \tag{6.105}$$

The threshold voltage is proportional to the density of trapped charge. For simple (two-level) kinetics of a single discrete trap level and a conductive level, with the time-derivative of free charge being proportional to the density of free charge, $\mathrm{d}p(t)/\mathrm{d}t=-p(t)/\tau$, a simple exponential behavior results (where τ is the relaxation time or 'characteristic time' of the transient). For traps distributed in energy, a convolution of relaxation times τ exists. As shown in Chapter 4, for exponential distributions, the power law emerges. Thus, under a constant gate bias, the threshold voltage evolves as

$$V_{\mathrm{T}}(t)=V_{\mathrm{g}}-(V_{\mathrm{g}}-V_{\mathrm{T0}})\left(\frac{t}{\tau}\right)^{-\alpha}. \tag{6.106}$$

In comparison, a stretched exponential [369] is of the form

$$\Delta V_{\mathrm{T}}(t)=\Delta V_{\mathrm{T}\infty}\left\{1-\exp\left[-\left(\frac{t}{\tau}\right)^{\beta}\right]\right\} \tag{6.107}$$

where β is between 0 and 1. For completeness sake it is also worth mentioning that Fortunato *et al.* derived another expression for the threshold voltage evolution based on the power-law trapped-charge

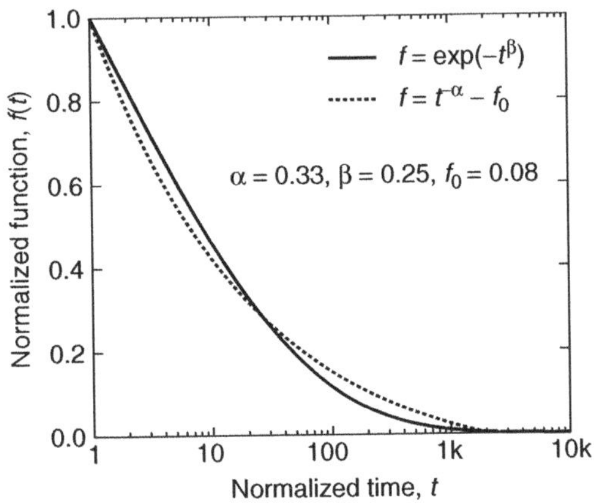

Figure 6.43 Difference between a stretched exponential ($\beta = 0.25$) and a power-law function ($\alpha = 0.33$); experimentally deciding between them is difficult

concept [357]. Experimentally deciding between the models is very difficult due to the fact that they diverge only in the tail of the transient (see Figure 6.43 for a comparison of the stretched exponential and power-law behavior), and due to the way the threshold voltage is measured; To measure the threshold voltage, the stressing gate bias has to be interrupted and the gate bias scanned to obtain a transfer curve from which the threshold voltage can be determined. During this interruption destressing or increased stressing takes place, thus distorting the stressing experiment. For this reason steady-bias experiments, such as current-transients, have preference.

The traps can be emptied by removing the gate bias. The threshold voltage will return to its natural value in a similar way and follow again a power law. This recovery of the device can be enhanced by the application of a reverse bias, for instance applying a positive bias to a hole-channel device. Alternatively, stress reversal can be induced by light [364].

In some cases stressing is observed in the opposite direction, i.e., a reduction of the threshold voltage upon prolonged application of bias [370]. This indicates that trapping might not be the only process involved in the stressing phenomenon. A movement of (water) ions has been suggested [371]. Also, the density of traps need not be constant. During operation new traps may be formed, as suggested by Lyon in SiO_2 devices [372].

6.10.5 Traps and Transients

In the above discussion it was assumed that thermal equilibrium exists at all times. The distribution of the charge over the conductive states and traps states is only determined by their relative abundance and the temperature. The history of the device plays no part. This is true for relatively fast traps. Fast in this case is relative to the measurement of a single data point, which is normally of the order of milliseconds. If trapping happens faster than that, the device can be considered in thermal equilibrium at all times.

When thermalization times are longer than the time it takes to measure a single data point, but faster than the time to measure a full curve (for instance transfer curve), the device can be in a different state at the end of the curve compared with the beginning of the curve. In this case hysteresis is expected. This means that each data point depends on its history inside the curve, for instance if we scan from low bias to high bias or reverse. However, repeating the scan will give identical results. A hysteresis curve (scanning a full loop, returning to the starting bias) repeatedly becomes a closed loop. In such a loop, the current has the tendency to drop because mobile charge is captured, resulting in 'clockwise' loops. Figure 6.44 shows an example of hysteresis observed in a TFT based on T6.

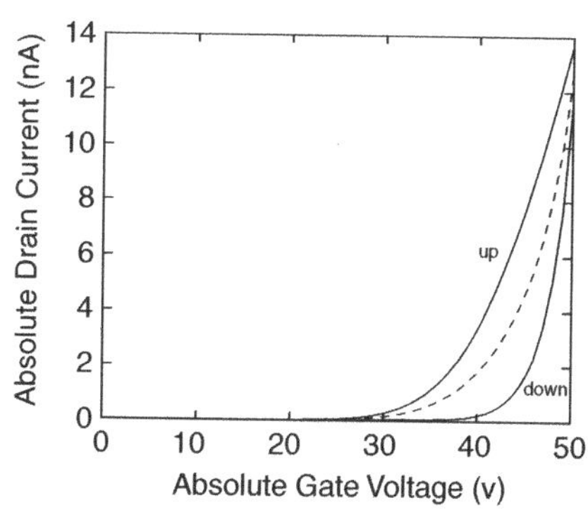

Figure 6.44 Example of hysteresis visible in a transfer curve of a T6 TFT. The dashed line shows the average of the up-scanning curve (top trace) and down-scanning curve (bottom trace). Traps with intermediate-range thermalization times are responsible for this effect. Reproduced with permission from Journal of Applied Physics, Electronic transport in field-effect transistors of sexithiophene by P. Stallinga, 96, 9 Copyright (2006) American Institute of Physics

When thermalization is even slower than the time it takes to measure a curve, the curves depend on the history of the device. Consequently, hysteresis curves are no longer closed loops. Twice measuring the same curve does not result in the same characteristics. The device at the end of the measurement is in a different state compared with the beginning. Any consecutive curve will fall below the previous one, unless the device is given enough time to relax. Because of the depth of the traps, and the associated long thermalization times, these effects can be on extremely long timescales; days, weeks or even longer are no exception.

Table 6.5 summarizes these effects for a measurement system that needs about 1 ms for a single data point and some seconds for measuring a curve. The systems with fastest thermalization have been described in the above sections. In this section the effects of slower traps are discussed.

The most complicated to model is the regime of intermediate thermalization timescales. Therefore it is better to avoid these situations. The hysteresis measurements only serve as an indication for the presence of traps and not as an analytical tool. When the presence of traps has been confirmed, it is better to go to the fast regime, i.e., measure as fast as possible, with pulsed techniques having a preference, or measure in the ultra-slow regime, with transient techniques in the timescale of interest. The complication arises that traps are normally distributed over all energy ranges and thus have highly dispersed thermalization times.

One can also measure the steady-state transient; instead of the repeated-scanning type described above, the device current is monitored at a fixed bias. The current transient that can then be expected is of the type described in the capture experiment of Section 4.2.2. It was shown there that the resulting transient is a power law when the

Table 6.5 Various manifestations of traps according to their depth and thermalization speed

Relative depth	Thermalization time	Effects
Shallow	<1 ms	Nonlinear transfer curves Nonlinear output curves Reduced effective mobility Meyer–Neldel Rule
Intermediate	1 ms–10 s	Hysteresis
Deep	>10 s	Stress

energetic levels are exponentially distributed in energy

$$I_{ds}(t) \propto t^{-\alpha} \tag{6.108}$$

where α depends on the energetic distribution of the conductive states, for instance the valence band, namely

$$\alpha = \frac{T}{T_1} \tag{6.109}$$

and $1/kT_1$ is the slope of the distribution of the conductive states in a logarithmic plot.

Finally, the transient behavior can also be studied by DLTS-like techniques (see Chapter 4), as shown by many authors [184, 373, 374].

6.10.6 The Origin of the Traps

It has been argued at the beginning of this chapter that only the first layer of the semiconductor is functional and conduction takes place there. It is then also logical to assume that the traps are located inside this first layer of the semiconductor. Yet, this conclusion might be too hasty. Traps might be located just below it, in the insulator, or the interface, or even just above in a nonactive layer.

The assumption that the insulator is a perfect insulator and the interface has a flat structure is not likely to be true. First of all, irregularities in the surface of the insulator might cause discontinuities in the first layer of the active layer deposited on top. Some parts of the active layer will be closer to the gate than others and these parts will therefore have lower energy. A charge that is transported from one electrode to the other preferentially resides in these lower-energy pockets, but to make it across, it has to be excited to the higher lying, continuous layer. To give an idea about the depth of these 'traps' we can do a small calculation. Imagine a TFT with -10 V at the gate. With an oxide thickness of 200 nm, this means about 25 meV per monolayer of 0.5 nm as a rough estimate, which is not negligible.

The interface can also be a significant source of electronic deep states. The insulator and the active material are fundamentally made of distinct materials (some exceptions aside, such as GaAs on top of low-temperature-grown GaAs). Lattice mismatches will cause, for example, dangling bonds. The surface (interface between material and vacuum) in silicon has a surface state density of the order equal to the atomic density, 10^{15} cm^{-2}, but by passivation by hydrogen to terminate the dangling bonds, the surface charge can be reduced to the order 10^{10} cm^{-2} [9]. For organic materials deposited on top of silicon oxide, surface treatment to passivate the dangling bonds has been proven to be essential and can easily increase the effective mobility by an order of magnitude [375]. Normal pre-deposition surface treatments are hexamethyldisilazane (HMDS), octadecyltrichlorosilane (OTS) [376], octadecyltrimethoxysilane (OTMS) [377], and O_2 plasma [378].

The insulator therefore plays an important role in the device performance and not only because it has to be as insulating as possible to prevent leakage currents. Alternatives for silicon oxide are sought [319, 379–383], especially organic alternatives are needed, since the goal is to make all-organic electronics [22, 257, 384–391]. For commercial applications the use of inorganic dielectrics such as SiO_2 grown at high temperatures is not of particular interest. A side-effect of using organics is that they possibly result in much higher carrier mobilities. The reason lies in the fact that silicon oxide and like materials have a large density of dangling bonds on silicon atoms (silicon atoms with coordination number equal to 3, where 4 is needed to pair off all covalent electrons), where organics have all atoms nicely covalently bonded. Such dangling bonds are notorious for introducing deep levels in the forbidden gap that are electrically active. In other words, a large density of slow traps can be expected on inorganic insulators. As demonstrated before throughout this chapter, these traps have severe detrimental effects on the behavior of the TFTs. Using organic materials for the insulator might possibly avoid these problems.

An additional interest is that high-K dielectrics can be developed that can lower the operating voltage [392–395]. Remember that the amount of charge induced by the gate is proportional to the oxide capacitance which, in turn, is linearly proportional to ε. For the same thickness and bias, more charge will be induced in the channel, or reversely, for the same charge and thickness a lower bias is needed. On the other hand, low-K insulators can have advantages in terms of increased mobility, reduced threshold voltage and lower hysteresis [396]. Low operating voltages can also be achieved by using ultra-thin insulating films [320], while obviously running the risk of short circuits and increased leakage current. Veres *et al.* wrote an extensive summary on gate dielectrics for organic TFTs [397].

Even with well-terminated dangling bonds, the surface has to be ultra-clean before deposition of the active layer. Any impurity will have a dramatic effect on the performance of the transistor. As an example, water is a known omnipresent impurity that is difficult to remove [398, 399], but its detrimental effect can also be used in a humidity sensor [36]. A direct link between water and the omnipresent phase transition at around 200 K [47] has been proven [48] and water seems to be a good candidate for the source of the traps. Confinement can lower the melting point much below 0°C, see the review article of Debenedetti on supercooled water [400].

The density of traps can also vary; new traps may be generated by the operation of the device. For SiO_2 the injection of electrons and, especially, holes into the oxide creates new interface states, as shown by Lyon [372]. The process of new trap generation has also been suggested to be responsible for the stressing in MOS-FETs [401], although there it seems to be related to current (hot carriers) rather than available carriers as in TFTs.

Even when the interface is of ideal quality, and no impurities are present, self-trapping at the interface via polaron formation can be a source of traps, as calculated by Mu noz Ramo *et al.* for perfect hafnium oxide insulators [402]. It can thus be concluded, as also stated by Veres *et al.* in their summary on gate insulators, that the choice of insulator is as important for high-quality OFET devices as the semiconductor itself [397].

No direct link can be made between the morphology and the transistor performance. The reason is that the performance is determined by the layer(s) close to the interface, whereas the morphology of the film is determined by its quality far away from the interface. Any relation between morphology and performance [324, 403, 404] is thus circumstantial.

6.10.7 Summary of the Effects of Traps on the TFT Characteristics

Summarizing, the presence of traps can manifest itself in many ways in the electrical characteristics of TFTs:

- Low carrier mobility.
- Nonlinear effects in the output curves, often erroneously attributed to the effects of the contacts.
- Nonlinear transfer curves (in the linear region, or nonquadratic transfer curves in the saturation region).
- Temperature dependence of as-measured mobility μ_{FET}.

- Temperature-and-bias dependence of mobility, resulting in the so-called Meyer–Neldel Rule. As such, the MN parameter can indeed be used as a figure-of-merit for the quality of TFT active layer, as suggested by Pichon *et al.* [405].
- Transient behavior. For intermediate traps this means that the measured curves show hysteresis, for the slowest traps, this means 'stressing', a parallel shift of the transfer curves along the voltage axis.

Analyzing the above, there exists thus a relation between the transient behavior and the thermal-activation behavior and the link between the two is traps. In trap-ridden materials, we can expect to have long-lasting transients (of the power-law type, as described in Chapter 4) and observe the MNR. This link has been observed by many people, see for example the work of Chen and Huang [99] or Crandall [85].

When the material quality is increased, for instance by increasing the purity or crystallinity, these effects are expected to disappear simultaneously. For instance, it is expected that when the mobility of the carrier increases, the 'contact effects' will disappear. This is something that would not be expected if these nonlinear effects were truly caused by the contacts. Since the effects described above are mostly unwanted, the solution is simple: increase the quality of the materials and devices. However, the main application of organic devices lies in the cheap-electronics area. Increasing the quality will inevitably induce an increase in cost. Therefore, it seems better to be able to use these traps and invent electronic ways to circumvent the effects. This, however, starts with a good understanding of the traps and their effects. The aim of this text was to give the experimental scientist some tools for the characterization of the materials and devices.

6.11 ADMITTANCE SPECTROSCOPY FOR THE DETERMINATION OF THE MOBILITY IN TFTs

While admittance spectroscopy is most adequate for two-terminal devices, it is interesting to see if it can also be applied to other architectures. Given the fact that the field effect mobility in TFTs is defined as the derivative of the transfer curve [Equation (6.36)], and the fact that admittance spectroscopy is measuring exactly that [Equation (3.2)], it is obvious that admittance spectroscopy is an excellent way of measuring the mobility. The as-measured mobility in TFTs can thus be redefined for admittance spectroscopy as

$$\mu_{\mathrm{FET}} \equiv \frac{L}{C_{\mathrm{ox}} V_{\mathrm{ds}} W} g_{\mathrm{m}} \tag{6.110}$$

where g_{m} is the trans-conductance

$$g_{\mathrm{m}} \equiv \frac{\partial I_{\mathrm{ds}}}{\partial V_{\mathrm{g}}} = \frac{I_{\mathrm{ac}}}{V_{\mathrm{ac}}} \tag{6.111}$$

and I_{ac} is the measured amplitude of the ac current and V_{ac} is the amplitude of the applied voltage signal. Figure 6.45 shows how to measure the TFT mobility in this way.

This technique was applied to a TFT of sexithiophene (T6) with parameters: $C_{\mathrm{ox}} = 192\,\mu\mathrm{F\ m^{-2}}$, $L = 5\,\mu\mathrm{m}$, $W = 1$ cm, $V_{\mathrm{ac}} = 40$ mV, $V_{\mathrm{g}} = -5$ V and $V_{\mathrm{ds}} = -1$ V. Figure 6.46 shows the result. The mobility strongly depends on the frequency of the AC probing voltage. This can be attributed to the presence of traps. For higher frequencies, the charges do not have enough time to thermalize and be trapped on the deep levels. This means that more and more charges remain available for conduction and this increases the effective mobility. The capacitance of the device can be roughly calculated as the area between the electrode, multiplied by the oxide capacitance density, $C = WLC_{\mathrm{ox}} = 9.6$ pF. The measured

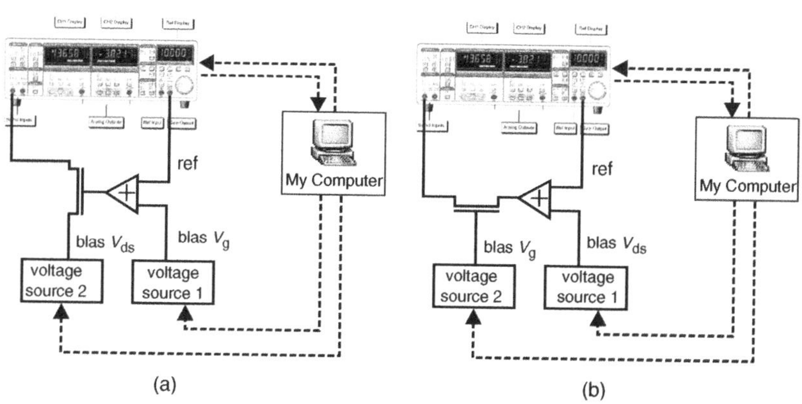

Figure 6.45 Experimental set-up for determining the TFT mobility by admittance spectroscopy based on a lock-in detector, two DC voltage sources for the gate and drain biases and a bias-T that mixes the DC bias with the AC voltage of the reference. The computer controls the data acquisition. (a) The way to measure the mobility via the trans-conductance. (b) An alternative way to measure the mobility via the modulation of the drain voltage

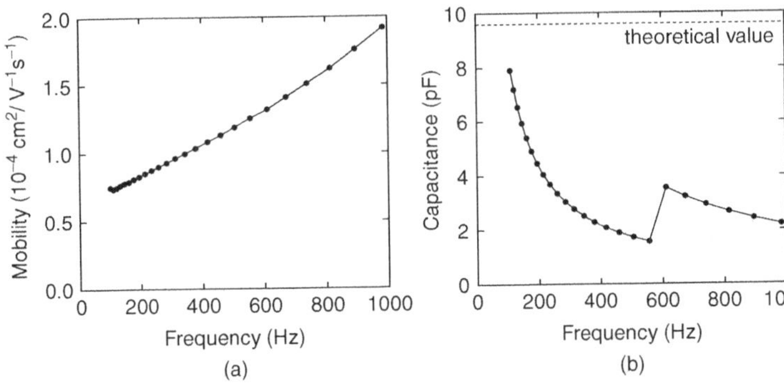

Figure 6.46 (a) Experimental result of determining mobility of a T6 TFT using admittance spectroscopy. The mobility strongly depends on probing frequency. (b) The accompanying capacitance of the device, which drops well below the predicted value of 9.6 pF for higher frequencies

admittance capacitance is well below that for most frequencies. This implies that ever less charge moves into the device during a voltage cycle for higher frequencies. Yet, the charge that does move in compensates by contributing much more to current.

The technique of measuring mobility via admittance spectroscopy has an additional advantage that the bias can be kept constant, compared with the necessity of making a bias scan in the DC methods. In this way, the evolution of the mobility can be evaluated without interrupting the constant bias in a stressing experiment. Figure 6.47 shows the result of measuring the mobility using steady-bias admittance spectroscopy.

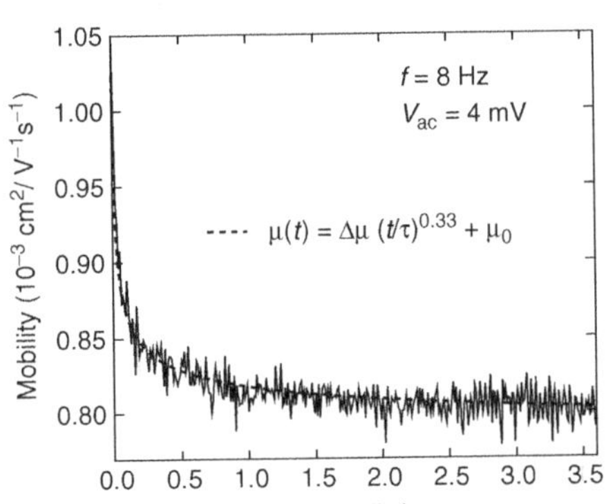

Figure 6.47 Transient measurement of mobility via admittance. The advantage of this method lies in the fact that a constant bias can be applied, which is important when performing so-called stress measurements

6.12 SUMMARY OF TFT MEASUREMENTS

In spite of the fact that a TFT has one more electrode compared with the two-terminal devices such as a resistor or a Schottky diode, its electrical representation is actually easier. This is connected to the fact that the charge (and current) are (assumed to be) governed by the out-of-plane field instead of the in-plane field of the resistor and diodes.

It has been shown here that the electrical characteristics are rather simple. Even the effects of the contacts are described rather straightforwardly, without complicated physical and mathematical models. Ambipolar behavior naturally emerges when ambipolar materials are assumed.

Some complication arises when the material (at the interface) is full of traps. It is shown here that the traps manifest themselves in many ways, including nonlinear effects in both output and transfer curves, temperature dependence of mobility and current, hysteresis effects and general transient effects such as stress, etc.

6.13 DIFFUSION TRANSISTOR

In the simple modeling above it was assumed that all current is drift current, see for example the basic drift-current equation [Equation (6.32)]. From other devices, such as p-n junctions, we know that diffusion currents can play a substantial role in the functioning of the device. These currents are proportional to the density gradients, for instance for hole-diffusion currents,

$$J_{\mathrm{p}} = -qD_{\mathrm{p}}\frac{\mathrm{d}p}{\mathrm{d}x} \qquad (6.112)$$

For a TFT the density gradients are never very large (compared with, for example, the p-n junction or the MOS-FET, where the densities are exponential functions). Diffusion currents are therefore not expected to play a major role. Because the diffusion coefficient and mobility are linked via Einstein's Relation ($D = q\mu/kT$), the contributions of diffusion are similar to effects of biases of the order of the thermal voltage, $V_{\mathrm{th}} = kT/q = 26$ mV (at room temperature); in most cases negligible.

In reality, diffusion current might become important when the mobility and diffusion coefficient are decoupled. Without going into detail about the origins of this phenomenon (see Chapter 5), it is interesting to describe what would happen to the electrical behavior of TFTs if they are indeed decoupled. The extreme case is the opposite of the only-drift-current situation described above, namely only diffusion current, or in other words, $\mu = 0$ and $D \neq 0$. In other words, the drift-current equation [Equation (6.32)] is replaced by the diffusion-current equation [Equation (6.112)], while maintaining the charge–voltage relation of the ideal capacitor [Equation (6.31)]. The differential equation for a diffusion transistor becomes

$$p(x) = C_{\mathrm{ox}}[V(x) - V_{\mathrm{g}}]/q$$
$$I_{\mathrm{x}}(x) = -qD_{\mathrm{p}}W\mathrm{d}p(x)/\mathrm{d}x \qquad (6.113)$$

A first observation is that the diffusion current and the drift current are in the same direction (going from high density to low density), where they are in opposing directions in p-n junctions (as in MOS-FETs). This finds its origin in the fact that the gradient is not maintained by an internal field, but by an external field. The former demands that, at zero bias, these currents are opposing the drift currents caused by the field, whereas the latter only demands that the currents are zero in the absence of bias. Another way of looking at it is saying that the external bias is responsible for maintaining the density gradient, rather than the built-in voltage.

A second observation is that, since the current is constant along the device, the gradient must be constant. Or, in other words, in case there is current, the density drops linearly from one side of the device to the other.

Figure 6.48(a) shows a general case for an n-channel device ($D_{\mathrm{p}} = 0$, $D_{\mathrm{n}} \neq 0$) with zero threshold voltage and a small drain bias compared with the larger gate bias. As can be seen, the solution of the differential equation is a linear drop in charge density. When the drain bias is increased, the charge density gradient is increased (Figure 6.48b) and the current grows linearly with V_{ds}. There are no saturation effects observable. This continues until the drain-bias is equal to the gate bias, which is the pinch-off voltage $V_{\mathrm{p}} = V_{\mathrm{g}}$. In this case the density of charge at the drain is zero. Further increases of V_{ds} will not increase the current and distribution. Similar to the case of pure-drift currents described earlier, the last voltage drop now occurs in an infinitesimal tiny region close to the drain. This is a singularity (and thus difficult to simulate), with an infinite electric field (any residual charge mobility will cause the current to be nonzero; $I_{\mathrm{x}} \propto n\mu E$). What is important is that in this case the current becomes independent of V_{ds}. To summarize, a diffusion transistor

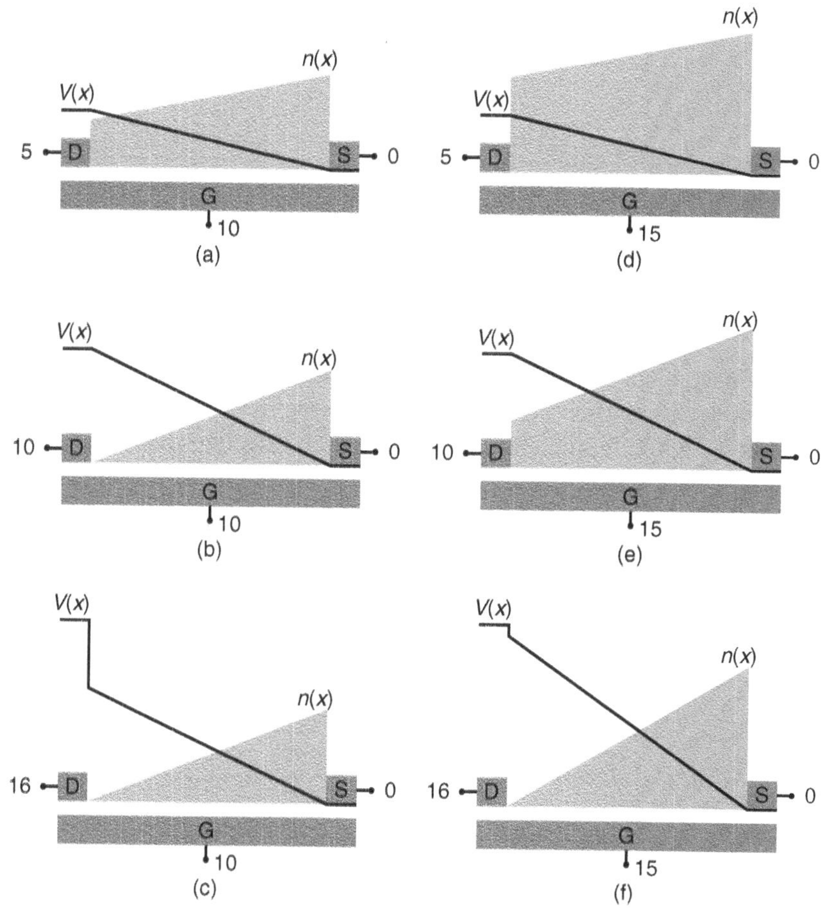

Figure 6.48 Schematic diagram of a TFT showing the voltage and charge density in an n-channel diffusion transistor

has a linearly rising I–V relation until abrupt saturation sets in. This is notably different from a drift-current TFT, where saturation effects are readily visible long before the actual saturation sets in, as evidenced by the quadratic curvature of the I–V curves.

Going back to the initial situation of small V_{ds} compared with V_g (Figure 6.48a), thus before saturation, it is obvious that changes in gate bias will offset the charge density equally along the entire channel and the gradient remains the same, see for example Figure 6.48(d). In this regime, the currents are therefore independent of gate-bias. In fact, the solution to Equation (6.113), with the usual boundary conditions,

$V(0) = 0$, $V(L) = V_{ds}$, and $I_x(x) = I_{ds}$, is

$$I_{ds}^{LIN} = C_{ox} D_p \frac{W}{L} V_{ds} \tag{6.114}$$

i.e., independent of gate bias.

In saturation, it is easily shown that the current depends linearly on V_g. This by substituting the pinch-off voltage into Equation (6.114),

$$I_{ds}^{SAT} = C_{ox} D_p \frac{W}{L} V_g \tag{6.115}$$

The effective field-effect mobility, defined via the derivative of the transfer curves, is thus zero for the linear regime and nonzero in the saturation regime. Somewhat unorthodox, an effective mobility can be obtained by analyzing the saturation regime with a linear-regime model [Equation (6.36)]. This yields $\mu_{FET} = D_p/V_{ds}$

Figure 6.49 shows the I–V and transfer curves of an ideal diffusion transistor. From this it can be seen that the difference between a 'drift transistor' and a 'diffusion transistor' is a one degree lower power law in the transfer curves:

	Drift	Diffusion
Linear	$\propto V_g V_{ds}$	$\propto V_{ds}$
Saturation	$\propto V_g^2$	$\propto V_g$

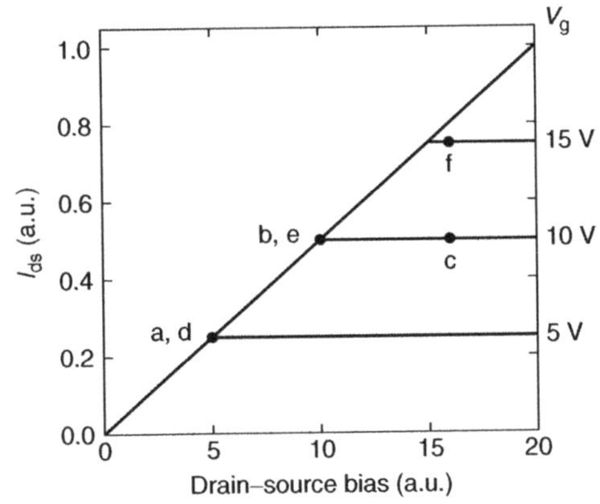

Figure 6.49 Output curves of a diffusion transistor for various gate biases, as indicated on the right-hand side. The situations for the cross-sections shown in Figure 6.48 are indicated

The characteristics become more complicated when both drift and diffusion currents are present. Combining the drift and diffusion currents results in a differential equation of the form

$$I_x(x) = -WC_{ox}\left(D_p + \mu_p\left[V(x) - V_g\right]\right)\frac{dV(x)}{dx} \qquad (6.116)$$

This shows that the diffusion current has an effect similar to a gate bias of $V_g = D_p/\mu_p$. If Einstein's Relation holds, this is 26 mV (at room temperature), and negligible in most cases. There has been some evidence, though, that highly disordered materials, such as organic semiconductors, can deviate from Einstein's Relation. As an academic exercise, the extreme case with $\mu = 0$ has been presented in this section.

A

Derivation of Equations (2.21), (2.25), (6.95) and (6.101)

To arrive at Equations (2.21) and (6.95) a density of states (DOS) exponentially decaying in energy is used,

$$N_{\mathrm{T}}(E) = N_{\mathrm{T}0} \exp\left(\frac{E_{\mathrm{V}} - E}{kT_2}\right) \tag{A.1}$$

where $N_{\mathrm{T}0}$ is the density of trap states at $E = E_{\mathrm{V}}$ and T_2 is the decay rate, parameters that describe the distribution. The dependence of N_{T}^{+} on the position of the Fermi level thus becomes

$$N_{\mathrm{T}}^{+}(E_{\mathrm{F}}) = \int_{-\infty}^{\infty} N_{\mathrm{T}}(E)\left[1 - f(E - E_{\mathrm{F}})\right] \mathrm{d}E \tag{A.2}$$

where f is the Fermi–Dirac distribution function,

$$f(E - E_{\mathrm{F}}) = \frac{1}{1 + \exp\left[(E - E_{\mathrm{F}})/kT\right]}. \tag{A.3}$$

The integral of Equation (A.2) converges when $T < T_2$. To a good approximation, the solution can be found by dividing the integral into two parts, see Figure A.1. In the first part, below E_{F}, the slope is $1/kT - 1/kT_2$ as a result of the difference of slopes in N_{T} and the exponential approximation for $1 - f$. Above E_{F}, $1 - f$ is considered

Electrical Characterization of Organic Electronic Materials and Devices Peter Stallinga

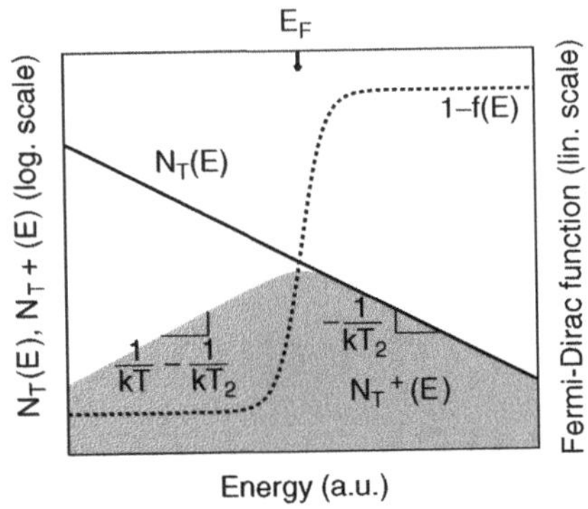

Figure A.1 Graphical schematic of the distribution in energy of trap states $N_T(E)$ (solid line) and charged trap states N_T^+ (shaded area). The latter is a result of a multiplication of the former by the Fermi–Dirac function $1 - f(E)$ (dashed line). This shows that the total trapped charge N_T^+ as a function of Fermi level, see the integral of Equation (A.4), can easily be approximated by dividing the integral into two parts. Reproduced with permission from Organic Electronics, Thin-film field-effect transistors: The effects of traps in the bias and temperature dependance of field-effect mobility, including the Meyer-Neldel rule by P. Stallinga and H. L. Gomes, 7, 6, 592–599 Copyright (2006) Elsevier

unity and the resulting slope is $1/kT_2$. With this help, it can easily be shown that the integral is equal to

$$N_T^+(E_F) = N_{TT}(T)\exp\left(\frac{E_V - E_F}{kT_2}\right) \tag{A.4}$$

where N_{TT} is

$$N_{TT}(T) = \alpha(T)N_{T0}\frac{(kT_2)^2}{kT_2 - kT} \tag{A.5}$$

in which $\alpha(T)$ is an ad-hoc correction factor that compensates for the error of integration; compare the rounded distribution of N_T^+ of Figure A.1 and the triangular integration described above. Numerical simulations show that $\alpha(T)$ oscillates between 1 and 0.8 in the temperature range $0–T_2$, see Figure A.2). This makes N_{TT} essentially temperature independent for T not very close to T_2. For $T \geq T_2$ the integral diverges.

If the conduction states are assumed to be discrete in energy, a standard valence band with N_V states at E_V, the density of holes follows

$$p(E_F) = N_V \exp\left(-\frac{E_F - E_V}{kT}\right). \tag{A.6}$$

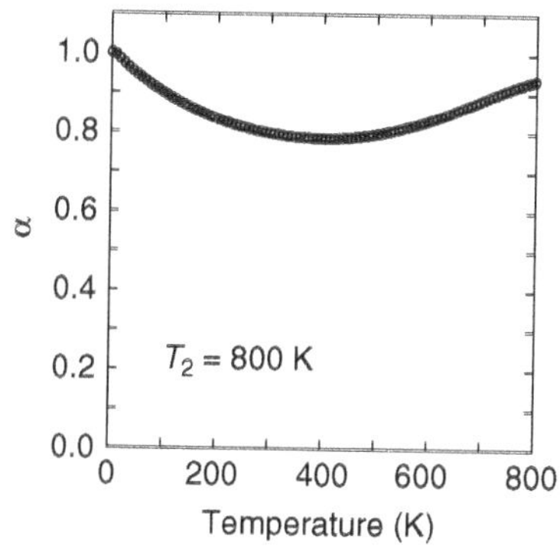

Figure A.2 The ad-hoc correction factor α as a function of temperature, calculated numerically. α is a slowly varying function of T and its contribution to the calculation is minimal. For the calculations, α is approximated by a third-order polynomial (solid line). Reproduced with permission from Organic Electronics, Thin-film field-effect transistors: The effects of traps in the bias and temperature dependance of field-effect mobility, including the Meyer-Neldel rule by P. Stallinga and H. L. Gomes, 7, 6, 592–599 Copyright (2006) Elsevier

These two densities of p and N_T^+ can be introduced into the appropriate total charge equation, ρ. Ignoring the tiny contribution of the free (and trapped) electron density n, the total charge is all positive charge, $\rho = p + N_T^+$, or

$$p + p^{T/T_2} \frac{N_{TT}(T)}{N_V^{T/T_2}} = \rho. \qquad (A.7)$$

For high densities of traps, the first term is negligible and the free hole density can then easily be determined as

$$p(\rho) = N_V \left[\frac{\rho}{qN_{TT}(T)} \right]^{T_2/T}. \qquad (A.8)$$

The current is proportional to the free charge and is thus a function of temperature. This is equal to Equations (2.21) and (6.95) if we substitute $\rho = \rho_0$ or $\rho = -C_{ox}V_g$, respectively.

To arrive at Equations (2.25) and (6.101), conduction states also exponentially distributed in energy are used:

$$N_V(E) = N_{V0} \exp\left(\frac{E_V - E}{kT_1}\right) \qquad (A.9)$$

where N_{V0} is the density of states at $E = E_V$, and T_1 is a parameter describing the distribution, as explained in the main text. Similar integration techniques result in

$$p(E_F) = N_{VT}(T)\exp\left(\frac{E_V - E_F}{kT_1}\right) \tag{A.10}$$

with

$$N_{VT}(T) = \beta(T)N_{V0}\frac{(kT_1)^2}{kT_1 - kT} \tag{A.11}$$

in which β is a function similar to α, but scaled with T_1 instead of T_2. Again, this can be substituted in the total charge equation and for high trap densities ($N_T^+ \gg p$), the free charge density follows

$$p(\rho) = N_{VT}(T)\left[\frac{\rho}{qN_{TT}(T)}\right]^{T_2/T_1} \tag{A.12}$$

If ρ represents the charge induced by the gate, $\rho = -C_{ox}V_g$, then the free charge density p and the current are nonlinear functions of V_g, as shown in Equation (6.101). The derivative of the function yields the mobility which can thus also be bias and temperature dependent. In DC measurements of two-terminal devices, a temperature-dependent current will be observed, as shown in Equation (2.25).

Bibliography

[1] L. G. Wade Jr. Organic Chemistry, Prentice Hall (1987).
[2] P. R. Wallace, Phys. Rev. **71**, 622 (1947).
[3] S. Ashley, Sci. Am. March, 14 (2009); A. K. Geim, K. S. Novoselov, Nat. Mater. **6**, 183 (2007). DOI: 10.1038/nmat1849.
[4] J. R. Sheats, J. Mater. Res. **19**, 1974 (2004).
[5] F. Garnier, Curr. Opin. Solid State Mater. Sci. **2**, 455 (1997).
[6] C. D. Dimitrakopoulos, P. R. L. Malenfant, Adv. Mater. **14**, 99 (2002).
[7] C. D. Dimitrakopoulos, D. J. Mascaro, IBM J. Res. Dev. **45**, 11 (2001).
[8] J. M. Shaw, P. F. Seidler, IBM J. Res. Dev. **45**, 3 (2001).
[9] S. M. Sze, Physics of Semiconductor Devices, 2nd Edn, John Wiley & Sons, Ltd (1981).
[10] M. Shur, Physics of Semiconductor Devices, Prentice Hall (1990).
[11] P. Blood, J. W. Orton, The Electrical Characterization of Semiconductors: Majority Carriers and Electron States, Techniques of Physics, Vol. 14. Academic Press (1992).
[12] H. Sirringhaus, T. Kawase, R. H. Friend, T. Shimoda, M. Inbasekaran, W. Wu, E. P. Woo, Science **290**, 2123 (2000).
[13] C. Clavijo Cedeño, J. Seekamp, A. P. Kam, T. Hoffmann, S. Zankovych, C. M. Sotomayor Torres, C. Menozzi, M. Cavallini, M. Murgia, G. Ruani, F. Biscarini, M. Behl, R. Zentel, J. Ahopelto, Microelectron. Eng. **61–62**, 25 (2002).
[14] J. Zaumseil, T. Someya, Z. Bao, Y.-L. Loo, R. Cirelli, J. A. Rogers, Appl. Phys. Lett. **82**, 793 (2003). DOI: 10.1063/1/1541941.
[15] P. Mach, S. J. Rodriguez, R. Nortrup, P. Wiltzius, J. A. Rogers, Appl. Phys. Lett. **78**, 3592 (2001). DOI: 10.1063/1.1377312.
[16] A. Knobloch, A. Manuelli, A. Bernds, W. Clemens, J. Appl. Phys. **96**, 2286 (2004). DOI: 10.1063/1.1767291.
[17] J. A. Rogers, Z. Bao, M. Meier, A. Dodabalapur, O. J. A. Schueller, G. M. Whitesides, Synth. Met. **115**, 5 (2000).
[18] P. Calvert, Chem. Mater. **13**, 3299 (2001). DOI: 10.1021/cm0101632.
[19] J. Z. Wang, Z. H. Zheng, H. W. Li, W. T. S. Huck, H. Sirringhaus, Nature Mater. **3**, 171 (2004). DOI: 10.1038/nmat1073.

Electrical Characterization of Organic Electronic Materials and Devices Peter Stallinga

[20] C. D. Sheraw, L. Zhou, J. R. Huang, D. J. Gundlach, T. N. Jackson, M. G. Kane, I. G. Hill, M. S. Hammond, J. Campi, B. K. Greening, J. Francl, J. West, Appl. Phys. Lett. **80**, 1088 (2002).

[21] L. A. Majewski, R. Schroeder, M. Voight, M. Grell, J. Phys. D: Appl. Phys. **37**, 3367 (2004). DOI: 10.1088/0022-3727/37/24/003.

[22] R. Parashkov, E. Becker, G. Ginev, T. Riedl, H.-H. Johannes, W. Kowalsky, Jpn. J. Appl. Phys. **43**, L130 (2004).

[23] G. H. Gelinck, H. E. A. Huitema, E. van Veenendaal, E. Cantatore, L. Schrijnemakers, J. B. P. H. van der Putten, T. C. T. Geuns, M. Beenhakkers, J. B. Giesbers, B.-H. Huisman, E. J. Meijer, E. Mena Benito, F. J. Touwslager, A. W. Marsman, B. J. E. van Rens, D. M. de Leeuw, Nature Mater. **3**, 106 (2004). DOI: 10.1038/nmat1061.

[24] J. B. Lee, V. Subramanian, IEEE Trans. Electron Devices **52**, 269 (2005). DOI: 10.1109/TED.2004.841331.

[25] H. Rost, J. Ficker, J. Sanchez Alonso, L. Leenders, I. McCulloch, Synth. Met. **145**, 83 (2004). DOI: 10.1016/j.synthmet.2004.04.008.

[26] S. H. Park, S. Cho, J. K. Lee, K. Lee, A. J. Heeger, Org. Electron. (2009). DOI: 10.1016/ j.orgel.2009.01.008.

[27] A. Aviram, M. A. Ratner, Chem. Phys. Lett. **29**, 277 (1974).

[28] K. Bradley, J.-C. P. Gabriel, G. Grüner, Nano Lett. **3**, 1353 (2003).

[29] P. Qi, A. Javey, M. Rolandi, Q. Wang, E. Yenilmez, H. Dai, J. Am. Chem. Soc. **126**, 11774 (2004). DOI: 10.1021/ja045900k.

[30] M. C. McAlpine, R. S. Friedman, S. Jin, K.-H. Lin, W. U. Wang, C. M. Lieber, Nano Lett. **3**, 1531 (2003). DOI: 10.1021/nl0346427.

[31] L. Torsi, A. Dodabalapur, L. Sabbatini, P. G. Zambonin, Sens. Actuators B **67**, 312 (2000); L. Torsi, A. Dodabalapur, N. Cioffi, L. Sabatini, P. G. Zambonin, Sens. Actuators B 77, 7 (2001); L. Torsi, M. C. Tanese, N. Cioffi, M. C. Gallazzi, L. Sabbatini, P. G. Zambonin, Sens. Actuators B **98**, 204 (2004). DOI: 10.1016/j.snb.2003.10.007; L. Torsi, M. C. Tanese, N. Cioffi, M. C. Gallazzi, L. Sabbatini, P. G. Zambonin, G. Raos, S. V. Meille, M. M. Giangregorio, J. Phys. Chem. B **107**, 7589 (2003). DOI: 10.1021/jp0344951; L. Torsi, A. J. Lovinger, B. Crone, T. Someya, A. Dodabalapur, H. E. Katz, A. Gelperin, J. Phys. Chem. B **106**, 12563 (2002). DOI: 10.1021/jp021473q.

[32] E. Bentes, H. L. Gomes, P. Stallinga, L. Moura, Proc. IEEE Sens. **2**, 766 (2004). DOI: 10.1109/ICSENS.2004.1426281.

[33] J. A. Covington, J. W. Gardner, D. Briand, N. F. de Rooij, Sens. Actuators B **77**, 155 (2001).

[34] C. Bartic, B. Palan, A. Campitelli, G. Borghs, Sens. Actuators B **83**, 115 (2002); C. Bartic, A. Campitelli, S. Borghs, Appl. Phys. Lett. **82**, 475 (2003). DOI: 10.1063/1.1527698.

[35] B. Crone, A. Dodabalapur, A. Gelperin, L. Torsi, H. E. Katz, A. J. Lovinger, Z. Bao, Appl. Phys. Lett. **78**, 2229 (2001).

[36] Z.-T. Zhu, J. T. Mason, R. Dieckmann, G. G. Malliaras, Appl. Phys. Lett. **81**, 4643 (2002). DOI: 10.1063/1.1527233.

[37] M. Leclerc, Adv. Mater. **11**, 1491 (1999).

[38] H.-A. Ho, M. Boissinot, M. G. Bergeron, G. Corbeil, K. Doré, D. Boudreau, M. Leclerc, Angew. Chem. Int. Ed. **41**, 1548 (2002).

[39] L. Looger, M. A. Dwyer, J. J. Smith, H. W. Hellinga, Nature **423**, 185 (2003).

[40] G. Maruccio, R. Cingolani, R. Rinaldi, J. Mater. Chem. **14**, 542 (2004).

[41] G. Maruccio, P. Visconti, A. Biasco, A. Bramanti, A. Della Torre, P. P. Pompa, V. Frascerra, V. Arima, E. D'Amone, R. Cingolani, R. Rinaldi, Electroanal. **16**, 1853 (2004).

[42] V. A. L. Roy, E. L.-M. Wong, B. C.-B. Ko, C.-T. Chao, S.-C. Yan, B. Yan, J.-C. Wu, C.-M. Che, Adv. Mater. **20**, 1258 (2008). DOI: 10.1002/adma.200701179.
[43] S. Damico, G. Maruccio, P. Visconti, E. Damone R. Cingolani, R. Rinaldi, S. Masiero, G. P. Spada, G. Gottarelli, Microelectron. J. **34**, 961 (2003).
[44] G. Caserta, B. Rispoli, A. Serra, Phys. Status Solidi **35**, 237 (1969).
[45] H. Bässler, Phys. Status Solidi B **175**, 15 (1993).
[46] A. T. Fromhold Jr. Quantum Mechanics for Applied Physics and for Engineering, Dover Publications (1981).
[47] P. Stallinga, H. L. Gomes, F. Biscarini, M. Murgia, D. M. de Leeuw, J. Appl. Phys. **96**, 5277 (2004).
[48] H. L. Gomes, P. Stallinga, M. Cölle, D. M. de Leeuw, F. Biscarini, Appl. Phys. Lett. **88**, 082101 (2006).
[49] T. Holstein, Ann. Phys. **8**, 325 (1959); T. Holstein, Ann. Phys. **8**, 343 (1959).
[50] S. Brazovskii, N. Kirova, Z. G. Yu, A. R. Bishop, A. Saxena, Opt. Mater. **9**, 502 (1998).
[51] M. P. Sarachik, P. Dai, Phys. Status Solidi B **230**, 205 (2002).
[52] V. Ambegaokar, B. I. Halperin, J. S. Langer, Phys. Rev. B **4**, 2612 (1971).
[53] S. D. Baranovskii, I. P. Zvyagin, H. Cordes, S. Yamasaki, P. Thomas, Phys. Status Solidi B **230**, 281 (2002).
[54] S. V. Rakhmanova, E. M. Conwell, Appl. Phys. Lett. **76**, 3822 (2000).
[55] K. Waragai, H. Akimichi, S. Hotta, H. Kano, Phys. Rev. B **52**, 1786 (1995).
[56] S. F. Nelson, Y.-Y. Lin, D. J. Gundlach, T. N. Jackson, Appl. Phys. Lett. **72**, 1854 (1998). DOI:10.1063/1.121205.
[57] D. C. Look, D. C. Walters, M. O. Manashreh, J. R. Sizelove, C. E. Stitz, K. R. Evans, Phys. Rev. B **42**, 3578 (1990).
[58] N. F. Mott, W. D. Twose, Adv. Phys. **10**, 107 (1961).
[59] B. I. Shklovskii, Sov. Phys. Semicond. **6**, 1053 (1973).
[60] H. A. Albuquerque, A. G. de Oliveira, G. M. Ribeiro, R. L. da Silva, R. M. Rubinger, J. Appl. Phys. **95**, 3553 (2004). DOI: 10.1063/1.1650530.
[61] G. Guillaud, R. Ben Chaabane, C. Jouve, M. Gamoudi, Thin Solid Films **258**, 279 (1995).
[62] L. Torsi, Microelectron. Reliab. **40**, 779 (2000).
[63] P. W. Atkins, Quanta, A Handbook of Concepts, 2nd Edn, University Press (1991).
[64] N. Karl, Adv. Sol. Stat. Phys. **14**, 261 (1974).
[65] J. Cornil, D. Beljonne, J.-P. Calbert, J.-L. Brédas, Adv. Mater. **13**, 1053 (2001).
[66] A. Troisi, G. Orlandi, J. Phys. Chem. B **109**, 1849 (2005). DOI: 10.1021/jp0457489.
[67] T. Rada, Q. Chen, N. V. Richardson, J. Phys.: Condens. Matter **15**, S2749 (2003).
[68] F. Garnier, Chem. Phys. **227**, 253 (1998).
[69] D. J. Gundlach, J. A. Nichols, L. Zhou, T. N. Jackson, Appl. Phys. Lett. **80**, 2925 (2002).
[70] D. J. Gundlach, Y. Y. Lin, T. N. Jackson, S. F. Nelson, D. G. Schlom, IEEE Electron Device Lett. **18**, 87 (1997).
[71] L. Torsi, A. Dodabalapur, L. J. Rothberg, A. W. P. Fung, H. E. Katz, Science **272**, 1462 (1996). DOI: 10.1126/science.272.5267.1462.
[72] J. S. Blakemore, Semiconductor Statistics, Dover Publications (1987).
[73] T. Minakata, I. Nagoya, M. Ozaki, J. Appl. Phys. **69**, 7354 (1991); T. Minakata, H. Imai, M. Ozaki, J. Appl. Phys. **72**, 4178 (1992); T. Minakata, M. Ozaki, J. Appl. Phys. **73**, 1819 (1993); T. Minakata, M. Ozaki, H. Imai, J. Appl. Phys. **74**, 1079 (1993).
[74] H.-J. Egelhaaf, L. Lüer, D. Oelkrug, G. Winter, P. Haisch, M. Hanack, Synth. Met. **84**, 897 (1997).

[75] J. Laubender, L. Chkoda, M. Sokolowski, E. Umbach, Synth. Met. **111**, 373 (2000).
[76] D. M. Taylor, H. L. Gomes, A. E. Underhill, S. Edge, P. I. Clemenson, J. Phys. D: Appl. Phys. **24**, 2032 (1991). DOI: 10.1088/0022-3727/24/11/019.
[77] S. Gasiorowicz, Quantum Physics, 2nd Edn, John Wiley & Sons, Ltd (1995).
[78] W. H. Press, S. A. Teukolsky, W. T. Vetterling, B. P. Flannery, Numerical Recipes in C, The Art of Scientific Computing, 2nd Edn Cambridge University Press (1992). This book is available for free on-line at http://www.nr.com/.
[79] M. Shur, M. Hack, J. Appl. Phys. **55**, 3831 (1984).
[80] N. Tessler, Y. Roichman, Org. Electron. **6**, 200 (2005).
[81] W. Meyer, H. Neldel, Z. Tech. **18**, 588 (1937) (in German).
[82] D. J. Fisher, The Meyer–Neldel Rule: The Compensation Effect (Defect and Diffusion Forum), Trans-tech Publications (2001).
[83] Y. Lubianiker, I. Balberg, J. Non-Cryst. Solids **227–230**, 180 (1998).
[84] M. Kondo, Y. Chida, A. Matsuda, J. Non-Cryst. Solids **198–200**, 178 (1996).
[85] R. S. Crandall, Phys. Rev. B **43**, 4057 (1991).
[86] S. K. Ram, S. Kumar, R. Vanderhaghen, P. Roca i Cabarrocas, J. Non-Cryst. Solids **299–302**, 411 (2002).
[87] L. Pichon, A. Mercha, R. Carin, T. Mohammed-Brahim, O. Bonnaud, Y. Helen, Solid-State Electron. **46**, 459 (2002).
[88] K. L. Ngai, Solid State Ionics **105**, 231 (1998).
[89] W. Bogusz, F. Krok, W. Piszczatowski, Solid State Ionics **119**, 165 (1999).
[90] K. Shimakawa, F. Abdel-Wahab, Appl. Phys. Lett. **70**, 652 (1997). DOI: 10.1063/1.118323.
[91] E. J. Meijer, M. Matters, P. T. Herwig, D. M. de Leeuw, T. M. Klapwijk, Appl. Phys. Lett. **76**, 3433 (2000).
[92] J. C. Wang, Y. F. Chen, Appl. Phys. Lett. **73**, 948 (1998).
[93] D. M. Goldie, J. Mater. Sci. **37**, 3323 (2002).
[94] E. J. Meijer, D. B. A. Rep, D. M. de Leeuw, M. Matters, P. T. Herwig, T. M. Klapwijk, Synth. Met. **121**, 1351 (2001).
[95] J. A. Merlo, C. D. Frisbie, J. Phys. Chem. B **108**, 19169 (2004).
[96] J. C. Dyre, J. Phys. C: Sol. Stat. Phys. **19**, 5655 (1986).
[97] R. Widenhorn, A. Rest, E. Bodegom, J. Appl. Phys. **91**, 6524 (2002).
[98] R. Widenhorn, M. Fitzgibbons, E. Bodegom, J. Appl. Phys. **96**, 7379 (2004). DOI: 10.1063/1.1818353.
[99] Y. F. Chen, S. F. Huang, Phys. Rev. B **44**, 13775 (1991).
[100] J. P. M. Serbena, I. A. Hümmelgen, T. Hadizad, Z. Y. Wang, Small **2**, 372 (2006). DOI: 10.1002/smll.200500305.
[101] M. S. Meruvia, A. R. Benvenho, I. A. Hümmelgen, A. A. Pasa, W. Schwarzacher, Appl. Phys. Lett. **86**, 263504 (2005).
[102] L. Ma, Y. Yang, Appl. Phys. Lett. **85**, 5084 (2004). DOI: 10.1063/1.1821629.
[103] Y.-C. Chao, H.-F. Meng, Appl. Phys. Lett. **88**, 223510 (2006).
[104] K. Kudo, D. X. Wang, M. Iizuka, S. Kuniyoshi, K. Tanaka, Thin Solid Films **331**, 51 (1998).
[105] M. S. Meruvia, A. R. Benvenho, I. A. Hümmelgen, J. A. Gomez, C. F. O. Graeff, R. W. C. Li, L. H. J. M. C. Aguiar, J. Gruber, Phys. Status Solidi A **202**, R158 (2005). DOI: 10.1002/pssa.200521270.
[106] P. Bröms, J. Birgersson, N. Johansson, M. Löglund, W. R. Salaneck, Synth. Met. **74**, 179 (1995).
[107] I. D. Parker, J. Appl. Phys. **75**, 1656 (1994).
[108] N. Zheludev, Nat. Photonics **1**, 189 (2007).
[109] C. W. Tang, S. A. VanSlyke, Appl. Phys. Lett. **51**, 913 (1987).

[110] J. H. Burroughes, D. D. C. Bradley, A. R. Brown, R. N. Marks, K. Mackay, R. H. Friend, P. L. Burns, A. B. Holmes, Nature **347**, 539 (1990). DOI: 10.1038/347539a0.

[111] R. H. Friend, R. W. Gymer, A. B. Holmes, J. H. Burroughes, R. N. Marks, C. Taliani, D. D. C. Bradley, D. A. Dos Santos, J. L. Brédas, M. Lögdlund, W. R. Salaneck, Nature **397**, 121 (1999).

[112] S. R. Forrest, Org. Electron. **4**, 45 (2003); S. R. Forrest, Nature **428**, 911 (2004).

[113] P. W. M. Blom, M. J. M. de Jong, C. T. H. F. Liedenbaum, Polym. Adv. Technol. **9**, 390 (1998).

[114] Z. Wu, H. Yang, Y. Duan, W. Xie, S. Liu, Y. Zhao, Semicond. Sci. Technol. **18**, L49 (2003).

[115] J. R. Lakowicz, I. Gryczynski, Y. Shien, J. Malicka, S. D'Auria, Z. Gryczynski in Fluoresence Spectroscopy, Imaging and Probes, New Tools in Chemical Physical and Life Sciences, Editors R. Kraayerhof, A. J. W. G. Visser, H. C. Gerritsen, Springer, 43 (2002).

[116] A. S. Riad, S. M. Khalil, S. Darwish, Thin Solid Films **249**, 219 (1994).

[117] T. G. Abdel-Malik, A. A. Ahmed, A. S. Riad, Phys. Status Solidi A **121**, 507 (1990).

[118] W. Shockley, H. J. Queisser, J. Appl. Phys. **32**, 510 (1961).

[119] R. H. Fowler, L. Nordheim, Proc. Phys. Soc. London, Sect. A **119**, 173 (1928). DOI: 10.1098/rspa.1928.0091.

[120] I. A. Hümmelgen, L. S. Roman, F. C. Nart, L. O. Péres, E. L. Sá, Appl. Phys. Lett. **68**, 3194 (1996).

[121] M. Koehler, I. A. Hümmelgen, Appl. Phys. Lett. **70**, 3254 (1997).

[122] A. Rose, Phys. Rev. **97**, 1538 (1955).

[123] H. C. Kao, W. Hwang, Electrical Transport in Solids with Particular Reference to Organic Semiconductors, Pergamon Press (1981).

[124] N. F. Mott, R. W. Gurney, Electronic Processes in Ionic Crystals, Oxford University Press (1940).

[125] M. A. Lampert, Phys. Rev. **103**, 1648 (1956).

[126] D. Natali, M. Sampietro, J. Appl. Phys. **92**, 5310 (2002).

[127] J. Lee, D. K. Hwang, C. H. Park, S. S. Kim, S. Im, Thin Solid Films **451–452**, 12 (2004).

[128] T. Y. Choi, H. S. Kang, D. H. Park, J. M. Koo, J. K. Lee, S. D. Ahn, J. Joo, Synth. Met. **137**, 929 (2003).

[129] P. Chattopadhyay, Semicond. Sci. Technol. **10**, 1099 (1995).

[130] L. J. van der Pauw, Philips Res. Rep. **13**, 1 (1958).

[131] S. B. Catalano, IEEE Trans. Electron Devices **10**, 185 (1963).

[132] J. Shao, G. T. Wright, Solid-State Electron. **3**, 291 (1961).

[133] P. Stallinga, H. L. Gomes, H. Rost, A. B. Holmes, M. G. Harrison, R. H. Friend, J. Appl. Phys. **89**, 1713 (2001).

[134] J. Santamaria, G. Gonzalez Diaz, E. Iborra, I. Martil, F. Sanchez-Quesada, J. Appl. Phys. **65**, 3236 (1989).

[135] D. M. Taylor, H. L. Gomes, J. Phys. D: Appl. Phys. **28**, 2554 (1995).

[136] E. H. Nicollian, A. Goetzberger, The Bell System Technical Journal **46**, 1055 (1967).

[137] H. C. Card, E. H. Rhoderick, Solid-State Electron. **15**, 993 (1972).

[138] P. Stallinga, H. L. Gomes, M. Murgia, K. Müllen, Org. Electron. **3**, 43 (2002).

[139] P. Gray, Phys. Rev. **140**, A179 (1965).

[140] H. C. F. Martens, J. N. Huiberts, P. W. M. Blom, Appl. Phys. Lett. **77**, 1852 (2000).

[141] H. C. F. Martens, H. B. Brom, P. W. M. Blom, H. F. M. Schoo, Phys. Status Solidi **218**, 283 (2000).

[142] S. W. Tsang, S. K. So, J. B. Xu, J. Appl. Phys. **99**, 013706 (2006).
[143] M. Schmeits, J. Appl. Phys. **101**, 084508 (2007). DOI: 10.1063/1.2719014.
[144] N. D. Nguyen, M. Schmeits, H. P. Loebl, Phys. Rev. B **75**, 075307 (2007).
[145] P. Stallinga, A. R. V. Benvenho, E. C. P. Smits. S. G. J. Mathijssen, M. Cölle, H. L. Gomes, D. M. de Leeuw, Org. Electron. **9**, 735 (2008).
[146] E. Smits, T. D. Anthopoulos, S. Setayesh, E. van Veenendaal, R. Coehoorn, P. W. M. Blom, B. de Boer, D. M. de Leeuw, Phys. Rev. B **73**, 205316 (2006).
[147] S. E. Guidoni, C. M. Aldao, Eur. J. Phys. **23**, 395 (2002).
[148] T. Speck, U. Seifert, Europhys. Lett. **74**, 391 (2006).
[149] J. Shewchun, M. A. Green, J. Appl. Phys. **46**, 5179 (1975).
[150] M. A. Green, J. Shewchun, J. Appl. Phys. **46**, 5185 (1975).
[151] A. Y. C. Yu, E. H. Snow, Solid-State Electron. **12**, 155 (1969).
[152] H. C. Card, E. H. Rhoderick, J. Phys. D: Appl. Phys. **4**, 1602 (1971).
[153] H. C. Card, E. H. Rhoderick, Solid-State Electron. **16**, 365 (1973).
[154] J. Shewchun, R. Singh, M. A. Green, J. Appl. Phys **48**, 765 (1977).
[155] T. Misawa, J. Phys. Soc. Jpn. **12**, 882 (1957).
[156] C. H. Champness, J. Pan, Can. J. Phys. **66**, 168 (1988).
[157] C. Lungenschmied, E. Ehrenfreund, N. S. Sariciftci, Org. Electron. **10**, 115 (2009).
[158] M. A. Green, J. Shewchun, Solid-State Electron. **16**, 1141 (1973).
[159] J. Werner, A. F. J. Levi, R. T. Tung, M. Anzlowar, M. Pinto, Phys. Rev. Lett. **60**, 53 (1987).
[160] S. Kar, W. E. Dahlke, Solid-State Electron. **15**, 221 (1972).
[161] D. Seghier, H. P. Gislason, J. Phys. D: Appl. Phys. **38**, 843 (2005).
[162] J. R. Kirtley, T. N. Theis, P. M. Mooney, S. L. Wright, J. Appl. Phys. **63**, 1541 (1988).
[163] A. K. Rice, K. J. Malloy, J. Appl. Phys. **87**, 7892 (2000).
[164] Z. Çelik-Butler, P. Vasina, N. Vibhavie Amarasinghe, IEEE Trans. Electron Devices **47**, 646 (2000).
[165] F. N. Hooge, Phys. Lett. A **29**, 139 (1969).
[166] F. N. Hooge, IEEE Trans. Electron Devices **41**, 1926 (1994).
[167] X. Y. Chen, C. Salm, F. N. Hooge, P. H. Woerlee, Solid-State Electron. **43**, 1715 (1999).
[168] S. Martin, A. Dodabalapur, Z. Bao, B. Crone, H. E. Katz, W. Li, A. Passner, J. A. Rogers, J. Appl. Phys. **87**, 3381 (2000).
[169] P. V. Necliudov, S. L. Rumyantsev, M. S. Shur, D. J. Gundlach, T. N. Jackson, J. Appl. Phys. **88**, 5395 (2000).
[170] L. K. J. Vandamme, R. Feyaerts, Gy. Trefan, C. Detcheverry, J. Appl. Phys. **91**, 719 (2002).
[171] B. K. Jones, IEEE Trans. Electron Devices **41**, 2188 (1994).
[172] L. K. J. Vandamme, IEEE Trans. Electron Devices **41**, 2176 (1994).
[173] R. Kohlrausch, Ann. Phys. Chem. **72**, 353 (1847).
[174] R. G. Palmer, D. L. Stein, E. Abrahams, P. W. Anderson, Phys. Rev. Lett. **53**, 958 (1984).
[175] M. Singh, Philos. Mag. **86**, 797 (2003).
[176] M. Silver, L. Cohen, Phys. Rev. B **15**, 3276 (1977).
[177] L. S. Gradshteyn, I. M. Ryzhik, Table of Integrals, Series and Products, 6th Edn, Editors A. Jeffrey, D. Zwillinger, Academic Press (2000).
[178] A. Hepp, N. von Malm, R. Schmechel, H. von Seggern, Synth. Met. **138**, 201 (2003). DOI: 10.1016/S0379-6779(02)01264-X.
[179] S. Karg, J. Steiger, H. von Seggern, Synth. Met. **111**, 277 (2000).
[180] J. Steiger, R. Schmechel, H. von Seggern, Synth. Met. **129**, 1 (2002).
[181] T. A. T. Cowell, J. Woods, Br. J. Appl. Phys. **18**, 1045 (1967). DOI: 10.1088/0508-3443/18/8/302.

[182] J. L. Castado, J. Garrido, J. Piqueras, J. Phys. D: Appl. Phys. **17**, 2047 (1984).

[183] H. Okushi, Y. Tokumaru, Jpn. J. Appl. Phys. **20**, L45 (1981).

[184] N. Sengouga, B. K. Jones, Solid-State Electron. **36**, 229 (1993); N. Sengouga, B. K. Jones, IEEE Trans. Electron Devices **40**, 471 (1993).

[185] P. Omling, E. R. Weber, L. Montelius, H. Alexander, J. Michel, Phys. Rev. B **32**, 6571 (1985).

[186] D. V. Lang, J. Appl. Phys. **45**, 3023 (1974).

[187] L. Dobaczewski, A. R. Peaker, K. Bonde Nielsen, J. Appl. Phys. **96**, 4689 (2004).

[188] O. Gaudin, R. B. Jackman, T.-P. Nguyen, P. Le Rendu, J. Appl. Phys. **90**, 4196 (2001).

[189] N. Karl, K.-H. Kraft, J. Marktanner, M. Münch, F. Schatz, R. Stehle, H.-M. Uhde, J. Vac. Sci. Technol. A **17**, 2318 (1999).

[190] N. Karl, Synth. Met. **133**, 649 (2003).

[191] V. Essex, P. E. Secker, Br. J. Appl. Phys. (J. Phys. D) **2**, 1107 (1969).

[192] P. Stallinga, H. L. Gomes, Synth. Met. **156**, 1305 (2006); P. Stallinga, H. L. Gomes, Synth. Met. **156**, 1316 (2006).

[193] E. Lebedev, Th. Dietrich, V. Petrova-Koch, S. Karg, W. Brütting, Appl. Phys. Lett. **71**, 2686 (1997).

[194] R. Österbacka, J. Juška, K. Arlauskas, H. Stubb, SPIE **3145**, 389 (1997).

[195] S. Forero, P. H. Nguyen, W. Brütting, M. Schwoerer, Phys. Chem. Phys. **1**, 1769 (1999).

[196] R. W. I. de Boer, M. Jochemsen, T. M. Klapwijk, A. F. Morpurgo, J. Niemax, A. K. Tripathi, J. Pflaum, J. Appl. Phys. **95**, 1196 (2004). DOI: 10.1063/1.1631079.

[197] G. Juška, K. Arlauskas, R. Österbacka, H. Stubb, Synth. Met. **109**, 173 (2000).

[198] P. W. M. Blom, M. C. J. M. Vissenberg, Phys. Rev. Lett. **80**, 3819 (1998).

[199] Q. Mohammad, S. S. Manoharan, J. Appl. Phys. **97**, 096101 (2005). DOI: 10.1063/1.1881791.

[200] M. Redecker, D. D. C. Bradley, M. Jandke, P. Strohriegl, Appl. Phys. Lett. **75**, 109 (1999). DOI:10.1063/1.124291.

[201] J. Bettenhausen, P. Strohriegl, W. Brütting, H. Tokihisa, T. Tsutsui, J. Appl. Phys. **82**, 4957 (1997).

[202] H. Scher, E. W. Montroll, Phys. Rev. B **12**, 2455 (1975).

[203] G. F. Ferreira, Phys. Rev. B **16**, 4719 (1977).

[204] M. Grünewald, B. Movaghar, B. Pohlmann, D. Würtz, Phys. Rev. B **32**, 8191 (1985).

[205] G. Pfister, Phys. Rev. Lett. **36**, 271 (1976).

[206] T. Tiedje, A. Rose, Solid State Commun. **37**, 49 (1980).

[207] F. W. Schmidlin, Phys. Rev. B **16**, 2362 (1977).

[208] W. Brütting, H. Riel, T. Beierlein, W. Riess, J. Appl. Phys. **89**, 1704 (2001).

[209] M. Kitamura, T. Imada, S. Kako, Y. Arakawa, Jpn. J. Appl. Phys. **43**, 2326 (2004).

[210] G. Juška, K. Genevičius, K. Arlauskas, R. Österbacka, H. Stubb, Phys. Rev. B **65**, 233208 (2002). DOI: 10.1103/PhysRevB.65.233208.

[211] I. H. Campbell, D. L. Smith, C. J. Neef, J. P. Ferraris, Appl. Phys. Lett. **74**, 2809 (1999).

[212] A. Einstein, Ann. Phys. **17**, 549 (1905) (in German).

[213] M. M. Perlman, S. Bamji, Appl. Phys. Lett. **33**, 581 (1978).

[214] W. D. Gill, J. Appl. Phys. **43**, 5033 (1972).

[215] S. M. Tuladhar, D. Poplavkyy, S. A. Choulis, J. R. Rurrant, D. C. Bradley, J. Nelson, Adv. Funct. Mater. **15**, 1171 (2005).

[216] L.-B. Lin, S. A. Jenekhe, P. M. Borsenberger, Appl. Phys. Lett. **69**, 3495 (1996).

[217] A. Hirao, T. Tsukamoto, H. Nishizawa, Phys. Rev. B **59**, 12991 (1999).

[218] M. E. Scharfe, Phys. Rev. B **2**, 5025 (1970).

[219] J. Noolandi, Phys. Rev. B **16**, 4466 (1977).
[220] A. Many, G. Rakavy, Phys. Rev. B **126**, 1980 (1962).
[221] B. Gross, Phys. Rev. **107**, 368 (1957).
[222] B. Gross, Phys. Rev. **110**, 337 (1958).
[223] B. Gross, M. M. Perlman, J. Appl. Phys. **43**, 853 (1972).
[224] J. Lindmayer, J. Appl. Phys. **36**, 196 (1965).
[225] G. Rosen, Phys. Rev. B **4**, 667 (1971) (and references therein).
[226] D. Basu, L. Wang, L. Dunn, B. Yoo, M. Heeney, I. McCulloch, Appl. Phys. Lett. **89**, 242104 (2006). DOI: 10.1063/1.2405378.
[227] W. Shockley, G. L. Pearson, Phys. Rev. **74**, 232 (1948).
[228] D. Kahng, IEEE Trans. Electron Devices **23**, 655 (1976) which includes the reference: P. K. Weimer, presented at the IRE-AIEE Devices Research Conference, Stanford (1961).
[229] P. G. Le Comber, W. E. Spear, A. Ghaith, Electron. Lett. **15**, 179 (1979).
[230] K. Weber, M. Grünewald, W. Fuhs, P. Thomas, Phys. Status Solidi B **110**, 133 (1982).
[231] G. Maruccio, A. Biasco, P. Visconti, A. Bramanti, P. P. Pompa, F. Calabi, R. Cingolani, R. Rinaldi, S. Corni, R. Di Felice, E. Molinari, M. P. Verbeet, G. W. Canters, Adv. Mater. **17**, 816 (2005). DOI: 10.1002/adma.200400628.
[232] J. A. Rogers, Z. Bao, K. Baldwin, A. Dodabalapur, B. Crone, V. R. Raju, V. Kuck, H. Katz, K. Amundson, J. Ewing, P. Drzaic, Proc. Natl Acad. Sci. USA **98**, 4835 (2001). DOI: 10.1073/pnas.09588098.
[233] D. Nilsson, T. Kugler, P.-O. Svensson, M. Berggren, Sens. Actuators B **86**, 193 (2002).
[234] Y.-H. Kim, D.-G. Moon, J.-I. Han, IEEE Electron Device Lett. **25**, 702 (2004). DOI: 10.1109/LED.2004.836502.
[235] C. Reese, M. Roberts, M.-M. Ling, Z. Bao, Mater. Today **7**, 20 (2004). DOI: 10.1016/S1369-7021(04)00398-0.
[236] C. R. Newman, C. D. Frisbie, D. A. da Silva Filho, J.-L. Brédas, P. C. Ewbank, K. R. Mann, Chem. Mater. **16**, 4436 (2004). DOI: 10.1021/cm049391x.
[237] H. Sirringhaus, N. Tessler, R. H. Friend, Synth. Met. **102**, 857 (1999).
[238] J.-L. Lin, W.-J. Sah, S.-C. Lee, IEEE Electron Device Lett. **12**, 120 (1991).
[239] H. Sirringhaus, N. Tessler, R. H. Friend, Science **280**, 1741 (1998).
[240] Y.-Y. Lin, D. J. Gundlach, S. F. Nelson, T. N. Jackson, IEEE Electron Device Lett. **18**, 606 (1997); Y.-Y. Lin, D. J. Gundlach, S. F. Nelson, T. N. Jackson, IEEE Trans. Electron Devices **44**, 1325 (1997).
[241] O. D. Jurchescu, J. Baas, T. T. M. Palstra, Appl. Phys. Lett. **84**, 3061 (2004). DOI: 10.1063/1.1704874.
[242] W. Clemens, W. Fix, J. Ficker, A. Knobloch, A. Ullmann, J. Mater. Res. **19**, 1963 (2004).
[243] B. Crone, A. Dodabalapur, Y.-Y. Lin, R. W. Filas, Z. Bao, A. LaDuca, R. Sarpeshkar, H. E. Katz, W. Li, Nature **403**, 521 (2000).
[244] W. Fix, A. Ullmann, J. Ficker, W. Clemens, Appl. Phys. Lett. **81**, 1735 (2002). DOI: 10.1063/1.1501450.
[245] B. A. Ridley, B. Nivi, J. M. Jacobson, Science **286**, 746 (1999).
[246] B. Sun, H. Sirringhaus, Nano Lett. **5**, 2408 (2005). DOI: 10.1021/nl051586w.
[247] Z.-X. Xu, V. A. L. Roy, P. Stallinga, M. Muccini, S. Toffanin, H.-F. Xiang, C.-M. Che, Appl. Phys. Lett. **90**, 223509 (2007). DOI: 10.1063/1.2740478.
[248] L.-L. Chua, J. Zaumseil, J.-F. Chang, E. C.-W. Ou, P. K.-H. Ho, H. Sirringhaus, R. Friend, Nature **434**, 194 (2005).
[249] A. Dodabalapur, Nature **434**, 151 (2005).
[250] E. Cantatore, E. J. Meijer, Proc. 29th Eur. Solid-State Circuits Conf. ESSCIRC '03, 29 (2003). DOI: 10.1109/ESSCIRC.2003.1257064.

[251] Y. Sakamoto, T. Suzuki, M. Kobayashi, Y. Gao, Y. Fukai, Y. Inoue, F. Sato, S. Tokito, J. Am. Chem. Soc. **126**, 8138 (2004). DOI: 10.1021/ja0476258.

[252] M. Ahles, R. Schmechel, H. von Seggern, Appl. Phys. Lett. **85**, 4499 (2004).

[253] G. Horowitz, J. Mater. Chem. **9**, 2021 (1999).

[254] A. Babel, S. A. Jenehke, Adv. Mater. **14**, 371 (2002).

[255] T. M. Pappenfus, R. J. Chesterfield, C. D. Frisbie, K. R. Mann, J. Casado, J. D. Raff, L. L. Miller, J. Am. Chem. Soc. **124**, 4184 (2002). DOI: 10.1021/ja025553j.

[256] J. A. Rogers, A. Dodabalapur, Z. Bao, H. E. Katz, Appl. Phys. Lett. **75**, 1010 (1999). DOI: 10.1063/1.124581.

[257] M. Matters, D. M. de Leeuw, M. J. C. M. Vissenberg, C. M. Hart, P. T. Herwig, T. Geuns, C. M. J. Mutsaers, C. J. Drury, Opt. Mater. **12**, 189 (1999).

[258] E. J. Meijer, D. M. de Leeuw, S. Setayesh, E. van Veenendaal, B.-H. Huisman, P. W. M. Blom, J. C. Hummelen, U. Scherf, T. M. Klapwijk, Nat. Mater. **2**, 678 (2003). DOI: 10.1038/nmat978.

[259] H. Sirringhaus, Nature Mater. **2**, 641 (2003).

[260] M. J. Powell, IEEE Trans. Electron Devices **36**, 2753 (1989).

[261] G. Horowitz, Adv. Mater. **10**, 365 (1998).

[262] G. Horowitz, R. Hajlaoui, R. Bourguiga, M. Hajlaoui, Synth. Met. **101**, 401 (1999).

[263] G. Horowitz, R. Hajlaoui, P. Delannoy, J. Phys. III France **5**, 355 (1995).

[264] A. R. Brown, Synth. Met. **88**, 37 (1997).

[265] A. R. Brown, A. Pomp, D. M. de Leeuw, D. B. M. Klaassen, E. E. Havinga, P. Herwig, K. Müllen, J. Appl. Phys. **79**, 2136 (1996).

[266] D. J. Gundlach, Y.-Y. Lin, T. N. Jackson, D. G. Schlom, Appl. Phys. Lett. **71**, 3853 (1997).

[267] A. J. Salih, D. M. Haynes, A. R. Hepburn, Synth. Met. **71**, 2257 (1995).

[268] F. Dinelli, M. Murgia, P. Levy, M. Cavallini, F. Biscarini, D. M. de Leeuw, Phys. Rev. Lett. **92**, 116802 (2004).

[269] M. Daraktchiev, A. von Mühlenen, F. Nüesch, M. Schaer, M. Brinkmann, M.-N. Bussac, L. Zuppiroli, New J. Phys. **7**, 133 (2005).

[270] T. Muck, V. Wagner, U. Bass, M. Leufgen, J. Geurts, L. W. Molenkamp, Synth. Met. **146**, 317 (2004). DOI: 10.1016/j.synthmet.2004.08.010.

[271] C. G. B. Garrett, W. H. Brattain, Phys. Rev. **99**, 376 (1955).

[272] G. Horowitz, R. Hajlaoui, H. Bouchriha, R. Bourguiga, M. Hajlaoui, Adv. Mater. **10**, 923 (1998).

[273] J. R. Brews, IEEE Trans. Electron Devices **ED-26**, 1282 (1979).

[274] G. Horowitz, J. Mater. Res. **19**, 1946 (2004).

[275] T. Yasuda, T. Goto, K. Fujita, T. Tsutsui, Appl. Phys. Lett. **85**, 2098 (2004).

[276] S. A. Choulis, Y. Kim, J. Nelson, D. D. C. Bradley, M. Giles, M. Shkunov, I. McCulloch, Appl. Phys. Lett. **85**, 3890 (2004).

[277] T. D. Anthopoulos, C. Tanase, S. Satayesh, E. J. Meijer, J. C. Hummelen, P. W. M. Blom, D. M. de Leeuw, Adv. Mater. **16**, 2174 (2004). DOI: 10.1002/adma.200400309.

[278] R. Chesterfield, C. R. Newman, T. M. Pappenfus, P. C. Ewbank, M. H. Haukaas, K. R. Mann, L. L. Miller, C. D. Frisbie, Adv. Mater. **15**, 1278 (2003). DOI: 10.1002/adma.200305200.

[279] A. Babel, J. D. Wind, S. A. Jenekhe, Adv. Funct. Mater. **14**, 891 (2004). DOI: 10.1002/adfm.200305180.

[280] T. Nishikawa, S.-I. Kobayashi, T. Nakanowatari, T. Mitani, T. Shimoda, Y. Kubozono, G. Yamamoto, H. Ishii, M. Niwano, Y. Iwasa, J. Appl. Phys. **97**, 104509 (2005). DOI: 10.1063/1.1903109.

[281] C. Santato, R. Capelli, M. A. Loi, M. Murgia, F. Cicoira, V. A. L. Roy, P. Stallinga, R. Zamboni, C. Rost, S. F. Karg, M. Muccini, Synth. Met. **146**, 329 (2004).

[282] M. Ahles, A. Hepp, R. Schmechel, H. von Seggern, Appl. Phys. Lett. **84**, 428 (2004).

[283] M. A. Loi, C. Rost-Bietsch, M. Murgia, S. Karg, W. Riess, M. Muccini, Adv. Funct. Mater. **16**, 41 (2006). DOI: 10.1002/adfm.200500424.

[284] C. Rost, S. karg, W. Riess, M. A. Loi, M. Murgia, M. Muccini, Appl. Phys. Lett. **85**, 1613 (2004). DOI: 10.1063/1.1785290.

[285] C. Santato, I. Manunza, A. Bonfiglio, F. Cicoira, P. Cosseddu, R. Zamboni, M. Muccini, Appl. Phys. Lett. **86**, 141106 (2005). DOI: 10.1063/1.1898429.

[286] R. J. Walters, G. I. Bourianoff, H. A. Atwater, Nat. Mater. **4**, 143 (2005). DOI: 10.1038/nmat1307.

[287] C. Rost, S. Karg, W. Riess, M. A. Loi, M. Murgia, M. Muccini, Synth. Met. **146**, 237 (2004).

[288] C. Rost, D. J. Gundlach, S. Karg, W. Rieß, J. Appl. Phys. **95**, 5782 (2004).

[289] R. A. Street, A. Salleo, Appl. Phys. Lett. **81**, 2887 (2002). DOI: 10.1063/1.1512950.

[290] I. Yagi, K. Tsukagoshi, Y. Aoyagi, Appl. Phys. Lett. **84**, 813 (2004).

[291] J. Zaumseil, K. W. Baldwin, J. A. Rogers, J. Appl. Phys. **93**, 6117 (2003).

[292] P. V. Necliudov, M. S. Shur, D. J. Gundlach, T. N. Jackson, Solid-State Electron. **47**, 259 (2003).

[293] H. Klauk, G. Schmid, W. Radlik, W. Weber, L. Zhou, C. D. Sheraw, J. A. Nichols, T. N. Jackson, Solid-State Electron. **47**, 297 (2003).

[294] G. B. Blanchet, C. R. Fincher, M. Lefenfeld, J. A. Rogers, Appl. Phys. Lett. **84**, 296 (2004).

[295] E. J. Meijer, G. H. Ghelinck, E. van Veenendaal, B.-H. Huisman, D. M. de Leeuw, T. M. Klapwijk, Appl. Phys. Lett. **82**, 4576 (2003).

[296] L. A. Majewski, R. Schroeder, M. Grell, Appl. Phys. Lett. **85**, 3620 (2004). DOI: 10.1063/1.1797540.

[297] J. A. Nichols, D. J. Gundlach, T. N. Jackson, Appl. Phys. Lett. **83**, 2366 (2003). DOI: 10.1063/1.1611278.

[298] D. Tanimura, M. Yano, W. Takashima, K. Kaneto, Curr. Appl. Phys. **5**, 159 (2005). DOI: 10.1016/j.cap.2004.06.008.

[299] H.-C. Lin, K.-L. Yeh, T.-Y. Huang, R.-G. Huang, S. M. Sze, IEEE Trans. Electron Devices **49**, 264 (2002).

[300] C. Wang, J. P. Snyder, J. R. Tucker, Appl. Phys. Lett. **74**, 1174 (1999).

[301] A. Bolognesi, A. Di Carlo, P. Lugli, Appl. Phys. Lett. **81**, 4646 (2002). DOI: 10.1063/1.1527983.

[302] S. Cherian, C. Donley, D. Mathine, L. LaRussa, W. Xia, N. Armstrong, J. Appl. Phys. **96**, 5638 (2004). DOI: 10.1063/1.1803945.

[303] A. Dodabalapur, J. Baumbach, K. Baldwin, H. E. Katz, Appl. Phys. Lett. **68**, 2246 (1996). DOI: 10.1063/1.115873.

[304] Y. Shen, M. W. Klein, D. B. Jacobs, J. C. Scott, G. G. Malliaras, Phys. Rev. Lett. **86**, 3867 (2001).

[305] E. M. Muller, J. A. Marohn, Adv. Mater. **17**, 1410 (2005). DOI: 10.1002/adma.200401174.

[306] G. Wang, Y. Luo, P. H. Beton, Appl. Phys. Lett. **83**, 3108 (2003). DOI: 10.1063/1.1617375.

[307] B. H. Hamadani, D. Natelson, Appl. Phys. Lett. **84**, 443 (2004).

[308] L. Bürgi, T. J. Richards, R. H. Friend, H. Sirringhaus, J. Appl. Phys. **94**, 6129 (2003). DOI: 10.1063/1.1613369.

[309] K. Seshadri, C. D. Frisbie, Appl. Phys. Lett. **78**, 993 (2001). DOI: 10.1063/1.1345805.

[310] K. P. Puntambekar, P. V. Pesavento, C. D. Frisbie, Appl. Phys. Lett. **83**, 5539 (2003). DOI:10.1063/1.1637443.

[311] H. Klauk, D. J. Gundlach, J. A. Nichols, T. N. Jackson, IEEE Trans. Electron Devices **46**, 1258 (1999).
[312] H. Klauk, M. Halik, U. Zschieschang, F. Eder, G. Schmid, C. Dehm, Appl. Phys. Lett. **82**, 4175 (2003). DOI: 10.1063/1.1579870.
[313] J. Lee, K. Kim, J. H. Kim, S. Im, D.-Y. Jung, Appl. Phys. Lett. **82**, 4169 (2003). DOI: 10.1063/1.1580993.
[314] J. Lee, J. H. Kim, S. Im, Appl. Phys. Lett. **83**, 2689 (2003). DOI: 10.1063/1.1613997.
[315] R. W. I. de Boer, T. M. Klapwijk, A. F. Morpurgo, arXiv:cond-mat/0307320v1 (2008).
[316] L. Torsi, N. Cioffi, C. Di Franco, L. Sabbatini, P. G. Zambonin, T. Bleve-Zacheo, Solid-State Electron. **45**, 1479 (2001).
[317] L. Torsi, A. Dodabalapur, H. E. Katz, J. Appl. Phys. **78** (1995).
[318] A. Babel, S. A. Jenehke, Synth. Met. **148**, 169 (2005). DOI: 10.1016/j.synthmet.2004.09.033.
[319] X. Peng, G. Horowitz, D. Fichou, F. Garnier, Appl. Phys. Lett. 57, 2013 (1990). DOI:10.1063/1.103994.
[320] J. Collet, O. Tharaud, A. Chapoton, D. Vuillaume, Appl. Phys. Lett. **76**, 1941 (2000). DOI: 10.1063/1.126219.
[321] B. A. Mattis, Y. Pei, V. Subramanian, Appl. Phys. Lett. **86**, 033113 (2005). DOI: 10.1063/1.1854217.
[322] A. Dodabalapur, J. Laquindanum, H. E. Katz, Z. Bao, Appl. Phys. Lett. **69**, 4227 (1996). DOI: 10.1063/1.116953.
[323] G. Horowitz, R. Hajlaoui, D. Fichou, A. Kassmi, J. Appl. Phys. **85**, 3202 (1999).
[324] F. Dinelli, M. Murgia, F. Biscarini, D. M. de Leeuw, Synth. Met. **146**, 373 (2004).
[325] G. Niu, J. D. Cressler, S. J. Mathew, S. Subbana, IEEE Trans. Electron Devices **47**, 648 (2000).
[326] M. J. Powell, J. W. Orton, Appl. Phys. Lett. **45**, 171 (1984).
[327] P. Stallinga, H. L. Gomes, Synth. Met. **156**, 1305 (2006).
[328] P. Checcoli, G. Conte, S. Salvatori, R. Paolesse, A. Bolognesi, M. Berliocchi, F. Brunetti, A. Damico, A. Di Carlo, P. Lugli, Synth. Met. **138**, 261 (2003). DOI:10.1016/S0379-6779(02)01308-5.
[329] S. Zhu, J. Chen, M.-F. Li, S. J. Lee, J. Singh, C. X. Zhu, A. Du, C. H. Tung, A. Chin, D. L. Kwong, IEEE Electron Device Lett. **25**, 565 (2004); J. Y. Choi, S. Ahmed, T. Dimitrova, J. T. C. Chen, D. K. Schroder, IEEE Trans. Electron Devices **51**, 1380 (2004); C. Wang, J. P. Snyder, J. R. Tucker, Appl. Phys. Lett. **74**, 1174 (1999); D. L. John, L. C. Castro, J. Clifford, D. L. Pulfrey, IEEE Trans. Nanotechnol. **2**, 175 (2003); B. Winstead, U. Ravaioli, IEEE Trans. Electron Devices **47**, 1241 (2000); S. Zhu, H. Y. Yu, S. J. Wang, J. H. Chen, C. Shen, C. Zhu, S. J. Lee, M. F. Li, D. S. H. Chan, W. J. Yoo, A. Du, C. H. Tung, J. Singh, A. Chin, D. L. Kwong, IEEE Electron Device Lett. **25**, 268 (2004); J. R. Lothian, F. Ren, J. M. Kuo, J. S. Weiner, Y. K. Chen, Solid-State Electron. **41**, 673 (1997).
[330] R. A. Vega, Schottky Field Effect Transistors and Schottky CMOS Circuitry, Thesis, Rochester Institute of Technology (2006).
[331] P. V. Necliudov, M. S. Shur, D. J. Gundlach, T. N. Jackson, J. Appl. Phys. **88**, 6594 (2000).
[332] G. Horowitz, Synth. Met. **138**, 101 (2003). DOI: 10.1016/S0379-6779(02) 01298-5.
[333] G. Horowitz, M. E. Hajlaoui, Adv. Mater. **12**, 1046 (2000); G. Horowitz, M. E. Hajlaoui, Synth. Met. **122**, 185 (2001).
[334] A. Bolognesi, M. Berliocchi, M. Manenti, A. Di Carlo, P. Lugli, K. Lmimouni, C. Dufour, IEEE Trans. Electron Devices **51**, 1997 (2004). DOI: 10.1109/TED.2004.838333.

[335] J. Levinson, F. R. Shepherd, P. J. Scanlon, W. D. Westwood, G. Este, M. Rider, J. Appl. Phys. **53**, 1193 (1982).
[336] R. A. Street, D. Knipp, A. R. Völkel, Appl. Phys. Lett. **80**, 1658 (2002).
[337] H. Sehil, N. M. Rahmani, F. Raoult, Mater. Chem. Phys. **42**, 101 (1995).
[338] R. Schroeder, L. A. Majewski, M. Grell, Appl. Phys. Lett. **84**, 1004 (2004).
[339] R. Schroeder, L. A. Majewski, M. Grell, Appl. Phys. Lett. **83**, 3201 (2003).
[340] J. R. Tucker, C. Wang, P. S. Carney, Appl. Phys. Lett. **65**, 618 (1994).
[341] L. Wang, D. Fine, T. Jung, D. Basu, H. von Seggern, A. Dodabalapur, Appl. Phys. Lett. **85**, 1772 (2004). DOI: 10.1063/1.1790033.
[342] P. Stallinga, H. L. Gomes, Org. Electron. **8**, 300 (2007). DOI: 10.1016/j.orgel.2006.11.004.
[343] L. Bürgi, H. Sirringhaus, R. H. Friend, Appl. Phys. Lett. **80**, 2913 (2002). DOI: 10.1063/1.1470702.
[344] P. Stallinga, H. L. Gomes, Synth. Met. **158**, 473 (2008).
[345] P. Stallinga, H. L. Gomes, Synth. Met. **156**, 1316 (2006).
[346] Wikipedia, accessed 1 March (2007).
[347] P. Stallinga, V. A. L. Roy, Z.-X. Xu, H.-F. Xiang, C.-M. Che, Adv. Mater. **20**, 2120 (2008).
[348] G. Horowitz, M. E. Hajlaoui, R. Hajlaoui, J. Appl. Phys. **87**, 4456 (2000).
[349] L. Bozano, S. A. Carter, J. C. Scott, G. G. Malliaras, P. J. Brock, Appl. Phys. Lett. **74**, 1132 (1999).
[350] P. Stallinga, H. L. Gomes, Org. Electron. **7**, 592 (2006).
[351] M. C. J. M. Vissenberg, M. Matters, Phys. Rev. B **57**, 12964 (1998).
[352] P. Stallinga, H. L. Gomes, Org. Electron. **6**, 137 (2005).
[353] F. R. Libsch, J. Kanicki, Appl. Phys. Lett. **62**, 1286 (1992).
[354] W. B. Jackson, J. M. Marshall, M. D. Moyer, Phys. Rev. B **39**, 1164 (1989).
[355] S. Paul, A. J. Flewitt, W. I. Milne, J. Robertson, Appl. Phys. Lett. **86**, 202110 (2005).
[356] A. Rahal, T. Mohammed-Brahim, H. Toutah, B. Tala-Ighil, Y. Helen, C. Prat, F. Raoult, Microelectron. Rel. **39**, 851 (1999).
[357] G. Fortunato, L. Mariucci, C. Reita, Appl. Phys. Lett. **59**, 826 (1991). DOI: 10.1063/1.105275.
[358] H. L. Gomes, P. Stallinga, F. Dinelli, M. Murgia, F. Biscarini, D. M. de Leeuw, T. Muck, J. Geurts, L. W. Molenkamp, V. Wagner, Appl. Phys. Lett. **84**, 3184 (2004); H. L. Gomes, P. Stallinga, D. M. de Leeuw, Mater. Sci. Forum **514**, 33 (2006).
[359] H. L. Gomes, P. Stallinga, D. M. de Leeuw, Mater. Sci. Forum **514–516**, 33 (2006).
[360] H. L. Gomes, P. Stallinga, F. Dinelli, M. Murgia, F. Biscarini, D. M. de Leeuw, M. Muccini, K. Müllen, Polym. Adv. Technol. **16**, 227 (2005). DOI: 10.1002/pat.558.
[361] H. L. Gomes, P. Stallinga, M. Murgia, F. Biscarini, T. Muck, V. Wagner, E. Smits, D. M. de Leeuw, Proc. SPIE **5940**, (2005). DOI: 10.1117/12.617862.
[362] W. A. Schoonveld, J. B. Oostinga, J. Vrijmoeth, T. M. Klapwijk, Synth. Met. **101**, 608 (1999).
[363] M. Matters, D. M. de Leeuw, P. T. Herwig, A. R. Brown, Synth. Met. **102**, 998 (1999).
[364] A. Salleo, R. A. Street, J. Appl. Phys. **94**, 471 (2003).
[365] R. A. Street, A. Salleo, M. L. Chabinyc, Phys. Rev. B **68**, 085316 (2003).
[366] J. Yuan, J. Zhang, J. Wang, D. Yan, W. Xu, Thin Solid Films **450**, 316 (2004).
[367] P. D'Angelo, P. Stoliar, T. Cramer, A. Cassinese, F. Zerbetto, F. Biscarini, Appl. Phys. A **95**, 55 (2009).
[368] U. Zschieschang, R. T. Weitz, K. Kern, H. Klauk, Appl. Phys. A **95**, 139 (2009).
[369] R. V. Chamberlin, G. Mozurkewich, R. Orbach, Phys. Rev. Lett. **52**, 867 (1984).

[370] S. J. Zilker, C. Detcheverry, E. Cantatore, D. M. de Leeuw, Appl. Phys. Lett. **79**, 1124 (2001).
[371] N. D. Young, A. Gill, Semicond. Sci. Technol. 7, 1103 (1992). DOI: 10.1088/0268-1242/7/8/013.
[372] S. A. Lyon, Appl. Surf. Sci. **39**, 552 (1989).
[373] B. K. Jones, M. A. Iqbal, Mater. Sci. Forum **258–263**, 933 (1997).
[374] M. G. Adlerstein, Electron. Lett. **12**, 297 (1976).
[375] N. A. Hastas, C. A. Dimitriadis, F. V. Farmakis, G. Kamarinos, Microelectron. Reliab. **43**, 671 (2003).
[376] M. Shtein, J. Mapel, J. B. Benziger, S. R. Forrest, Appl. Phys. Lett. **81**, 268 (2002). DOI: 10.1063/1.1491009.
[377] J.-M. Kim, J.-W. Lee, J.-K. Kim, B.-K. Ju, J.-S. Kim, Y.-H. Lee, M.-H. Oh, Appl. Phys. Lett. **85**, 6368 (2004).
[378] S. C. Lim, S. H. Kim, J. H. Lee, M. K. Kim, D. J. Kim, T. Zyunga, Synth. Met. **148**, 75 (2005). DOI: 10.1016/j-synthmet.2004.08.034.
[379] T. Higuchi, T. Murayama, E. Itoh, K. Miyairi, Thin Solid Films **499**, 374 (2006).
[380] G. Wang, D. Moses, A. J. Heeger, H.-M. Zhang, M. Narasimhan, R. E. Demaray, J. Appl. Phys. **95**, 316 (2004).
[381] A. I. Kingon, J.-P. Maria, S. K. Streiffer, Nature **406**, 1032 (2000).
[382] C. D. Dimitrakopoulos, B. K. Furman, T. Graham, S. Hedge, S. Purushothaman, Synth. Met. **92**, 47 (1998).
[383] F.-C. Chen, C.-W. Chu, J. He, Y. Yang, J.-L. Lin, Appl. Phys. Lett. **85**, 3295 (2004). DOI: 10.1063/1.126219.
[384] H. Klauk, M. Halik, U. Zschieschang, G. Schmid, W. Radlik, J. Appl. Phys. **92**, 5259 (2002). DOI: 10.1063/1.1511826.
[385] J. Puigdollers, C. Voz, A. Orpella, R. Quidant, I. Martin, M. Vetter, R. Alcubilla, Org. Electron. 5, 67 (2004). DOI: 10.1016/ j.orgel.2003.10.002.
[386] E. Becker, R. Parashkov, G. Ginev, D. Schneider, S. Hartmann, F. Brunetti, T. Dobbertin, D. Metzdorf, T. Riedl, H.-H. Johannes, W. Kowalsky, Appl. Phys. Lett. **83**, 4044 (2003). DOI: 10.1063/1.1623951.
[387] R. Parashkov, E. Becker, S. Hartmann, G. Ginev, D. Schneider, H. Krautwald, T. Dobbertin, D. Metzdorf, F. Brunetti, C. Schildknecht, A. Kammoun, M. Brandes, T. Riedl, H.-H. Johannes, W. Kowalsky, Appl. Phys. Lett. **82**, 4579 (2003). DOI: 10.1063/1.1584786.
[388] G. H. Gelinck, T. C. T. Geuns, D. M. de Leeuw, Appl. Phys. Lett. 77, 1487 (2000).
[389] M. Halik, H. Klauk, U. Zschieschang, T. Kriem, G. Schmid, W. Radlik, K. Wussow, Appl. Phys. Lett. **81**, 289 (2002). DOI: 10.1063/1.1491604.
[390] M. Halik, H. Klauk, U. Zschieschang, G. Schmid, C. Dehm, M. Schütz, S. Maisch, F. Effenberger, M. Brunnbauer, F. Stellacci, Nature **431**, 963 (2004). DOI: 10.1038/nature02987.
[391] J. Park, S. Y. Park, S.-O. Shim, H. Kang, H. H. Lee, Appl. Phys. Lett. **85**, 3283 (2004). DOI: 10.1063/1.1805703.
[392] C. Bartic, H. Jansen, A. Campitelli, S. Borghs, Org. Electron. **3**, 65 (2002).
[393] G. Velu, C. Legrand, O. Tharaud, A. Chapoton, D. Remiens, G. Horowitz, Appl. Phys. Lett. **79**, 659 (2001).
[394] C. D. Dimitrakopoulos, S. Purushothaman, J. Kymissis, A. Callegari, J. M. Shaw, Science **283**, 822 (1999).
[395] R. Schroeder, L. A. Majewski, M. Grell, Adv. Mater. **17**, 1535 (2005). DOI: 10.1002/adma.200401398.
[396] J. Veres, S. D. Ogier, S. W. Leeming, D. C. Cupertino, S. M. Khaffaf, Adv. Funct. Mater. **13**, 199 (2003).
[397] J. Veres, S. Ogier, G. Lloyd, D. de Leeuw, Chem. Mater. **16**, 4543 (2004). DOI: 10.1021/cm049598q.

[398] A. C. Mayer, R. Ruiz, R. L. Headrick, A. Kazimirov, G. G. Malliaras, Org. Electron. **5**, 257 (2004).
[399] K. Shin, X. Hu, X. Zheng, M. H. Rafailovich, J. Sokolov, V. Zaitsev, S. A. Schwarz, Macromolecules **34**, 4993 (2001).
[400] P. G. Debenedetti, J. Phys.: Condens. Matter **15**, R1669 (2003).
[401] S. Mahapatra, C. D. Parikh, V. Ramgopal Rao, C. R. Viswanathan, J. Vasi, IEEE Trans. Electron Devices **47**, 171 (2000).
[402] D. Muñoz Ramo, A. L. Shluger, J. L. Gavartin, G. Bersuker, Phys. Rev. Lett. **99**, 155504 (2007). DOI: 10.1103/PhysRevLett.99.155504.
[403] P. Ostoja, P. Maccagnani, M. Gazzano, M. Cavallini, J. C. Kengne, R. Kshirsagar, F. Biscarini, M. Melucci, M. Zambianchi, G. Barbarella, Synth. Met. **146**, 243 (2004).
[404] P. C. Chang, J. Lee, D. Huang, V. Subramanian, A. R. Murphy, J. M. J. Frechet, Chem. Mater. **16**, 4783 (2004). DOI: 10.1021/cm0496570.
[405] L. Pichon, A. Mercha, J. M. Routore, R. Carin, O. Bonnaud, T. Mohammed-Brahim, Thin Solid Films **427**, 350 (2003).

Index

Electrical Characterization of Organic Electronic Materials and Devices Peter Stallinga

Printed and bound by CPI Group (UK) Ltd, Croydon, CR0 4YY
16/09/2026
14963471-0003